# Instruments and Methods for Cyclotron Produced Radioisotopes

# Instruments and Methods for Cyclotron Produced Radioisotopes

Special Issue Editors

**Saverio Braccini**
**Francisco Alves**

MDPI • Basel • Beijing • Wuhan • Barcelona • Belgrade

MDPI

*Special Issue Editors*

Saverio Braccini
University of Bern
Switzerland

Francisco Alves
University of Coimbra
Portugal

*Editorial Office*
MDPI
St. Alban-Anlage 66
4052 Basel, Switzerland

This is a reprint of articles from the Special Issue published online in the open access journal *Instruments* (ISSN 2410-390X) from 2018 to 2019 (available at: https://www.mdpi.com/journal/instruments/special_issues/Cyclotron_Radioisotopes).

For citation purposes, cite each article independently as indicated on the article page online and as indicated below:

LastName, A.A.; LastName, B.B.; LastName, C.C. Article Title. *Journal Name* **Year**, *Article Number*, Page Range.

**ISBN 978-3-03928-202-9 (Pbk)**
**ISBN 978-3-03928-203-6 (PDF)**

Cover image courtesy of Saverio Braccini.

# Contents

# About the Special Issue Editors

**Saverio Braccini** is professor of experimental physics at the Laboratory for High Energy Physics (LHEP) of the University of Bern, where he leads research on the medical applications of particle physics. He is co-chair of CYCLEUR, the European Network of Cyclotron Research Centres. He proposed the realisation of the medical cyclotron laboratory at the Bern University Hospital, an innovative facility for radioisotope production and multi-disciplinary research. He teaches general physics, medical radiation physics and radiation protection. His main research interest is the development of innovative accelerators and detectors for medical applications. In particular, he is developing novel irradiation techniques for the production of radio-metals with solid targets. In fundamental high-energy physics, he contributed to experiments with the Large Electron Positron Collider (LEP) and the Large Hadron Collider (LHC) at CERN.

**Francisco Alves** holds a Ph.D. in Biomedical Sciences and is professor of radiation physics, radiation protection and biophysics at Coimbra Health School. He is co-chair of CYCLEUR, the European Network of Cyclotron Research Centres. He lectures at the European School of Nuclear Medicine and serves as an expert for the International Atomic Energy Agency (IAEA). He contributed to the design of the cyclotron laboratory of ICNAS, the Institute for Nuclear Sciences Applied to Health, of the University of Coimbra where he leads research, operation and maintenance activities. ICNAS is the first research and clinical positron emission tomography (PET) centre in Portugal. His main research interest is the production of radio-metals using cyclotrons and liquid targets. In particular, he pioneered the development of liquid targets for the production of $^{68}$Ga, in quality and quantity, for routine clinical applications.

*instruments*

MDPI

*Editorial*

# Special Issue "Instruments and Methods for Cyclotron Produced Radioisotopes"

**Saverio Braccini [1,\*] and Francisco Alves [2,\*]**

1    Albert Einstein Center for Fundamental Physics, Laboratory for High Energy Physics, University of Bern, Sidlerstrasse 5, CH-3012 Bern, Switzerland
2    ICNAS—Institute for Nuclear Sciences Applied to Health, University of Coimbra, Pólo das Ciências da Saúde, Azinhaga de Santa Comba, 3000-548 Coimbra, Portugal
\*    Correspondence: Saverio.Braccini@lhep.unibe.ch (S.B.); franciscoalves@uc.pt (F.A.)

Received: 5 November 2019; Accepted: 6 November 2019; Published: 8 November 2019

**Abstract:** The 17th Workshop on Targets and Target Chemistry (WTTC17) was held in Coimbra (Portugal) on 27–31 August 2018. A few months before, the 13th Workshop of the European Cyclotron Network (CYCLEUR) took place in Lisbon (Portugal) on 23–24 November 2017. These two events reassembled major experts in the field of radioisotope production, targets, target chemistry and cyclotrons. In the last few years, significant advances have been obtained in these fields with direct implications for science and society. Instruments and methods, originally developed for nuclear and particle physics, played a crucial role and remarkable developments are on-going. The production of novel radioisotopes for both diagnostics and therapy is expected to produce a breakthrough in nuclear medicine in the next years, paving the way towards theranostics and personalized medicine. This Special Issue presents a collection of original scientific contributions on the latest developments on instruments and methods for medical and research cyclotrons as well as on target and target chemistry for the production of radioisotopes.

**Keywords:** cyclotrons; targets; target chemistry; radioisotopes; theranostics

---

## 1. Introduction

Translational research is fundamental for the development of modern medicine. On the basis of the findings of basic science and of the technology developed to obtain them, novel medical applications can be conceived and put into practice with a direct benefit for the society. A sound example of this virtuous process is represented by cyclotrons. Originally conceived for nuclear and particle physics, they are nowadays fundamental for the supply of medical radioisotopes.

In the last ten years, the number of facilities based on compact medical cyclotrons largely increased, mainly to match the constantly growing demand of radioisotopes for Positron Emission Tomography (PET) imaging. These accelerators are often in operation in hospitals and provide proton beams in the energy range 15–25 MeV and in the intensity range 10–500 µA. Deuteron beams are sometimes also available. The production of radioisotopes is performed also with larger cyclotrons providing 30 MeV proton beams with intesities of the order of 1 mA. They are mostly installed in laboratories or radio-pharmaceutical industries. Furthermore, a few large research facilities operate 70 MeV proton cyclotrons. 30 MeV and 70 MeV proton cyclotrons are in some cases able to accelerate also $\alpha$ particles or other ions.

Compact medical cyclotrons are mainly used for the production of $^{18}$F, which is presently the most common PET radioisotope. In the recent years, several novel PET radioisotopes are studied to widen the portfolio of radio-labelled bio-molecules to investigate specific diseases. Along this line, positron emitting radio-metals have a prominent role since some of them could be used in combination with with a beta-minus emitting partner to label the same molecule for therapeutic purposes. These two radioisotopes form a so-called theranostic pair which allows the combination of therapy and diagnostics, paving the way towards personalized medicine.

The path towards novel radio-labelled tracers and therapeutic agents is like a relay race, where physics plays a crucial role in optimizing the irradiation methodologies and in developing novel targets and chemistry is essential to provide effective methods to manipulate the irradiated target material and label the compounds. More in general, this field joins multi-disciplinary efforts not only from physics and chemistry but also from engineering, pharmacy and medicine. Furthermore, a close connection with industry is essential to bring the results of scientific research to the patients.

This special issue collects 15 research papers, 4 communications, 2 technical notes and one review. It represents a comprehensive summary of the most recent advances in the fields of medical cyclotrons, targets, radio-chemistry and non-convectional medical radioisotopes.

## 2. Cyclotrons and Related Developments

Medical cyclotrons are usually installed in hospitals and research centres. Some of them are operated by radio-pharmaceutical companies, that have sometimes their own production facilities. These accelerators are characterized by a large scientific potential that may extend beyond radioisotope production, especially if they are equipped with external beam transfer lines. A large number of new facilities are under construction or planned worldwide. It is important to remark that an accurate planning phase is crucial to reach the goals of such complex installations. An excellent example of a state-of-the art facility for the production of medical radioisoptopes is the new Center for Radiopharmaceutical Cancer Research at the Helmholtz-Zentrum Dresden-Rossendorf [1], where a variable energy (18–30 MeV) cyclotron was recently installed. This cyclotron is equipped with two beamlines, two target selectors and several liquid, gas and solid target stations to produce a very large variety of research radioisotopes.

For an optimal production of radioisotopes either in quality or in quantity, an accurate knowledge of the production cross-sections and of the features of the accelerator are mandatory. In particular, the energy of the pristine beam is crucial if solid targets are bombarded. Methods for the measurement of the beam energy of a medical cyclotron were developed by the University of Bern [2] using a multi-leaf Faraday cup and by the University of Coimbra [3] using stacks of natural titanium foils interleaved by niobium degraders.

For an efficient use of a medical cyclotron, regular preventive maintenance is of paramount importance. The wear and the lifetime of components of the accelerator, as the ion source, are key features. Along this line, a new kind of ion source filament was studied and tested at TRIUMF [4].

## 3. Targets and Related Developments

Targets can be classified according to the form of the bombarded material: gas, liquid or solid. Commercial solutions are available for compact medical cyclotrons, although research is ongoing especially on solid target stations. For an efficient exploitation of large high-energy and high-power cyclotrons, specific targets have to be developed. This is the case of the 70 MeV cyclotron in operation at iThemba LABS in South Africa [5] or of the thorium metal target for the production of $^{225}$Ac developed at TRIUMF [6].

The control of the temperature of both the target and of the cooling system is of paramount importance to avoid problems that may cause damage to the target and to the equipment with potential radiation protection implications. For this purpose, a system to measure the temperature in the cyclotron targets cooling water during bombardment was developed at the University of Coimbra [7].

Solid targets are used to produce non-conventional radioisotopes, radio-metals in particular. They present several critical issues as the release of the irradiated target followed by the transfer into a radio-chemistry laboratory. This is accomplished using different methodologies which are the subject of continuous improvements. A novel quick-release target system aimed at decreasing the retrieval time of the irradiated target to less than one minute was developed at TRIUMF [8], allowing to reduce the radiation dose to the operators. The bombardment of a solid target is followed by target dissolution and chemical separation, often implying complex logistics and potential radiation protection hazards. To simplify this process, W.Z. Gelbart and R.R. Johnson [9] proposed a system encompassing a solid target with in-situ dissolution. Cost-effective methods for solid target construction were developed. In particular, 3D printing was used at the Cyclotron Facility in Perth [10].

$^{11}$C is a PET radioisotope that, due to its short half-life of about 20 minutes, cannot be transported far away from the production site and is of interest only for hospital based facilities where the PET scanner and the cyclotron are located at very short distance. Despite this disadvantage, $^{11}$C is used to label relevant medical compounds and novel targets are under study, as the one presented in the paper by J. Peeples et al. [11] based on boron nitride nanotubes (BNNTs).

$^{68}$Ga is an emerging PET radioisotope that is usually produced by means of Ge/Ga generators. Cyclotron production is challenging and several research groups are focusing on liquid or solid target irradiation techniques. A novel method based on a fused zinc target was investigated at TRIUMF [12].

In the last years, accelerator production of $^{99m}$Tc was investigated to cope with the potential crisis of the production of Mo/Tc generators. The preparation of $^{100}$Mo targets is difficult since molybdenum metal cannot be electroplated. To overcome this difficulty, W.Z. Gelbart and R.R. Johnson [13] proposed a method to prepare targets that uses a specific cladding process. To realize $^{100}$Mo and $^{nat}$Y solid targets for cyclotron production of $^{99m}$Tc and $^{89}$Zr, magnetron sputtering was proposed by H. Skliarova et al. [14]. Novel ideas were also put forward, as the powder-in-gas target proposed by G. Lange [15].

## 4. Radio Chemistry Developments

The availability of radio-metals is fundamental for the development of theranostics in nuclear medicine. Reliable and efficient methods for the separation and the purification are under study. The University of Coimbra [16] developed an automated process based on a commercially available module suitable for $^{68}$Ga, $^{64}$Cu and $^{61}$Cu obtained through irradiation of liquid targets.

The production of medical radioisotopes is very often performed by irradiating rare and expensive isotope-enriched target materials that have to be recovered and reused. Although quite standardized, the production processes of $^{18}$F can be improved. In particular, the $^{18}$O enriched water has to be recovered and recycled. A method for optimized treatment and recovery of $^{18}$O enriched irradiated water was developed by the Ruhr University Bochum [17]. For recycling highly $^{100}$Mo-enriched target material for cyclotron production of $^{99m}$Tc, a closed-loop solution was developed, as reported in the paper by H. Skliarova et al. [18].

## 5. Non-Conventional Medical Radioisotopes

Non-conventional medical radioisotopes are the focus of research activities by several groups worldwide. In particular, radio-metals can be used to label peptides and proteins. Scandium is an interesting case, since $^{43}$Sc/$^{47}$Sc and $^{44}$Sc/$^{47}$Sc represent promising theranostic pairs. For an optimal production of scandium isotopes with the desired purity, accurate knowledge of the production cross-sections is mandatory. The measurement of the cross-sections is a complex process and the possibility to derive them from Thick Target Yield (TTY) measurements in the case of scandium was studied by M. Sitarz et al. [19]. Other radio-metals of interest are $^{52}$Mn and $^{165}$Er studied in Orleans [20] and $^{45}$Ti, which was proposed for PET, as reported in the review by P. Costa et al. [21].

Although not commonly available, $\alpha$ particle beams can be used for the production of medical radioisotopes. This is the case of $^{97}$Ru, which is a potential radioisotope for Single Photon Emission Computed Tomography (SPECT). Its production was studied with the ARRONAX [22] multi-particle cyclotron via the reaction $^{nat}$Mo($\alpha$,X).

## 6. Outlook

The production of radio-isotopes by means of cyclotrons is an expanding field of scientific research. Some of the most recent developments—such as theranostics in nuclear medicine—are still in their infancy and a large number of findings and advances is expected in the near future. This Special Issue represents a summary for experts active in the field as well as a guideline for students and young scientists.

**Conflicts of Interest:** The authors declare no conflict of interest.

## References

1. Kreller, M.; Pietzsch, H.J.; Walther, M.; Tietze, H.; Kaever, P.; Knieß, T.; Füchtner, F.; Steinbach, J.; Preusche, S. Introduction of the New Center for Radiopharmaceutical Cancer Research at Helmholtz-Zentrum Dresden-Rossendorf. *Instruments* **2019**, *3*, 9, doi:10.3390/instruments3010009.
2. Nesteruk, K.P.; Ramseyer, L.; Carzaniga, T.S.; Braccini, S. Measurement of the Beam Energy Distribution of a Medical Cyclotron with a Multi-Leaf Faraday Cup. *Instruments* **2019**, *3*, 4, doi:10.3390/instruments3010004.
3. Do Carmo, S.J.; de Oliveira, P.M.; Alves, F. Simple, Immediate and Calibration-Free Cyclotron Proton Beam Energy Determination Using Commercial Targets. *Instruments* **2019**, *3*, 20, doi:10.3390/instruments3010020.
4. Prevost, D.; Jayamanna, K.; Graham, L.; Varah, S.; Hoehr, C. New Ion Source Filament for Prolonged Ion Source Operation on A Medical Cyclotron. *Instruments* **2019**, *3*, 5, doi:10.3390/instruments3010005.
5. Steyn, G.F.; Anthony, L.S.; Azaiez, F.; Baard, S.; Bark, R.A.; Barnard, A.H.; Beukes, P.; Broodryk, J.I.; Conradie, J.L.; Cornell, J.C.; et al. Development of New Target Stations for the South African Isotope Facility. *Instruments* **2018**, *2*, 29, doi:10.3390/instruments2040029.
6. Robertson, A.K.; Lobbezoo, A.; Moskven, L.; Schaffer, P.; Hoehr, C. Design of a Thorium Metal Target for 225Ac Production at TRIUMF. *Instruments* **2019**, *3*, 18, doi:10.3390/instruments3010018.
7. Do Carmo, S.J.C.; De Oliveira, P.M.; Alves, F. A Target-Temperature Monitoring System for Cyclotron Targets: Safety Device and Tool to Experimentally Validate Targetry Studies. *Instruments* **2018**, *2*, 9, doi:10.3390/instruments2030009.
8. Zeisler, S.; Clarke, B.; Kumlin, J.; Hook, B.; Varah, S.; Hoehr, C. A Compact Quick-Release Solid Target System for the TRIUMF TR13 Cyclotron. *Instruments* **2019**, *3*, 16, doi:10.3390/instruments3010016.
9. Gelbart, W.Z.; Johnson, R.R. Solid Target System with In-Situ Target Dissolution. *Instruments* **2019**, *3*, 14, doi:10.3390/instruments3010014.
10. Chan, S.; Cryer, D.; Price, R.I. Enhancement and Validation of a 3D-Printed Solid Target Holder at a Cyclotron Facility in Perth, Australia. *Instruments* **2019**, *3*, 12, doi:10.3390/instruments3010012.
11. Peeples, J.; Chu, S.H.; O'Neil, J.P.; Janabi, M.; Wieland, B.; Stokely, M. Boron Nitride Nanotube Cyclotron Targets for Recoil Escape Production of Carbon-11. *Instruments* **2019**, *3*, 8, doi:10.3390/instruments3010008.
12. Zeisler, S.; Limoges, A.; Kumlin, J.; Siikanen, J.; Hoehr, C. Fused Zinc Target for the Production of Gallium Radioisotopes. *Instruments* **2019**, *3*, 10, doi:10.3390/instruments3010010.
13. Gelbart, W.Z.; Johnson, R.R. Molybdenum Sinter-Cladding of Solid Radioisotope Targets. *Instruments* **2019**, *3*, 11, doi:10.3390/instruments3010011.
14. Skliarova, H.; Cisternino, S.; Cicoria, G.; Marengo, M.; Cazzola, E.; Gorgoni, G.; Palmieri, V. Medical Cyclotron Solid Target Preparation by Ultrathick Film Magnetron Sputtering Deposition. *Instruments* **2019**, *3*, 21, doi:10.3390/instruments3010021.
15. Lange, G. Vortex Target: A New Design for a Powder-in-Gas Target for Large-Scale Radionuclide Production. *Instruments* **2019**, *3*, 24, doi:10.3390/instruments3020024.
16. Alves, V.H.; Do Carmo, S.J.C.; Alves, F.; Abrunhosa, A.J. Automated Purification of Radiometals Produced by Liquid Targets. *Instruments* **2018**, *2*, 17, doi:10.3390/instruments2030017.
17. Uhlending, A.; Henneken, H.; Hugenberg, V.; Burchert, W. Optimized Treatment and Recovery of Irradiated [18O]-Water in the Production of [18F]-Fluoride. *Instruments* **2018**, *2*, 12, doi:10.3390/instruments2030012.
18. Skliarova, H.; Buso, P.; Carturan, S.; Rossi Alvarez, C.; Cisternino, S.; Martini, P.; Boschi, A.; Esposito, J. Recovery of Molybdenum Precursor Material in the Cyclotron-Based Technetium-99m Production Cycle. *Instruments* **2019**, *3*, 17, doi:10.3390/instruments3010017.

*Instruments* **2019**, *3*, 60

19. Sitarz, M.; Jastrzębski, J.; Haddad, F.; Matulewicz, T.; Szkliniarz, K.; Zipper, W. Can We Extract Production Cross-Sections from Thick Target Yield Measurements? A Case Study Using Scandium Radioisotopes. *Instruments* **2019**, *3*, 29, doi:10.3390/instruments3020029.

20. Vaudon, J.; Frealle, L.; Audiger, G.; Dutillly, E.; Gervais, M.; Sursin, E.; Ruggeri, C.; Duval, F.; Bouchetou, M.L.; Bombard, A.; et al. First Steps at the Cyclotron of Orléans in the Radiochemistry of Radiometals: 52Mn and 165Er. *Instruments* **2018**, *2*, 15, doi:10.3390/instruments2030015.

21. Costa, P.; Metello, L.F.; Alves, F.; Duarte Naia, M. Cyclotron Production of Unconventional Radionuclides for PET Imaging: the Example of Titanium-45 and Its Applications. *Instruments* **2018**, *2*, 8, doi:10.3390/instruments2020008.

22. Sitarz, M.; Nigron, E.; Guertin, A.; Haddad, F.; Matulewicz, T. New Cross-Sections for $^{nat}$Mo($\alpha$,x) Reactions and Medical $^{97}$Ru Production Estimations with Radionuclide Yield Calculator. *Instruments* **2019**, *3*, 7, doi:10.3390/instruments3010007.

![instruments logo]

*instruments*

MDPI

*Article*

# A Target-Temperature Monitoring System for Cyclotron Targets: Safety Device and Tool to Experimentally Validate Targetry Studies

**Sergio J. C. do Carmo [1,\*], Pedro M. de Oliveira [1] and Francisco Alves [2,3]**

[1]    ICNAS—Produção, University of Coimbra, Pólo das Ciências da Saúde, Azinhaga de Santa Comba,
       3000-548 Coimbra, Portugal; p.de.oliveira20@gmail.com
[2]    Institute for Nuclear Sciences Applied to Health (ICNAS), University of Coimbra,
       Pólo das Ciências da Saúde, Azinhaga de Santa Comba, 3000-548 Coimbra, Portugal; franciscoalves@uc.pt
[3]    Instituto Politécnico de Coimbra (IPC), Coimbra Health School, 3046-854 Coimbra, Portugal
\*    Correspondence: sergiocarmo@uc.pt; Tel.: +351-239-488-510

Received: 17 May 2018; Accepted: 19 June 2018; Published: 21 June 2018

**Abstract:** The present work describes an experimental system enabling temperature measurement in cyclotron targets' cooling water during bombardment. The developed system provides sensible and immediate response to variations of irradiation conditions during bombardment and enables quantification of the temperature rise in the cooling water due to beam interaction with the irradiated target and with its collimator. Such a system finds application either as a monitoring safety device to instantaneously detect and register abnormal alterations in target conditions to anticipate thermal-related incidents and as a tool to experimentally validate cyclotron targetry optimization studies and thermal simulations.

**Keywords:** cyclotron; targetry; thermal study

## 1. Introduction

As the production yield of radionuclides in cyclotrons depends on the interaction of the beam with the target material, a large amount of studies have been conducted over the last decades to study beam/target interfaces [1–3], regardless of gas, liquid, or solid phase of target material. These studies range from beam characterization and target design to thermal modelling; commonly aiming at improving and optimizing the production processes. A shared concern consists of maintaining the target temperature relatively low despite the considerable temperature rise generated by the beam interaction to maximize the possible target current and in order to guarantee that the impact of the impinging particles have no repercussion on the radionuclide production process and/or damage the target. For instance, several authors reported distinct problems arising from the very high temperature increment within the target during irradiation, such as melting or even evaporation of the target material in solid targets (in most cases, cost-prohibitive enriched material), sometimes resulting in a degradation of high-vacuum or damage on the backing target components [4]. There is therefore an interest in monitoring the target temperature for safety reasons, to anticipate and avoid temperature-related incidents in order to obtain the desirable production yields. However, while temperature monitoring is well established for the solid target technique, little empirical information is available concerning liquid or gas targets [5,6]. Besides, since combined thermal and fluid simulations are widely used in designing and optimizing cyclotron targets, experimental temperature monitoring devices are also of great importance in assessing target performance and to experimentally validate the results of thermal simulation studies [7].

The present work describes an experimental system used to measure the temperature of the water used to cool liquid targets during irradiation when it thermally interacts with the irradiated liquid-target arrangement. The developed system is not only sensitive to beam variations but it also quantifies the temperature increments due to beam interaction within the target, so it can be used either for safety monitoring applications or as a tool to experimentally validate results from thermal simulation studies.

## 2. Materials and Methods

Three thermistors were implemented in the cooling system arrangement of a commercial IBA conical-shaped Nirta C8® niobium liquid target [8,9]; adapted to measure the temperature of the cooling water at the back of the niobium insert, i.e., the inlet temperature, and at the exits of the target and collimator. Two thermistors (screw type pipe probes from Semitec Corporation (Tokyo, Japan) [10]) were placed at the exits of the target and collimator internal water channels. The third transducer was installed at the back of the niobium insert, also placed inside the water cooling tunnel, and was custom-made to fit inside the existing geometry without interfering the water flow and so the target performance. This is a 1.5 mm diameter and 100 mm long rod manufactured by USSensor Corporation (Orange, CA, USA) [11], thermally insulated up its extremity since it is totally inserted in the cooling channel. The distance separating the back of the conical shaped target and the thermistor, typically 1 mm, can be adjusted and minimized down to contact by screwing the back end of the thermistor to the adapted diffuser of the refrigerated water. As a result, cooling water temperature can be measured exactly before interacting with the back of the niobium target insert and immediately after exiting the target and the collimator, thus making it possible to quantify the temperature increments in the cooling water due to the beam interaction in both the target and the collimator separately. The thermistors placed within the liquid target setup were connected outside the cyclotron vault to voltage dividers in order to determine their temperature-dependent resistance during irradiation. The measured voltages were converted to temperature measurements using an Arduino-based interface [12] registering the data thanks to a PC-based data logger terminal [13].

## 3. Results

Figure 1 presents the temperature of the refrigerated water at the back of the target insert, at the exit of the target and at the exit of the collimator arrangements as a function of time in a typical irradiation of enriched $^{18}$O-water for the production of $^{18}$F. It also shows quantification of the rise in temperature from the target and the collimator during bombardment. Figure 1 confirms that the three measured temperatures are identical whenever there is no beam on the target and that the temperature of the inlet cooled water at the back of the target insert is as expected always similar to the temperature of the refrigerated water supplied to the cyclotron vault. Figure 1 also shows that although the temperature of the refrigerated water only stabilizes after a long period (due to slow thermal stabilization in the different vast sub-systems cyclotron; such as coils of the magnetic field, radiofrequency structure, etc.), rising temperature occurs and stabilizes instantaneously in the target and the collimator, as already reported by Steinbach et al. [14]. Moreover, Figure 1 indicates that the temperature increments seem to be almost independent of the refrigerated water temperature since these are nearly constant during the irradiation.

In order to quantify the increment in temperature of the cooling water due to the target + collimator arrangement only, overall thermal stabilization of the cyclotron was reached before irradiating the target. The referred temperatures were registered as the several distinct sub-systems of cyclotron were switched on individually; waiting each time for thermal stabilization before switching on the next sub-system. As can be seen in Figure 2, the heat load due to target irradiation is not macroscopically relevant when compared to the overall heat exchange from the accelerator since it does not result in an additional temperature increment in the cooling water.

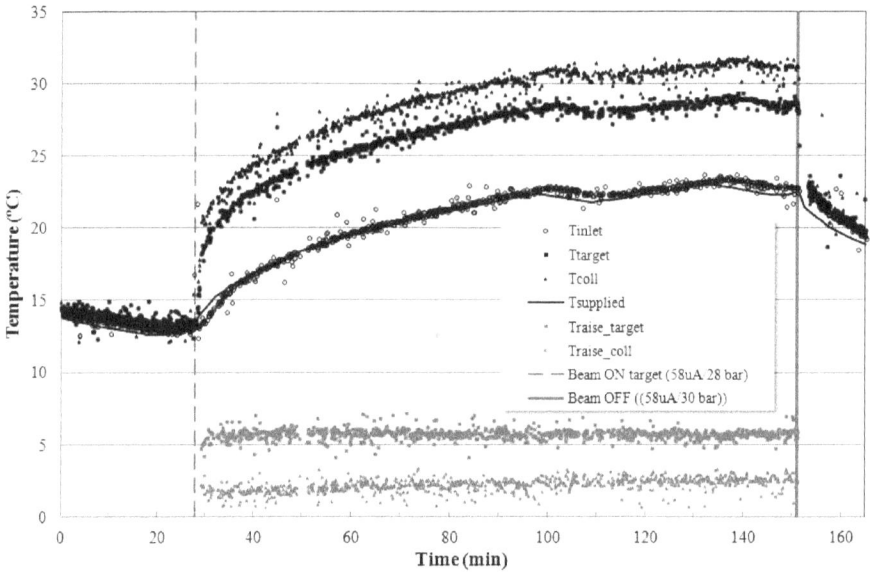

**Figure 1.** Temperature of the refrigerated water (black symbols) at the back of the target insert (Tinlet), at the exit of the target (Ttarget), and at the exit of the collimator (Tcoll) and temperature of the supplied cooling water (Tsupplied) as a function of time. The resulting temperature increases (grey symbols) inside the target and the collimator are also represented.

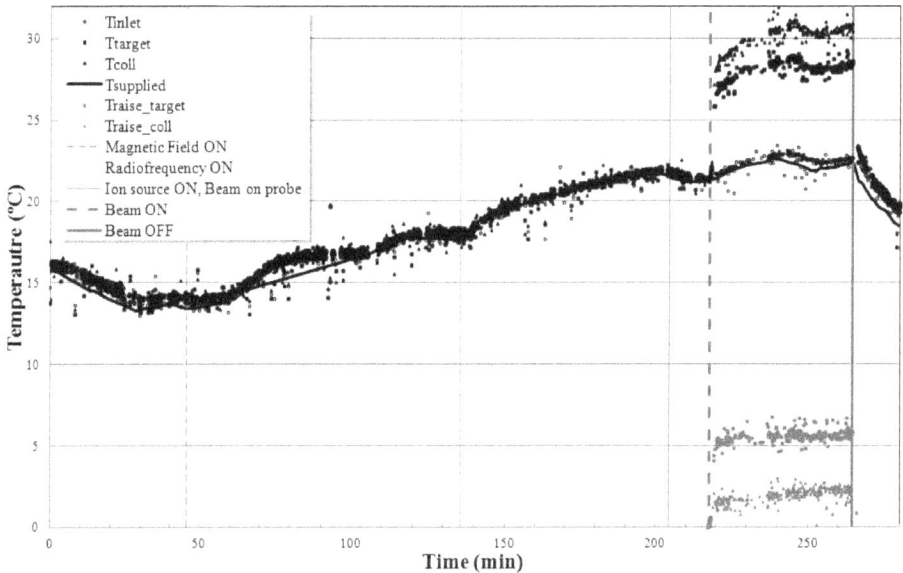

**Figure 2.** Temperature of the refrigerated water (black symbols) at the back of the target insert (Tinlet), at the exit of the target (Ttarget), at the exit of the collimator (Tcoll) and temperature increases inside the target and the collimator (grey symbols) as a function of time when each cyclotron thermally-relevant sub-system is switched on one by one after thermal stabilization is reached.

The temperature increments as a function of the beam current on target were also investigated. For that purpose, thermal stabilization of the cyclotron was previously reached and maintained by irradiating another identical target long enough to reach thermal equilibrium. Figure 3 confirms that thermal equilibrium was previously reached as the measured temperatures are stable and the only variations registered were due to, and follow, cyclic variations in the temperature of the cooling water in the primary cooling circuit, arising from operating cycles of the chiller unit. Only then was the target of interest irradiated, with no resulting global thermal change. Figure 3 shows that the temperature increments in the target and the collimator depend on the beam current and that the developed system is sensitive to such thermal changes, while the global system registers no macroscopic temperature change since thermal equilibrium was previously guaranteed. The technique is sensitive to such temperature changes and registers these alterations and allows their quantification; as illustrated in Figure 4 where the temperature increases in both the target and the collimator are represented for distinct beam currents. Such beam current dependence was expected since, at thermal equilibrium (i.e., time-independent), Newton´s law of cooling shows that the rate of convection heat transfer and the difference in temperatures between the liquid target and its surroundings are proportionally related through the equation

$$\frac{dQ}{dt} = I_{target}\Delta E = hA\left(T_{target} - T_w\right) \tag{1}$$

where $Q$ is the thermal energy (in J), $I_{target}$ is the target beam current, $\Delta E$ is the energy loss within the liquid target (i.e., 18 MeV), $h$ is the convective heat-transfer coefficient (in $W/(m^2 \cdot K)$ and assumed to be independent of the temperature), $A$ is the heat transfer surface (i.e., the inner surface of the niobium cavity (in $m^2$)), $T_{target}$ is the temperature of the liquid target water (assumed to be constant), and $T_w$ is the temperature of the refrigerated water. Figure 5 illustrates and confirms the expected proportion between the temperature increments in the cooling water and the beam current.

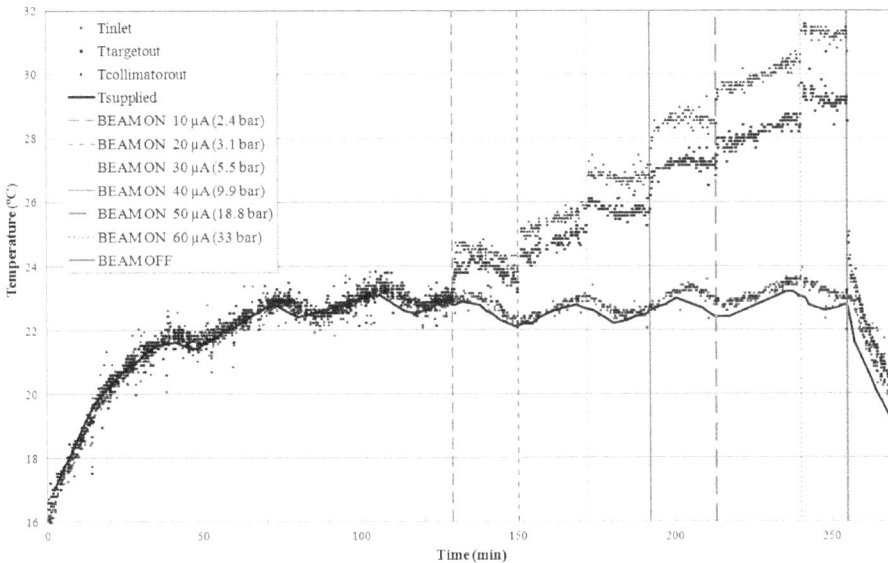

**Figure 3.** Temperature of the refrigerated water at the back of the target insert (Tinlet), exits of the target (Ttargetout), and exit of the collimator (Tcollimatorout) as a function of time after thermal stabilization and for distinct beam currents on the target.

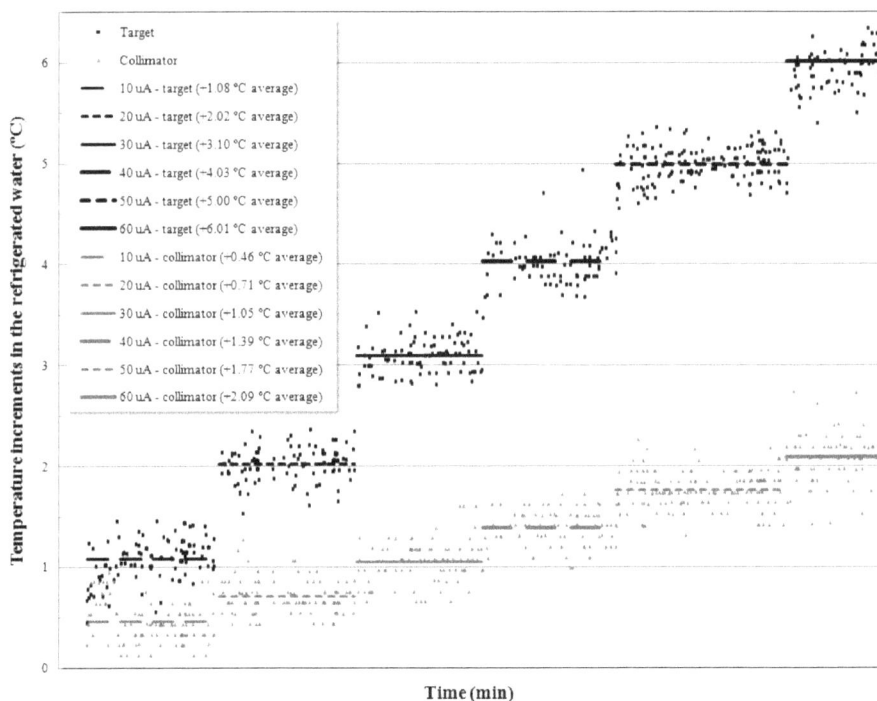

**Figure 4.** Temperature increases in the target and the collimator with respect to the temperature of the refrigerated water at the back of the target insert (Tinlet) as a function of time after thermal stabilization and for distinct beam currents on the target.

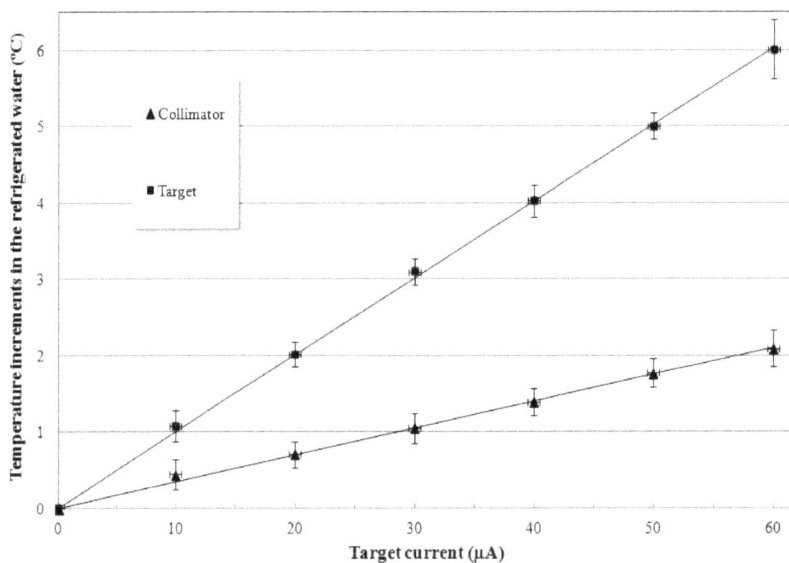

**Figure 5.** Temperature increases in the target and the collimator as a function of the beam current on target.

As stated by Equation (1), the heat transfer depends on the temperature differential; i.e., on the temperature of the refrigerated cooling water. As a result, there was also interest in registering the evolution of the referred temperatures of interest over a typical irradiation, during which the temperature of the cooling water was altered. For that purpose, the technique previously described was adopted, i.e., another identical target was irradiated long enough to reach thermal equilibrium. Figure 6 shows that even if each individual measured temperature is particularly dependent of the temperature of the refrigerated water, the temperatures increments are almost identical for three broadly distinct water cooling temperatures. The temperature increments indeed increase as the temperature of the cooling water decreases (i.e., as the temperature differential increases, as expected from Equation (1)) but Figure 6 also shows that these variations are marginal. This result confirms that heat exchange is mostly dependent on the liquid-target surface available and on its capacity to effectively remove the heat generated in the liquid from the beam interaction, as pointed out by Steyn et al. [6].

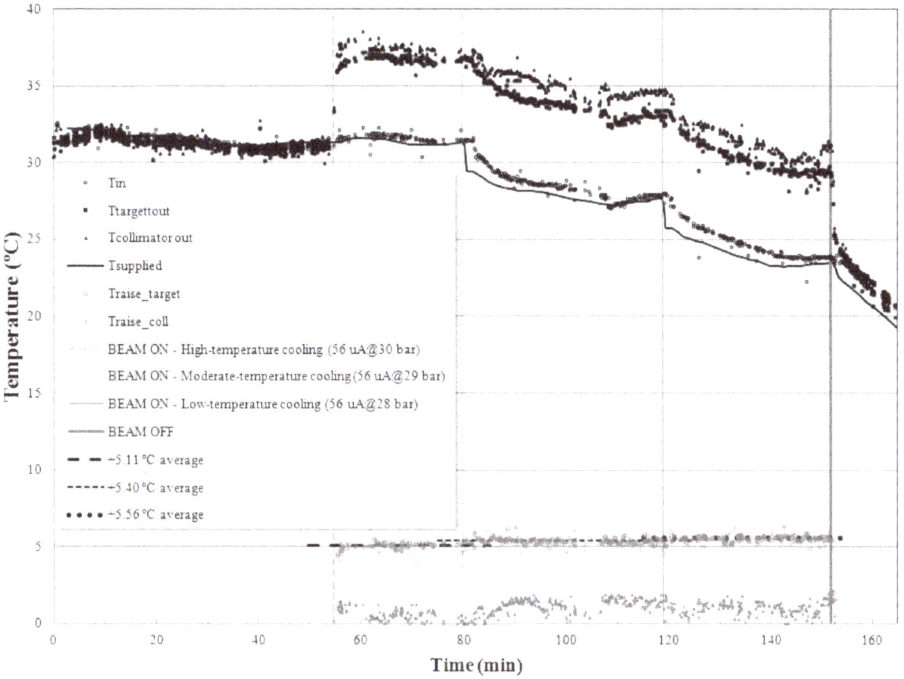

**Figure 6.** Temperatures of the refrigerated water (black symbols) at the back of the target insert (Tin), at the exits of the target (Ttargetout), at the exit of the collimator (Tcollimatorout) and temperature increases inside the target and the collimator (grey symbols) as a function of time after thermal stabilization and with the supplied refrigerated water at 3 distinct temperatures during irradiation.

## 4. Conclusions

The present work describes an experimental set-up used to measure the temperature of the cooling water in a liquid or gas target under irradiation. The system is sensitive to variations of the conditions of irradiation during bombardment, providing an immediate and quantitative response to the rise in temperature in the refrigerated water. The measurements confirmed the relationship between the temperature increments in the cooling water and the beam current. These also demonstrate that the initial temperature of the refrigerated water is not relevant to the thermal exchange, and so too the target performance, since the temperature of the liquid target inside the cavity is greatly superior. This particular result confirms that the target performance depends mostly on the liquid-target surface available and its capacity to effectively remove the heat generated. The technique developed can be used to experimentally validate cyclotron-targetry optimization studies and thermal simulations. On the other hand, it is also useful as a safety interlock to instantaneously detect and register abnormal alterations in the irradiation conditions during bombardment in order to anticipate thermal-related incidents.

**Author Contributions:** Conceptualization, S.J.C.d.C.; Methodology, S.J.C.d.C.; Software, S.J.C.d.C.; Validation, S.J.C.d.C.; Formal Analysis, P.M.d.O.; Investigation, S.J.C.d.C.; Writing-Original Draft Preparation, S.J.C.d.C.; Writing-Review & Editing, F.A.; Visualization, S.J.C.d.C.; Supervision, F.A.; Project Administration, F.A.

**Funding:** This research received no external funding.

**Acknowledgments:** The authors want to acknowledge Daniel Schiemann from SEMITEC Corporation for generously donating thermistors used in the experiments.

**Conflicts of Interest:** The authors declare no conflict of interest.

## References

1. Nortier, F.M.; Stevenson, N.R.; Gelbart, W.Z. Investigation of the thermal performance of solid targets for radioisotope production. *Nucl. Instrum. Methods* **1995**, *A355*, 236–241. [CrossRef]
2. Heselius, S.-J.; Lindblom, P.; Solin, O. Optical studies of the influence of an intense ion beam on high-pressure gas targets. *Int. J. Appl. Radiat. Isot.* **1982**, *33*, 653–659. [CrossRef]
3. Helus, F.; Uhlir, V.; Gasper, H. Contribution to cyclotron targetry: V. Measurement of the temperature in a gas target during irradiation. *J. Radioanal. Nucl. Chem.* **1996**, *210*, 233–235. [CrossRef]
4. McCarthy, D.W.; Shefer, R.E.; Klinkowstein, R.E.; Bass, L.A.; Margeneau, W.H.; Cutler, C.S.; Anderson, C.J.; Welch, M.J. Efficient production of high specific activity 64Cu using a biomedical cyclotron. *Nucl. Med. Biol.* **1997**, *24*, 35–43. [CrossRef]
5. Steyn, G.F.; Vermeulen, C. A Saturation Boiling Model for an Elongated Water Target Operating at a High Pressure during $^{18}$F Production Bombardments. Journal of Physics: Conference Series 2014. Available online: http://events.saip.org.za/getFile.py/access?contribId=329&sessionId=30&resId=1&materialId=paper&confId=14 (accessed on 9 May 2018).
6. Steyn, G.F.; Vermeulen, C. Saturation conditions in elongated single-cavity boiling water targets. In Proceedings of the 15th international Worshop on Targetry and Target Chemistry—WTTC, Prague, Czech Republic, 18–21 August 2014; pp. 149–150.
7. Gagnon, K.; Wilson, J.S.; Quarrie, S.A. Thermal modelling of a solid cyclotron target using finite element analysis: An experimental validation. In Proceedings of the 13th international Worshop on Targetry and Target Chemistry—WTTC, Roskilde, Denmark, 26–28 July 2010; p. 11.
8. IBA, Louvain-la-Neuve, Belgium. Available online: https://www.iba-radiopharmasolutions.com/ (accessed on 20 June 2018).
9. Devillet, F.; Geets, J.-M.; Ghyoot, M.; Kral, E.; Natergal, B.; Mooij, R.; Vosjan, M. Performance of IBA new conical shaped niobium [$^{18}$O] water targets. In Proceedings of the 15th international Worshop on Targetry and Target Chemistry—WTTC, Prague, Czech Republic, 18–21 August 2014; pp. 145–148.
10. SEMITEC Corporation, Japan. Available online: http://www.semitec.co.jp/ (accessed on 20 June 2018).
11. USSENSOR CORP., USA. Available online: http://www.ussensor.com/ (accessed on 20 June 2018).
12. Arduino. Available online: http://www.arduino.cc/ (accessed on 20 June 2018).
13. CoolTermWin Software Package. Available online: http://freeware.the-meiers.org/ (accessed on 20 June 2018).
14. Steinbach, J.; Guenther, K.; Loesel, E.; Grunwald, G.; Mikecz, P.; Ando, L.; Szelecsenyi, F.; Beyer, G.J. Temperature Course in small volume [$^{18}$O] water targets for [$^{18}$F] F-production. *Appl. Radiat. Isot.* **1990**, *41*, 753–756. [CrossRef]

*instruments*

MDPI

*Article*

# Measurement of the Beam Energy Distribution of a Medical Cyclotron with a Multi-Leaf Faraday Cup

**Konrad P. Nesteruk** *,†, **Luca Ramseyer, Tommaso S. Carzaniga and Saverio Braccini**

Albert Einstein Center for Fundamental Physics (AEC), Laboratory for High Energy Physics (LHEP), University of Bern, Sidlerstrasse 5, CH-3012 Bern, Switzerland; luca.ramseyer@gmx.ch (L.R.); tommaso.carzaniga@lhep.unibe.ch (T.S.C.); saverio.braccini@lhep.unibe.ch (S.B.)
* Correspondence: konrad.nesteruk@psi.ch
† Current address: Paul Scherrer Institut, Forschungsstrasse 111, 5232 Villigen PSI, Switzerland.

Received: 18 December 2018; Accepted: 1 January 2019; Published: 4 January 2019

**Abstract:** Accurate knowledge of the beam energy distribution is crucial for particle accelerators, compact medical cyclotrons for the production of radioisotopes in particular. For this purpose, a compact instrument was developed, based on a multi-leaf Faraday cup made of thin aluminum foils interleaved with plastic absorbers. The protons stopping in the aluminum foils produce a measurable current that is used to determine the range distribution of the proton beam. On the basis of the proton range distribution, the beam energy distribution is assessed by means of stopping-power Monte Carlo simulations. In this paper, we report on the design, construction, and testing of this apparatus, as well as on the first measurements performed with the IBA Cyclone 18-MeV medical cyclotron in operation at the Bern University Hospital.

**Keywords:** beam energy; ion accelerator; medical cyclotron

---

## 1. Introduction

Beam energy is a key parameter for all particle accelerators. In the case of medical cyclotrons for radioisotope production, the yield and the purity of the produced radioisotopes strongly depend on the cross-section of both the desired radioisotope and the possible impurities. This is particularly crucial in the case of the bombardment of thin targets by means of solid target stations.

Compact medical cyclotrons for the production of radioisotopes have an energy in the range of 15–25 MeV and are characterized by a considerable potential for fundamental and applied research, especially if they are equipped with a beam transfer with independent access to the beam area [1]. They are usually installed in hospitals, where space constraints severely limit this possibility. For all activities beyond the production of $^{18}$F—the main radioisotope for Positron Emission Tomography (PET)—accurate knowledge of the beam energy distribution is often crucial.

Along this line, a compact, easy-to-use, and cost-effective apparatus for the measurement of the beam energy distribution of the compact medical cyclotron has been designed, built, and tested. In this paper, we report on the first measurements obtained with this instrument performed with the Bern medical cyclotron.

## 2. Materials and Methods

The laboratory at the Bern University Hospital (Inselspital) [2] features an IBA Cyclone 18/18 high-current cyclotron and two bunkers with independent access. This accelerator is used daily for the production of $^{18}$F for PET diagnostics. The beam is brought to the second bunker by means of a 6 m-long Beam Transport Line (BTL), which is used for multi-disciplinary research activities. The BTL was used for the measurements presented in this paper.

The core of the experimental setup is a multi-leaf Faraday cup composed of aluminum foils interleaved with a non-conductive material. Protons have a range within this Faraday cup according to their energy and stop either in a foil or in the material. The number of protons stopping within one aluminum foil is assessed by measuring the current they produce by means of an electrometer. A similar, but more complex apparatus was developed to measure the energy distribution of a 68-MeV proton beam for eye melanoma hadrontherapy [3].

The multi-leaf Faraday cup we designed and constructed allows assessing the proton beam energy distribution by measuring the current collected by successive foils. The resolution of the measurement depends on the thickness of the absorber material and of the aluminum foils. The thickness and number of layers were chosen according to the outcome of a simulation performed with the SRIM [4] software, with which the transport of ions in matter can be accurately calculated. A previous measurement of the beam energy and energy spread of the Bern medical cyclotron [5] was used as input to the simulation to calculate the proton range within the multi-leaf Faraday cup. On the basis of simulations and material availability, 50 μm ± 10%-thick aluminum sheets and 25.4 μm ± 10%-thick Mylar® polyester foils were chosen. Following the SRIM calculations, the protons in the range of 16.5–20.5 MeV were able to be measured by interleaving 11 aluminum foils with 11 Mylar sheets and preceding such a stack by a 1.5 mm-thick aluminum absorber, as shown in Figure 1.

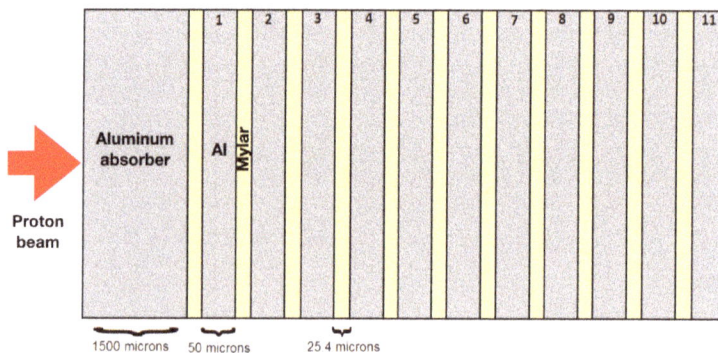

**Figure 1.** Scheme of the multi-leaf Faraday cup for the beam energy measurements of the Bern medical cyclotron.

The complete apparatus installed on the BTL is shown in Figure 2. The beam collimator, visible in the figure, was designed to shape the beam, and its surrounding disk was conceived of to reduce undesirable effects caused by single off-aperture scattered protons.

The measurements were performed by recording the mean current induced by protons in each aluminum foil. Each foil was connected to a switch by means of a coaxial cable. The switch was placed outside the BTL bunker and connected to an electrometer (Keysight B2987A) that read out the 11 channels in sequence. The fast sampling mode of the electrometer was used in order to reduce possible beam instabilities occurring during the read-out of the 11 channels to a negligible level in the order of statistical fluctuations.

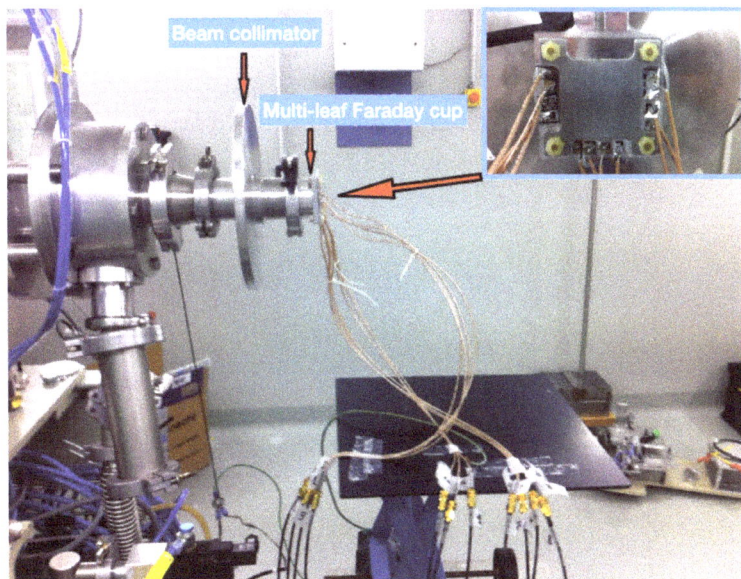

**Figure 2.** The multi-leaf Faraday cup installed on the Beam Transport Line (BTL) of the Bern cyclotron. A close-up is shown in the inset.

## 3. Results

The measurements were taken in two different conditions of cyclotron operation. The first mode of operation corresponds to the optimal isochronism condition and is determined by maximizing the transmission of the beam from the ion source to the stripper. This mode of operation is used for radioisotope production and for some experiments requiring a high beam current. For most of the research activities performed with the BTL, the cyclotron is operated in the regime of non-optimal isochronism in order to obtain stable beam currents from the nA to the pA range [6].

The first series of measurements was performed for the optimal isochronism condition. The ion source arc current was always set to a value corresponding to induced currents in the nA range in all the channels of the multi-leaf Faraday cup. In this current range, the electrometer shows the most stable readings. The measured beam energy distribution in the condition of the optimal isochronism is shown in Figure 3. The probability density is expressed in units of normalized induced current $I/I_{max}$, where $I_{max}$ corresponds to the channel with the highest measured current. The distribution presents one peak at a beam energy of $(18.68 \pm 0.13)$ MeV. The measurement was repeated for two slightly different currents set in the main coil of the cyclotron, but remaining in the vicinity of the optimal isochronism. A good agreement was found, and the peak was localized at the same beam energy value within the uncertainty. In order to compare the obtained results with the measurement performed with a different method and reported in [5], the mean energy and RMS of the distribution were also evaluated and found to be $(18.7 \pm 0.3)$ MeV and $(0.9 \pm 0.2)$ MeV, respectively. The values reported in [5] were $(18.3 \pm 0.3)$ MeV and $(0.4 \pm 0.2)$ MeV for the beam energy and RMS, respectively. The beam energy values are therefore in agreement within $1\sigma$ and the RMS values within $1.8\sigma$. The peak energy of $(18.76 \pm 0.02)$ MeV, found in [5], is in a good agreement with the results presented in this paper. It has to be noted that the two compared methods are characterized by a different resolution and were performed in slightly different conditions. The region of the optimal isochronism, corresponding to our definition based on a maximum beam transmission, spans a certain range of main coil settings, and the resulting energy distributions can vary to some extent. Another possible reason for the observed difference is the fact that the method reported in [5] gives the probabilities of finding beam energy within histogram bins, while the measurements described in this paper characterize the distribution by the evaluation of discrete points. Furthermore, the main coil of the cyclotron warms up during operation, making the operation conditions slightly different.

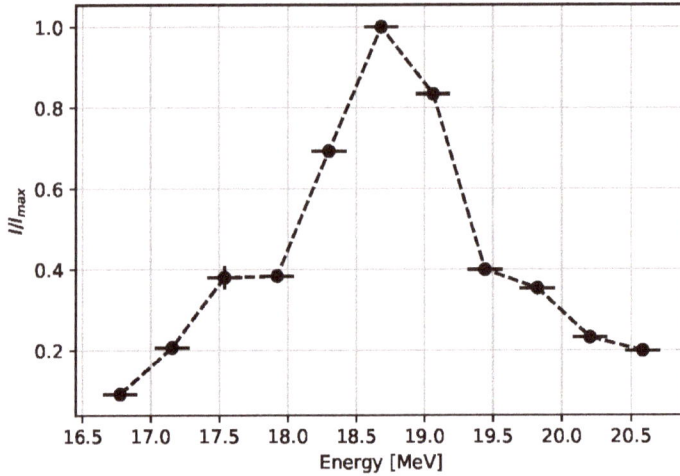

**Figure 3.** Beam energy distribution measured for the optimal isochronism condition of the Bern medical cyclotron. The dashed line is a guide to the eye.

The second series of measurements was conducted in the non-optimal isochronism regime. The main coil current was set to a value corresponding to a low transmission of the beam from the ion source to the stripper. The ion source arc current was again adjusted to provide electrometer readings in the nA range. The obtained beam energy distribution is shown in Figure 4. This time, two peaks are visible. The first one corresponds to a beam energy of $(18.68 \pm 0.13)$ MeV, while the second to an energy of $(19.82 \pm 0.13)$ MeV. Also for the non-optimal isochronism condition, the distribution was measured for two different values of the main coil currents, and a good agreement was found.

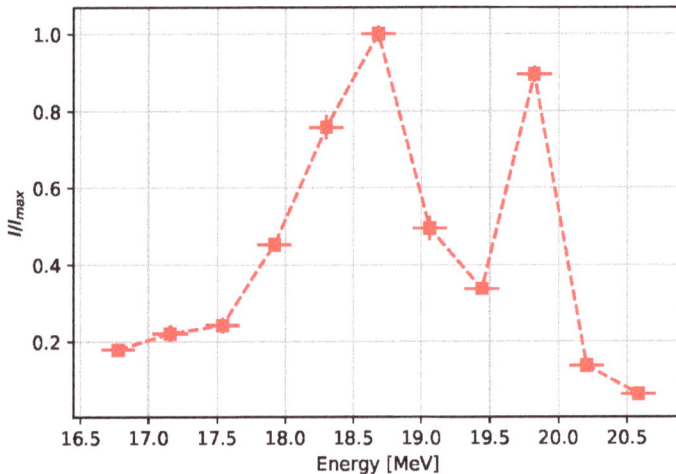

**Figure 4.** Beam energy distribution measured for the non-optimal isochronism condition of the Bern medical cyclotron. The dashed line is a guide to the eye.

The results obtained in both series for the optimal and non-optimal isochronism conditions present a peak at the beam energy of $(18.68 \pm 0.13)$ MeV. An additional peak of the distribution corresponding to the non-optimal isochronism condition can be explained by the fact that the beam size and shape on the stripper are optimized for the isochronous condition such that the majority of the accelerated beam is stripped in one turn. Operation of the machine far from the optimum can lead to an effect in which a significant number of ions are stripped after making extra turns in the cyclotron, which is manifested by the second peak in the energy distribution corresponding to a 1.1-MeV higher beam energy. This energy difference is equivalent to about 18 extra turns. Since the radial separation between particle orbits at the energies over 18 MeV is very small (below 1 mm) and there is no optimal horizontal focusing provided in this mode of operation, a large beam spot at the stripper location is likely, which leads to those extra turns. Steering effects in the quadrupoles along the BTL might also play a role in making the transport efficiency energy dependent.

## 4. Conclusions

In this paper, we presented a simple and efficient method to measure the energy distribution of an ion beam extracted from a medical cyclotron based on a multi-leaf Faraday cup. On the basis of simulations, a specific apparatus was designed, built, and tested at the Bern medical cyclotron laboratory to assess the energy distribution at the end of the Beam Transport Line (BTL) connected to an IBA Cyclone 18-MeV cyclotron. Measurements were performed for two modes of cyclotron operation. The first one corresponds to the optimal isochronism condition and applies to radioisotope production and to experiments requiring a high beam current. The second mode is used for most of the research activities with currents in the nA and pA range and corresponds to the regime of non-optimal isochronism. As expected, the energy distributions of the extracted proton beam for the two modes of operation were found to be different. Although for both modes, a peak at an energy of $(18.68 \pm 0.13)$ MeV is observed, the distribution corresponding to the non-optimal isochronism shows an additional peak at $(19.82 \pm 0.13)$ MeV. This effect is mostly due to the beam extraction by stripping. For the optimal isochronism, the results were compared to measurements previously performed, obtained with a different method, and a good agreement was found. The obtained results are higher with respect to the nominal beam energy of 18 MeV. This is consistent with the fact that for the BTL, the stripper is located at a radius ~5 mm larger with respect to the nominal one for beam transport optimization purposes, making the beam energy higher [2].

The proposed method can be applied to any cyclotron to assess the beam energy precisely for scientific and industrial purposes, in particular to optimize radioisotope production with solid target stations.

**Author Contributions:** S.B. and K.P.N. conceived of and designed the experiment; K.P.N., T.S.C., and L.R. performed the experiment; K.P.N. and L.R. analyzed the data; K.P.N. and S.B. wrote and revised the paper; S.B. coordinated the project.

**Acknowledgments:** We acknowledge contributions from the LHEP engineering and technical staff. One of the authors (T.S.C.) was financially supported by the Swiss National Science Foundation (Grant CR23I2_156852).

**Conflicts of Interest:** The authors declare that there is no conflict of interest.

## References

1. Braccini, S. Compact medical cyclotrons and their use for radioisotope production and multi-disciplinary research. In Proceedings of the Cyclotrons2016 TUD01, Zurich, Switzerland, 11–16 September 2016; p. 229.
2. Braccini, S. The new Bern PET cyclotron, its research beam line, and the development of an innovative beam monitor detector. *AIP Conf. Proc.* **2013**, *1525*, 144. [CrossRef]
3. Seidel, S.; Bundesmann, J.; Damerow, T.; Denker, A.; Kunert, C.; Weber, A. A multi-leaf Faraday cup especially for proton therapy of ocular tumors. In Proceedings of the Cyclotrons2016, Zurich, Switzerland, 11–16 September 2016; Volume MOE02, pp. 118–120.

4.    Ziegler, J.F.; Ziegler, M.D.; Biersack, J.P. SRIM—The stopping and range of ions in matter. *Nucl. Instrum. Methods* **2010**, *268*, 1818–1823. [CrossRef]

5.    Nesteruk, K.P.; Auger, M.; Braccini, S.; Carzaniga, T.S.; Ereditato, A.; Scampoli, P. A system for online beam emittance measurements and proton beam characterization. *J. Instrum.* **2018**, *13*, P01011. [CrossRef]

6.    Auger, M.; Braccini, S.; Ereditato, A.; Nesteruk, K.P.; Scampoli, P. Low current performance of the Bern medical cyclotron down to the pA range. *Meas. Sci. Technol.* **2015**, *26*, 094006. [CrossRef]

instruments

MDPI

Article

# Simple, Immediate and Calibration-Free Cyclotron Proton Beam Energy Determination Using Commercial Targets

Sergio J.C. do Carmo [1,*], Pedro M. de Oliveira [1] and Francisco Alves [2,3]

[1]  ICNAS—Produção, University of Coimbra, Pólo das Ciências da Saúde, Azinhaga de Santa Comba, 3000-548 Coimbra, Portugal; p.de.oliveira20@gmail.com
[2]  Institute for Nuclear Sciences Applied to Health (ICNAS), University of Coimbra, Pólo das Ciências da Saúde, Azinhaga de Santa Comba, 3000-548 Coimbra, Portugal; franciscoalves@uc.pt
[3]  Instituto Politécnico de Coimbra (IPC), Coimbra Health School, 3046-854 Coimbra, Portugal
*  Correspondence: sergiocarmo@uc.pt

Received: 5 January 2019; Accepted: 28 February 2019; Published: 5 March 2019

**Abstract:** This work presents a simple method for determining the energy of the proton beam in biomedical cyclotrons, using no additional experimental set-up and only materials from radioisotope routine productions that are therefore available on-site. The developed method requires neither absolute efficiency calibration nor beam current measurements, thus avoiding two major sources of uncertainty. Two stacks composed of natural titanium thin foils, separated by an energy degrader of niobium, were mounted in a commercial target and irradiated. The resulting activities of $^{48}$V were assessed by a HPGe spectrometer.

**Keywords:** cyclotron; stack-foil; monitor reaction

## 1. Introduction

Accurate knowledge of the incident beam energy is fundamental for the production of medical radioisotopes; either to optimize production yields or to prevent the co-production of undesired radioimpurities. However, since biomedical cyclotrons are not properly equipped for energy measurements, several indirect measurement techniques have been studied and reported over the years. These methods are commonly based on activity measurements of radioisotopes produced via well-documented and recommended monitor-reactions, requiring both beam current and activity measurements. In order to avoid the difficulty arising from beam current measurements, several authors measured activity ratios from distinct radioisotopes produced simultaneously, either in a single monitor foil or in a stack of target foils, and compared the results to calculated ratios from the recommended cross-sections in the published data [1–4]. However, as these methods rely on the determination of absolute activities for two distinct radioisotopes through γ-spectrometry, the results are highly influenced by uncertainties in the absolute efficiency calibration. In order to surmount this drawback, Burrage et al. [5] suggested the determination of activity ratios for a single radioisotope. Because a unique photopeak is characterized, the technique presents the advantage of requiring neither direct beam-current measurement nor problematic γ-spectroscopy absolute efficiency calibration. Burrage et al. [5] implemented the method by characterizing the production of $^{65}$Zn in a stack of copper foils; with the technique later improved by Asad et al. [6]. Gagnon et al. [7] also made use of this technique with only two foils of copper separated by an energy degrader of adequate thickness. These latter methods make use of the fact that each monitor excitation function presents a unique shape so that the activity profile vs. depth, i.e., vs. foil, is specific of the monitor reaction but also dependent on the incident energy. Since the cross-sections and the stopping-power can be estimated for each foil, it is possible to compute and predict the activity ratios between foils for several distinct initial energies and to determine which computed energy best fits the experimental data. As previously pointed out by Gagnon et al. [7], uncertainties in the cross-sections do not influence the ratio of activities between foils because it is the profile of the excitation function that determines the activity ratio.

The present work describes an improved method based on the stacked-foils technique, in an experimental configuration similar to the useful work of Burrage et al. [5] and Asad et al. [6]. The use of several foils in a stack configuration enables to experimentally determine several activity ratios instead of elaborating the result of an entire experiment on a single ratio as is the case in the work developed by Gagnon et al. [7]. The developed method was used to determine the proton beam energy in a IBA Cyclone 18/9 cyclotron [8], accelerating protons to 18 MeV. The beam energy was measured at several exit ports and for distinct high-voltages for the radio-frequency. Both the materials used and the experimental arrangement were chosen so that the technique can be immediately performed, exploiting only materials from routine productions, therefore available on-site and later reusable, and without any set-up amendment.

## 2. Materials and Methods

For proton energies up to about 20 MeV the well established $^{nat}Cu(p,x)^{63,65}Zn$ reactions represented in Figure 1 have been commonly used as monitor reactions [9]. Figure 1 also shows that the shapes of these monitor reactions only show significant variations in the 5–10 and 14–20 MeV energy ranges. As a result, when a stack is used in the overall energy range as in the work of Asad et al. [6], part of the experiment contains little information because several foils present similar activities. In the present work, two stacks were exclusively distributed in the two energy ranges showing significant variations, whereas a beam degrader was used around the inadequate 10–14 MeV energy range, following the strategy adopted in the work reported by Gagnon et al. [7]. Moreover, in order to achieve more significant activity differences between foils so that the "method signal" is more significant, we ought to exploit monitor reactions providing more pronounced variations in the energy ranges of interest. As illustrated in Figure 1, where the absolute of the derivates of the monitor reactions considered are also represented in the energy ranges of interest, the $^{nat}Ti(p,x)^{48}V$ monitor reaction presents more accentuate absolute variations, in particular in the low-energy region, and was thus chosen as monitor reaction for the present study. This advantageous characteristic is combined with the practicality arising from the fact that 12.5 μm titanium foils are commonly used as vacuum windows in commercial liquid target arrangements in IBA cyclotrons; so that these foils are not only immediately and easily available on site but can also be reused in routine production afterwards. Besides, $^{48}V$ presents an adequate long half-life of 16 days enabling measurements several days after bombardment. As the stacks were meant for narrower energy ranges, thinner and/or fewer foils are more suitable. Thinner foils present the advantage of providing a smaller and thus more defined energy loss in each foil; an improvement also due to the choice of titanium instead of copper because of its smaller atomic number. Such improved characteristic over the 25–100 μm thick foils of copper used in previous methods is also achieved by using the 12.5 μm thick titanium foils.

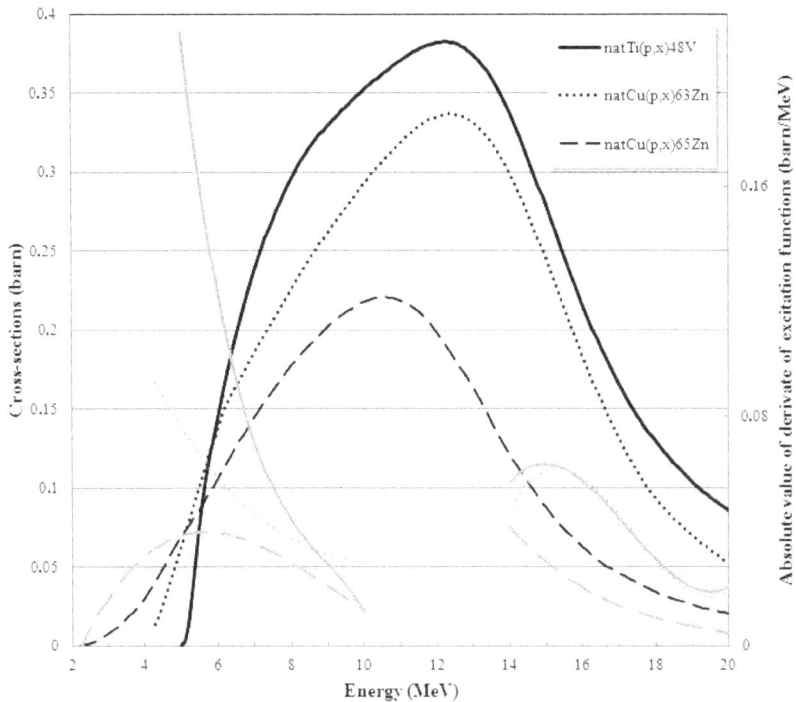

**Figure 1.** Excitation functions of the monitor reactions of interest (black lines) and absolute values of their derivates in the energy ranges of interest (grey lines).

The thickness of each of the 99.6% pure and 12.5 μm thick titanium circular foils used was determined by weight determination. Thickness differences between foils are of no concern as the expected activity of each foil are determined, taking into account the experimentally determined thicknesses. The degrader used in the present work was also made from material available from routine productions; namely two 250 μm thick niobium disks commonly used as target windows for the production of radiometals in liquid targets [10–12]. The stack foil arrangement consists of two stacks of ten titanium foils each separated by the niobium degrader. The total thickness, i.e., the number of titanium foils in each stack, was calculated so that the exit beam energy remains slightly higher than the 5.0 MeV threshold of the excitation function. Such consideration also enables one to avoid the larger uncertainties in the proton stopping-power, and therefore in energy, at lower energies. The stack arrangement was mounted in a standard liquid target system with no modifications, precisely at the place where a 12.5 μm titanium foil is usually placed as vacuum window. The rest of the target assembly remained as for routine productions, with the liquid target filled with ultra-pure water. Such an arrangement means that the experimental set-up can be immediately used in any cyclotron target and at any exit port with no additional material and/or modification required; while also benefiting from the continuous helium cooling flux available at the end side of the stack. Irradiations of the liquid target containing the stack were performed at 1 μA and during 5–10 min so that the foil activities remain inferior to about 100 kBq at End-Of-Bombardment (EOB). The first foil of the second stack crossed, i.e., the foil just after the Nb energy degrader, is the foil expected to present the higher activity as illustrated in Figure 2. Such maximum activity, and thus the maximum irradiation time, was determined so that the foil activities could be determined immediately after proper cooling time taking into account the particular geometry of the HPGe set-up used. The irradiated stack was allowed to cool down for at least one day to minimize the presence of numerous undesired radionuclide in the spectra. Activity measurements were carried out using a high purity germanium (HPGe) spectrometer (model GEM30P4-70 from Canberra) with a dead-time inferior to 4%. The relevant 944.1, 983.5 and 1312.1 keV characteristic $\gamma$-lines can all be used to identify and quantify $^{48}$V. Although the 983.5 and 1312.1 keV $\gamma$-lines are also characteristic of $^{48}$Sc, the $^{nat}$Ti(p,x)$^{48}$Sc reaction is relevant only for proton energies higher than 18 MeV [13]. Even if unnecessary, as only relative activities were necessary, the HPGe spectrometer was calibrated in absolute efficiency. Activity measurements can alternatively be performed using a dose calibrator, as considered by Gagnon et al. [7].

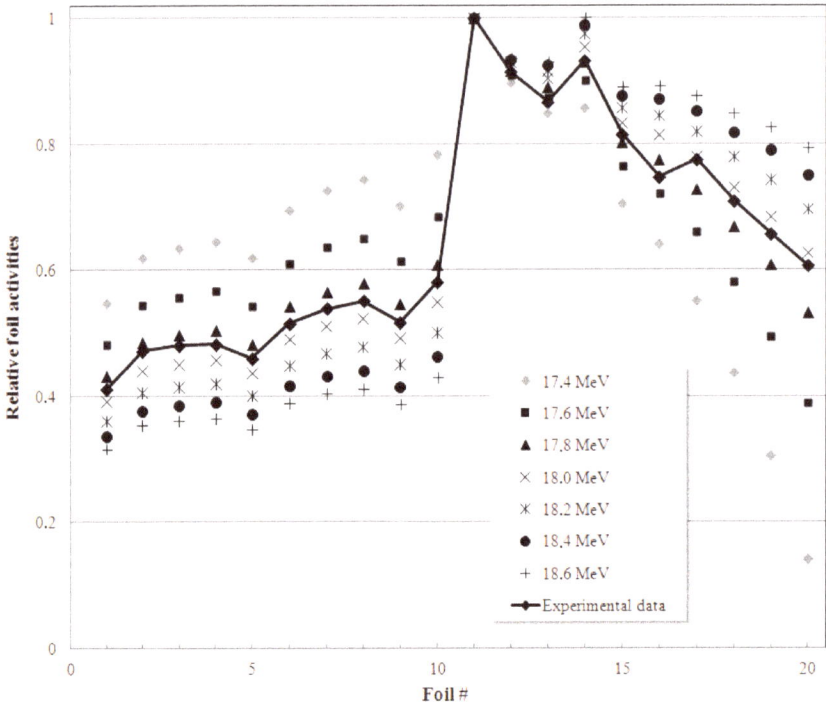

**Figure 2.** Calculated relative activities in the 20 foils of the stack, for several impinging energies for a typical experiment (symbols without line) and respective experimental data (full line). The beam crosses the stacks travelling from foil #1 to foil #20. The experimentally determined beam energy in this particular case was 17.900 MeV.

Stopping power for protons in $^{nat}$Ti and $^{nat}$Nb were obtained using the SRIM software [14] and used to determine continuous polynomial function fits. The IAEA recommended cross-sections for the $^{nat}$Ti(p,x)$^{48}$V reaction [9] were also fitted to two distinct continuous polynomial functions, for the high 14–20 and low 5–10 MeV energy ranges. These continuous functions enable the computation of the activities of each foil taking into account the experimentally deduced thicknesses, using small increments of 0.5 μm. The procedure was repeated for several initial impinging energies in the 17.4–18.6 MeV energy range as the nominal energy is 18 MeV. Although the beam current and the irradiation time considered in the calculations match the ones used in typical irradiations; these are not important because the calculated activities are only used to determine relative intensities. Figure 2 presents typical calculated relative activity profiles, determined for different impinging energies, together with an activity profile determined experimentally. Figure 2 illustrates the fact that the activity profile vs. depth depends on the initial beam energy.

In order to determine the computed activity profile that best matches the experimental data, the experimentally obtained relative activities were compared to the computed relative activities by calculating their residual for each foil. The residuals were then squared and summed for the several initial energies considered in the calculations to be used in an iterative least-squares minimization technique to adjust the experimental data, as described in Asad et al. [6]. Figure 3 presents a typical example of the sum of squared residuals as a function of the initial energy considered in the computation; illustrating the fact that there is a matching computed initial energy providing minimized squared residuals.

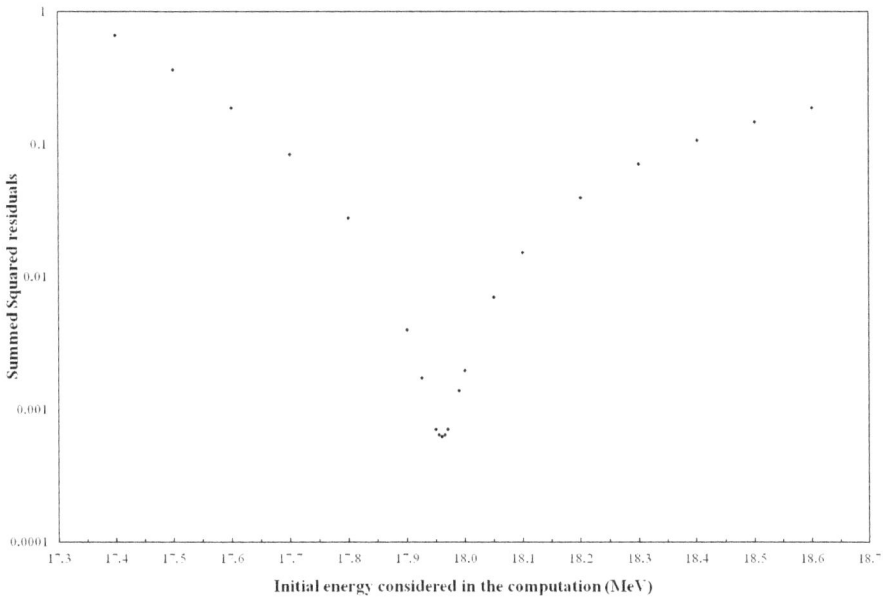

**Figure 3.** Summed squared residuals between the experimental data and calculated activities as a function of the initial energy considered in the calculations. The experimentally determined beam energy in this particular case was 17.960 MeV.

The procedure described was repeated for distinct high-voltages in the radiofrequency system and using a same exit port and also at different exit ports while maintaining constant 32 kV in the radio-frequency system. Table 1 shows that the radio-frequency voltage affects the beam energy on target; an expected result as the voltage alters the condition of acceleration. Two of the distinct exit ports used were also intentionally chosen as diametrically opposed in order to evaluate the influence of the last acceleration stage between exit ports. The experimentally obtained beam energy for these two diametrically opposed exits were $17.900 \pm 11.6\%$ MeV and $17.960 \pm 11.6\%$ MeV; a result in agreement with the fact that the maximum energy gain is 32 kV between the two exits considered.

**Table 1.** Experimentally determined proton beam energy for a fixed exit port and at distinct high voltages for the radio-frequency.

| High-voltage used | 28 kV | 32 kV | 36 kV |
|---|---|---|---|
| Beam energy ($\pm 11.6\%$) (MeV) | 17.905 | 17.960 | 17.930 |

The technique provides precise results as the matching computed energy is experimentally determined considering the combined activities in 20 foils. As illustrated in Figure 3, slightly different computed energies indeed result in distinct sums of residuals. Such consistency is in agreement with the fact that, given a set of computed foil activity, the result is only determined by the experimentally measured foil activities for which the counting statistical error was determined to be not superior to 1% (as the efficiency calibration is not relevant because only activity ratios are considered, the accuracy in the determination of the foil activities is only governed by the statistic in counting events in the Gaussian peak).

One has to point out that the results are obtained bearing in mind that the incident beam energy is not fully monoenergetic. Indeed, as the particles are not all exactly centered with the geometrical center of the cyclotron during the accelerating revolutions, the beam shape alters in between accelerations and results in a certain beam width with consequent energy spread. Additionally, the stripping process at the end of acceleration of the H$^-$ ions also slightly increases the final energy spread. The present method only determines average incident beam energies.

Even if identical experiments lead to precise consistency in the computed results, this characteristic is unfortunately not related to the accuracy of the technique. As several aspects, external to the computing technique, are inevitably involved in the method, a discussion concerning the accuracy of the method must arise. For instance, the beam also suffers beam straggling when crossing the stack, leading to an energy spread influencing the results, which was previously evaluated to be of 2.7% [6] In addition, besides the referred counting statistical error, the calculations of the foil activities are based on the knowledge of the thickness of each foil and inevitably rely on recommended excitation functions and stopping powers from databases. As the errors of these parameters were estimated to be of 1, 10 and 5%, respectively, the uncertainty of the technique was estimated to be of 11.6%; a typical limitation for energy determination techniques based on stacks.

## 3. Conclusions

The present work describes a technique for indirect measurement of proton beam energy. The method needs no beam current measurement nor absolute efficiency calibration. The technique was projected to exploit only materials available from routine production and enabling their reuse while simultaneously requiring no additional set-up, as a commercial target is sufficient to establish the required experimental arrangement. As a result, this experiment for beam energy measurement can be performed immediately in any biomedical cyclotron with ease.

**Author Contributions:** Conceptualization, S.J.C.d.C.; methodology, S.J.C.d.C.; validation, S.J.C.d.C.; formal analysis, S.J.C.d.C.; investigation, S.J.C.d.C.; data curation, P.O.; writing—original draft preparation, S.J.C.d.C.; writing—review and editing, F.A.; visualization, S.J.C.d.C.; supervision, F.A.

**Funding:** This research received no external funding.

**Conflicts of Interest:** The authors declare no conflict of interest.

## References

1. Kopecky, P. Proton beam monitoring via the Cu(p, x) $^{58}$Co, $^{63}$Cu(p, 2n) $^{62}$Zn and $^{65}$Cu(p, n) $^{65}$Zn reactions in copper. *Int. J. Appl. Radiat. Isot.* **1985**, *36*, 657–661. [CrossRef]
2. Kim, J.H.; Park, H.; Kim, S.; Lee, J.S.; Chun, K.S. Proton beam energy measurement with the stacked Cu foil technique for medical radioisotope production. *J. Korean Phys. Soc.* **2006**, *48*, 755–758.
3. Avila-Rodriguez, M.A.; Wilson, J.S.; Schueller, M.J.; McQuarrie, S.A. Measurement of the activation cross section for the (p,xn) reactions in niobium with potential applications as monitor reactions. *Nucl. Instrum. Methods Phys. Res. B* **2008**, *266*, 3353–3358. [CrossRef]
4. Avila_Rodriguez, M.A.; Rajander, J.; Lill, J.-O.; Gagnon, K.; Schlesinger, J.; Wilson, J.S.; McQuarrie, S.A.; Solin, O. Proton energy determination using activated yttrium foils and ionization chambers for activity assay. *Nucl. Instrum. Methods Phys. Res. B* **2008**, *267*, 1867–1872. [CrossRef]
5. Burrage, J.W.; Asad, A.H.; Fox, R.A.; Price, R.I.; Campbell, A.M.; Siddiqui, S. A simple method to measure proton beam energy in a standard medical cyclotron. *Aust. Phys. Eng. Sci. Med.* **2009**, *32*, 92–97. [CrossRef]

6. Asad, A.H.; Chan, S.; Cryer, D.; Burrage, J.W.; Siddiqui, S.A.; Price, R.I. A new, simple and precise method for measuring cyclotron proton beam energies using the activity vs. depth profile of zinc-65 in a thick target of stacked copper foils. *Appl. Radiat. Isot.* **2015**, *105*, 20–25. [CrossRef] [PubMed]
7. Gagnon, K.; Jensen, M.; Thisgaard, H.; Publicover, J.; Lapi, S.; McQuarrie, S.A.; Ruth, T.J. A new and simple calibration-independent method for measuring the beam energy of a cyclotron. *Appl. Radiat. Isot.* **2011**, *69*, 247–253. [CrossRef] [PubMed]
8. Ion Beam Applications, Chemin du Cyclotron, 1348 Louvain-La-Neuve, Belgium. Available online: https://www.iba-radiopharmasolutions.com/ (accessed on 5 March 2019).
9. Monitor Reactions 2017. Available online: https://www-nds.iaea.org/medical/monitor_reactions.html (accessed on 5 January 2019).
10. Alves, F.; Alves, V.H.; Neves, A.C.B.; do Carmo, S.J.C.; Nactergal, B.; Hellas, V.; Kral, E.; Gonçalves-Gameiro, C.; Abrunhosa, A.J. Cyclotron production of Ga-68 for human use from liquid targets: From theory to practice. *AIP Conf. Proc.* **2017**, *1845*, 020001.
11. Alves, F.; Alves, V.H.P.; do Carmo, S.J.C.; Neves, A.C.B.; Silva, M.; Abrunhosa, A.J. Production of copper-64 and gallium-68 with a medical cyclotron using liquid targets. *Mod. Phys. Lett. A* **2017**, *32*, 1740013. [CrossRef]
12. do Carmo, S.J.C.; Alves, V.H.P.; Alves, F.; Abrunhosa, A.J. Fast and cost-effective cyclotron production of $^{61}$Cu using a $^{nat}$Zn liquid target: An opportunity for radiopharmaceutical production and R&D. *Dalton Trans.* **2017**, *46*, 14556–14560. [PubMed]
13. Khandaker, M.U.; Kim, K.; Lee, M.W.; Kim, K.S.; Cho, Y.S.; Lee, Y.O. Investigations of the $^{nat}$Ti(p,x)$^{43,44m,44g,46,47,48}$Sc,$^{48}$V nuclear processes up to 40 MeV. *Appl. Radiat. Isot.* **2009**, *67*, 1348–1354. [CrossRef] [PubMed]
14. The Stopping Power and Range of Ions in Matter (SRIM Code, Version 2013). Available online: http://www.srim.org (accessed on 5 January 2019).

*instruments*

MDPI

*Article*

# New Ion Source Filament for Prolonged Ion Source Operation on A Medical Cyclotron

Dave Prevost [1], Keerthi Jayamanna [1], Linda Graham [1], Sam Varah [2] and Cornelia Hoehr [2,*]

[1] Accelerator Division, TRIUMF, Vancouver, BC V6T 2A3, Canada; dprevost@triumf.ca (D.P.); keerthi@triumf.ca (K.J.); lindag@triumf.ca (L.G.)

[2] Life Sciences Division, TRIUMF, Vancouver, BC V6T 2A3, Canada; svarah@triumf.ca

* Correspondence: choehr@triumf.ca; Tel.: +1-604-222-1047

Received: 7 December 2018; Accepted: 14 January 2019; Published: 16 January 2019

**Abstract:** Cyclotrons are an important tool for accelerator sciences including the production of medical isotopes for imaging and therapy. For their successful and cost-efficient operation, the planned and unplanned down time of the cyclotron needs to be kept at a minimum without compromising reliability. One of the often required maintenance activities is the replacement of the filament in the ion source. Here, we are reporting on a new ion source filament tested on a medical cyclotron and its prolonging effect on the ion source operation.

**Keywords:** ion source; filament; cyclotron; medical isotopes

## 1. Introduction

Since their introduction in the 1930's, cyclotrons play an important part in the production of radionuclides in a variety of applications, from basic physics, to agriculture and medicine [1]. At TRUMF, several cyclotrons and beam lines are dedicated to the production of medical isotopes for imaging and therapy [2]. The Life Sciences division employs the TR13 cyclotron [3,4], a 13 MeV negative-hydrogen machine, to produce most of their isotopes in gaseous, liquids and solid targets (e.g. $^{11}$C, $^{13}$N, $^{18}$F, $^{44}$Sc, $^{52}$Mn, $^{55}$Co, $^{61/64}$Cu, $^{68}$Ga, $^{86}$Y, $^{89}$Zr, $^{94m}$Tc, $^{117/118/119}$Sb, $^{192}$Ir, $^{203}$Pb) [5].

One of the main components of a cyclotron is the ion source where the projectile beam is created. While this can be protons, negative hydrogen ions, or deuterons, negative hydrogen ions are favoured [6] as the beam can be extracted from the cyclotron via a stripping foil in the acceleration plane. The ions pass through a carbon stripper foil, changing their charge from −1 to +1 and therefore changing their trajectory in the magnetic field of the cyclotron [7] and leading to beam extraction onto a target. The TR13 cyclotron operates with an external multi-cusp negative ion source. This causes less activation and consequently lower personnel dose and radioactive waste during maintenance and decommissioning of the cyclotron than with an internal ion source [8]. General information about ion sources can be found in [9].

To create negative hydrogen ions, ultra-high purity hydrogen gas is flowing into the ion source chamber, where the ions are created, which are then accelerated into the plane of the cyclotron. To achieve this, a tantalum filament is heated via a high current to create electrons. A bias voltage is applied between the filament and the chamber wall to control the ionization. These electrons form a plasma, which is held in place by rare-earth magnets placed in several rings in the chamber (see Figure 1). A fast-feedback loop constantly monitors in real-time the value of the arc current between the filament and the chamber of the wall, and adjusts the filament current to maintain the arc current at a constant pre-set level. The H$^-$ ions are formed in the plasma. Several mechanisms are competing for the H$^-$ production and destruction (ν vibrational sate) [7]:

H⁻ Production:

| | | |
|---|---|---|
| $H_2^*$ ($v \geq 4$) + e (~0.5 eV) → H⁻ + H | Dissociative electron attachment, | $\sigma = \sim 1.6 \times 10^{-16}$ cm$^2$ |
| $H_2$ + e (~3.7 eV) → H⁻ + H* | Dissociative electron attachment, | $\sigma = \sim 1.6 \times 10^{-21}$ cm$^2$ |
| $H_2$ ($v = 0$) + e (~38 eV) H⁻ + H⁺ + e | Polar dissociation, | $\sigma = \sim 1.6 \times 10^{-20}$ cm$^2$ |

H⁻ Destruction:

| | | |
|---|---|---|
| H⁻ + $H_2$ → $H_2$ + H | Charge transfer, | $\sigma = 2.5 \times 10^{-13}$ cm$^2$ |
| H⁻ + e ($\geq$15 eV) → H + 2e | Collisional detachment, | $\sigma = 4 \times 10^{-15}$ cm$^2$ |
| H⁻ + $H_2$ → $H_2$ + H | Collisional detachment, | $\sigma = 1 \times 10^{-15}$ cm$^2$ |

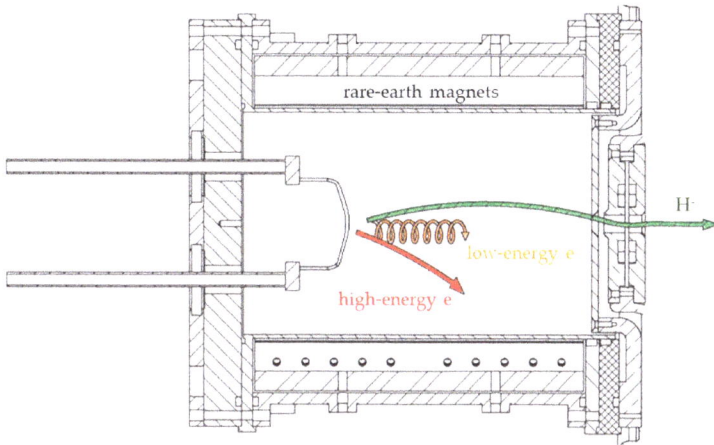

**Figure 1.** Sketch of the TR13 ion source.

To successfully produce H⁻, low and higher energy electrons from the ion source are therefore separated by the magnetic field arrangement, see [7].

The filament design in the ion source of the TR13 cyclotron at TRIUMF has been in use for many years and is made of tantalum wire shaped like a half circle (Filament I), see Figure 2a. The filament is mounted in duplicate on the removable end plate of the ion source for easy access. Over time, the tantalum wire of the filament loses material and wears thin, see Figure 2b.

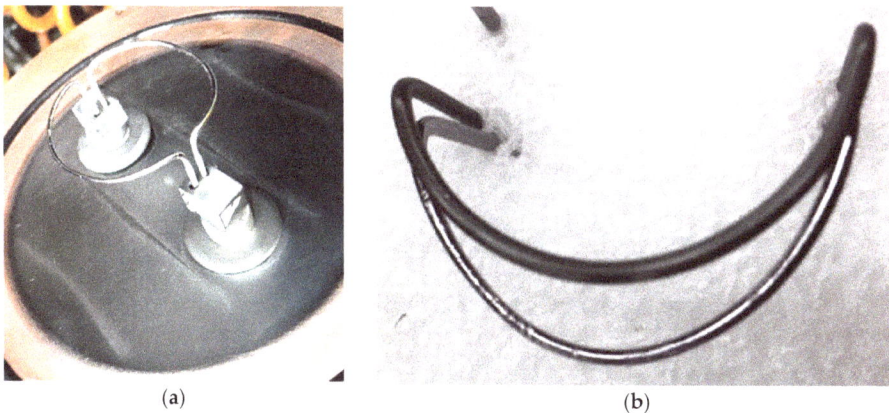

(a)          (b)

**Figure 2.** (a) Ion source back plate with two of the Filaments I mounted. The discoloration under the two filaments is a coating of worn-off filament material. (b) Used Filament I (bottom) and new Filament I (top). It can clearly be seen, that the filament wears thinner during operation.

This is reflected in a dropping current flowing through Filament I during routine operation, see Figure 3. To avoid filament breakage and therefore loss of negative hydrogen beam, the Filament I pair is being replaced when the current drops to about 100 A at the TR13. At normal operation Filament I in the ion source needs to be replaced on average every three months to ensure beam delivery, which causes the cyclotron to be off for a minimum of a day each time as the ion source chamber needs to be vented, shielding needs to be removed, services need to be disconnected, the backplate of the ion source needs to be removed and the filament needs to be replaced. The reverse is then necessary before operation can resume. To significantly shorten this down time, a new spiral filament, Filament II, designed for the 520 MeV cyclotron at TRIUMF [5], was tested at the TR13 cyclotron.

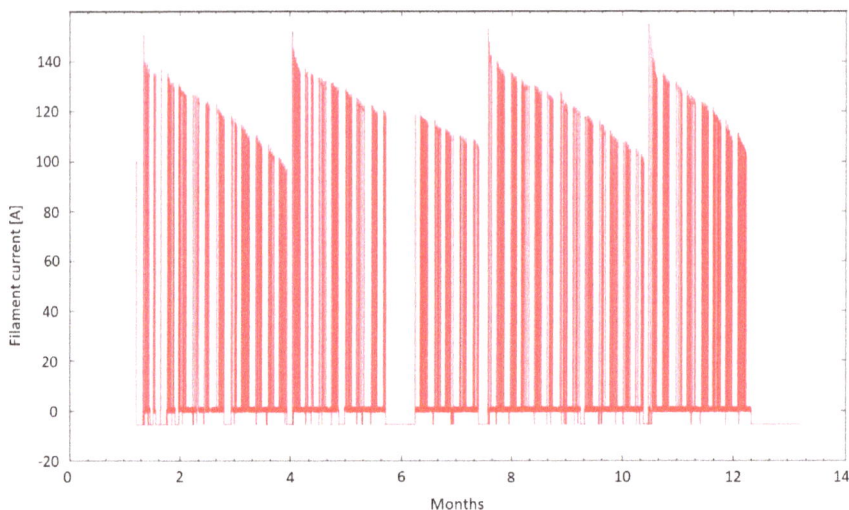

**Figure 3.** Filament current during operation of Filament I over time. The TR13 cyclotron is not operated 24 hours a day but is turned off after daily operation and for the weekend. The longer pause around month six is the annual shutdown during the Christmas holiday.

## 2. Materials and Methods

The TR13 cyclotron, the work horse of the Life Sciences Division at TRIUMF, is a 13 MeV cyclotron with an external ion source, accelerating negative-hydrogen ions. For more details of the design and operation see [3,4]. The ion source relies on a filament for the production of negative hydrogen ions which are then transported into the plane of the cyclotron for acceleration and bombardment of medical isotope targets. The new Filament II was originally designed for the external ion source at the 520 MeV cyclotron at TRUMF and has already been described in [7]. While a tantalum filament produces very bright H$^-$ beams, it is also known to degrade fast [7]. The new Filament II is made from a tungsten alloy, which degrades slower. The Filament II is shaped in house. Several lengths and shapes were studied as mentioned in [7] and the spiral filament in Figure 4a with an outer spiral diameter of 10 mm and six windings achieved the longest lifetime. As this new Filament II is slightly thicker (3.0 mm versus 1.5 mm for Filament I), a higher current is needed to give the same surface temperature and electron emission. To hold the new filament wire, the mounting stands on the ion-source back plate (see Figure 4) had to be modified to accept the 3.0 mm diameter. Due to the thicker wire, a higher filament current was necessary, 230 A at the beginning versus 150 A for Filament I. While our power supply (Xantrex, XKW, 12V and 250A) was able to provide this extra current, this should be taken into account when considering upgrading filaments. No other modification to the ion-source chamber, including the magnetic field, was necessary.

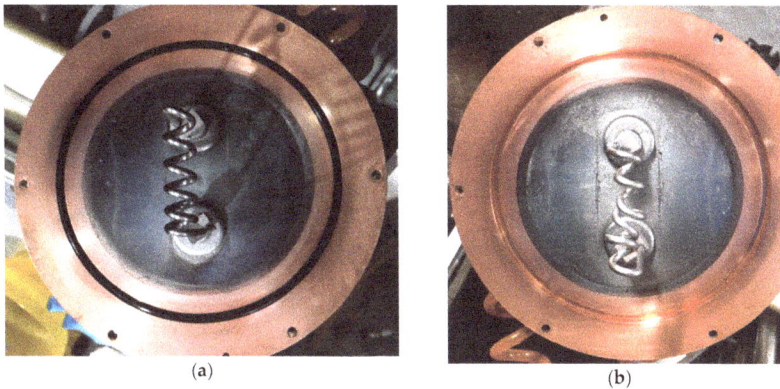

**Figure 4.** (**a**) Ion source back plate with new Filament II. (**b**) Ion source back plate with 6 months old Filament II. The operation deformed the spiral to the extent that neighboring windings touched and created an electrical short.

## 3. Results and Discussion

The operation of Filament II is the same as of the old Filament I. The automated routine tracking the currents and voltages was able to operate the new Filament II without any changes necessary. Figure 5. shows the filament current of the new Filament II over time. At the beginning of this graph, Filament I was still in use with its start-up current of ~150 μA and a fast current-drop. Shortly before month 8, the new Filament II was installed. While the higher start-up current of ~230 A is obvious, it is also very clear that the filament current declines significantly slower than for Filament I. We estimate from the observed drop that Filament I could be in use for about two years before it needs to be replaced.

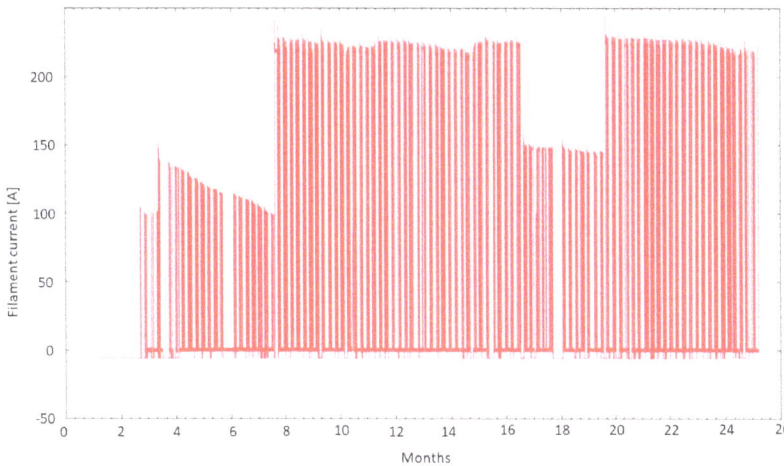

**Figure 5.** Filament current versus time. Until month eight, the old Filament I was in use with its typical current decline from 150 A down to about 100 A when the filament needs to be replaced to ensure ongoing ion source operation. After month 8, the new Filament II was installed. While it requires a higher current to create the plasma in the ion source (230 A), it has a significantly lower degradation time scale as reflected in the smaller slope. A short in the spiral of Filament II between month 17 and 20, see Figure 4b, resulted in an effective shorter length of the filament and therefore a lower filament current.

29

Also shown in Figure 5 at around month 17, a sudden filament current drop down to 150 A was observed. This sudden drop was caused by the deformation of the spiral to the point that two neighboring windings touched and shorted, effectively shortening the filament length, see Figure 4b. While this reduced the current output of the ions source due to the shorter active length and made tuning for higher beam currents difficult and tedious, it should be noted that it did not hinder routine isotope production. In month 20, the Filament II was replaced and the filament current consequently recovered to the previously observed 230 A.

Due to the higher current necessary to operate Filament II, a temperature rise in the electrical cables feeding into the ion source was observed, see Figure 6. During operation, the cable supplying the new spiral filament rose to 47.0 °C. It should be noted that although our power supply is rated to provide the increased current as it is rated for up to 250 A, we plan on upgrading the supply in the future.

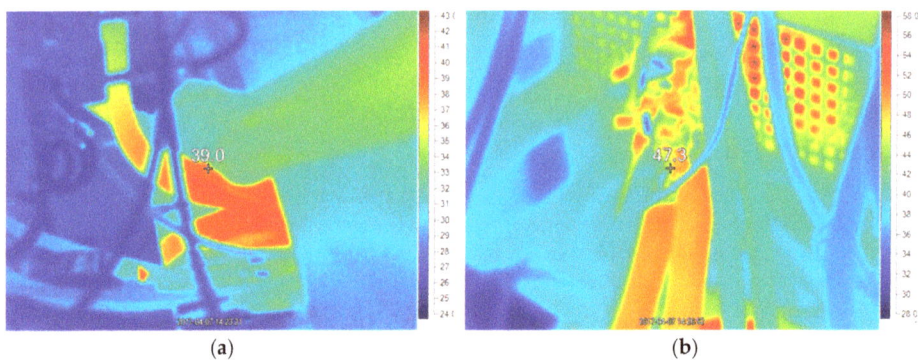

(a)            (b)

**Figure 6.** (**a**) Thermal image of the electrical cables supplying the new Filament II with current where it connects to the ion source back plate. (**b**) Thermal image of the same electrical cables connecting to the power supply.

## 4. Conclusions

In this paper, we presented the operational experience of a new spiral filament [7] on our 13 MeV medical cyclotron. The switch from our old half-circle Filament I to the new spiral Filament II required only minimum mechanical modifications and no alteration in operation. The only significant change is the higher current necessary to ensure a similar electron emission as with the previous Filament I. While our power supply was able to accommodate this current rise, this could potentially lead to the need to purchase an upgraded power supply for other sites. With the new Filament II, it is now not necessary to change the ion source filament approximately every three months, and we estimate that during routine operation it should last over two years, greatly reducing our planned down time. A newer version of the Filament II tested at TRIUMF is resistant to sagging and deformation and will be tested at the TR13 in the near future. Pending final negotiations, the filament will be licensed to D-Pace [10].

**Author Contributions:** Conceptualization, K.J. and C.H.; Formal analysis, D.P. and K.J.; Funding acquisition, D.P.; Investigation, D.P., K.J., L.G. and S.V.; Methodology, D.P. and K.J.; Project administration, D.P.; Resources, D.P. and K.J.; Supervision, D.P.; Visualization, K.J. and C.H.; Writing—original draft, C.H.; Writing—review & editing, D.P.

**Funding:** TRIUMF receives federal funding via a contribution agreement with the National Research Council of Canada.

**Conflicts of Interest:** The authors declare no conflict of interest.

*Instruments* **2019**, 3, 5

## References

1. Ruth, T.J. The production of radionuclides for radiotracers in nuclear medicine. *Rev. Accel. Sci. Technol.* **2009**, *2*, 17–33. [CrossRef]
2. Hoehr, C.; Bénard, F.; Buckley, K.; Crawford, J.; Gottberg, A.; Hanemaayer, V.; Kunz, P.; Ladouceur, K.; Radchenko, V.; Ramogida, C.; et al. Medical isotope production at TRIUMF—From imaging to treatment. *Phys. Procedia* **2017**, *90*, 200–208. [CrossRef]
3. Buckley, K.R.; Huser, J.; Jivan, S.; Chun, S.K.; Ruth, T.J. $^{11}$C-methane production in small volume, high pressure gas targets. *Radiochim. Acta* **2000**, *88*, 201–205. [CrossRef]
4. Laxdal, R.E.; Altman, A.; Kuo, T. Beam measurements on a small commercial cyclotron. In Proceedings of the EPAC 94, London, UK, 27 June–1 July 1994; pp. 545–547.
5. Infantino, A.; Oehlke, E.; Mostacci, D.; Schaffer, P.; Trinczek, M.; Hoehr, C. Assessment of the production of medial isotope using the Monte Carlo code FLUKA: Simulations against experimental measurements. *Nucl. Instrum. Methods Phys. Res. B* **2016**, *366*, 117–123. [CrossRef]
6. Sluyters, T.; Prelec, K. Will negative hydrogen ion sources soon replace proton sources in high energy accelerators? In Proceedings of the IXth International Conference on High Energy Accelerators SLAC, Stanford, CA, USA, 2–7 May 1974; p. 536.
7. Jayamanna, K.; Ames, F.; Bylinskii, I.; Lovera, M.; Minato, B. A 60 mA DC H$^-$ multi cusp ion source developed at TRIUMF. *Nucl. Instrum. Methods Phys. Res. A* **2018**, *895*, 150–157. [CrossRef]
8. Korenev, S. Critical analysis of negative hydrogen ion sources for cyclotrons. In Proceedings of the 20th International Conference on Cyclotrons and their Applications, Vancouver, BC, Canada, 16–20 September 2013; p. 192.
9. Ion Sources. Available online: https://inis.iaea.org/collection/NCLCollectionStore/_Public/26/001/26001458.pdf (accessed on 15 January 2019).
10. D-Pace. Available online: http://www.d-pace.com/ (accessed on 15 January 2019).

*instruments*

MDPI

Communication

# Can We Extract Production Cross-Sections from Thick Target Yield Measurements? A Case Study Using Scandium Radioisotopes

Mateusz Sitarz [1,2,3,*], Jerzy Jastrzębski [1], Férid Haddad [3], Tomasz Matulewicz [2], Katarzyna Szkliniarz [4] and Wiktor Zipper [4]

[1]   Heavy Ion Laboratory, University of Warsaw, 02-093 Warszawa, Poland; jastj@slcj.uw.edu.pl
[2]   Faculty of Physics, University of Warsaw, 02-093 Warszawa, Poland; Tomasz.Matulewicz@fuw.edu.pl
[3]   Groupement d'Intérêt Public ARRONAX, 44817 Saint-Herblain Cedex, France; haddad@arronax-nantes.fr
[4]   Institute of Physics, Department of Nuclear Physics and Its Applications, University of Silesia,
      41-500 Chorzów, Poland; kasia.tworek@op.pl (K.S.); wiktor.zipper@us.edu.pl (W.Z.)
*   Correspondence: msitarz@slcj.uw.edu.pl

Received: 27 March 2019; Accepted: 11 May 2019; Published: 14 May 2019

**Abstract:** In this work, we present an attempt to estimate the reaction excitation function based on the measurements of thick target yield. We fit a function to experimental data points and then use three fitting parameters to calculate the cross-section. We applied our approach to $^{43}Ca(p,n)^{43}Sc$, $^{44}Ca(p,n)^{44g}Sc$, $^{44}Ca(p,n)^{44m}Sc$, $^{48}Ca(p,2n)^{47}Sc$ and $^{48}Ca(p,n)^{48}Sc$ reactions. A general agreement was observed between the reconstructions and the available cross-section data. The algorithm described here can be used to roughly estimate cross-section values, but it requires improvements.

**Keywords:** medical Sc radioisotopes; radioisotope production; thick target yield measurements; cross-section reconstruction; numerical analysis

## 1. Introduction

The interest in three scandium radioisotopes, $^{43}Sc$, $^{44g/m}Sc$ and $^{47}Sc$, in nuclear medicine has already been acknowledged and discussed in [1–19] (the selected properties of these radioisotopes are summarized in Table 1). Both positron emitters $^{43}Sc$ and $^{44g}Sc$ are promising PET radioisotopes that can compete with the commonly used $^{68}Ga$ [1–5], while $^{44g}Sc$ offers unique possibilities in the three-photon PET technique [6–8]. Additionally, $^{44m}Sc$ can be used as a $^{44m}Sc/^{44g}Sc$ long-lived in-vivo generator as it decays mainly by a low energy transition to the ground state [9–12]. Meanwhile, $^{47}Sc$ is a β-emitter suitable for both therapeutic purposes and SPECT imaging [13], which is emphasized also within the IAEA Coordinated Research Project [14,15]. As mentioned in [16,17], this radioisotope is a matched pair for diagnostic $^{43}Sc$ and $^{44g}Sc$ radioisotopes.

In our recent papers [19,20], we have reported on the production routes of medical scandium radioisotopes as well as extending this data with scandium formed in natural and enriched thick $CaCO_3$ targets (from around 50 up to 1000 mg/cm$^2$) irradiated with α particles up to 30 MeV, deuterons up to 8 MeV and protons up to 30 MeV. The thick targets were used because we found that it was not feasible to prepare thinner (in the order of 1 mg/cm$^2$) self-supporting $CaCO_3$ as a homogeneously thick target for our experimental set-up. The significant stopping-power of our targets allowed us to obtain experimental thick target yield (TTY) values for scandium production.

**Table 1.** Nuclear data [18] of medically interesting scandium radioisotopes ($^{43}$Sc, $^{44g/m}$Sc, $^{47}$Sc). $^{48}$Sc, as a radioactive impurity, is also listed here with reference to the analysis in this paper.

| Radio-Nuclide | $T_{1/2}$ | $E_{average}$ β- or β+ | Branching or Transition | Main γ-Lines [keV] and Intensities |
|---|---|---|---|---|
| $^{43}$Sc | 3.89 h | β+ 476 keV | β+ 88% | 373 (22.5%) |
| $^{44g}$Sc | 3.97 h | β+ 632 keV | β+ 95% | 1157 (99.9%) |
| $^{44m}$Sc | 58.61 h | N/A | IT 99% | 271 (86.7%), 1002 (1.2%), 1126 (1.2%), 1157 (1.2%) |
| $^{47}$Sc | 3.35 d | β- 162 keV | β- 100% | 159 (68.3%) |
| $^{48}$Sc | 43.67 h | β- 220 keV | β- 100% | 175 (7.5%), 984 (100%), 1038 (97.6%), 1213 (2.4%), 1312 (100%) |

In this work, we want to complement our research by evaluating the $^{43}$Ca(p,n)$^{43}$Sc, $^{44}$Ca(p,n)$^{44g}$Sc, $^{44}$Ca(p,n)$^{44m}$Sc, $^{48}$Ca(p,2n)$^{47}$Sc and $^{48}$Ca(p,n)$^{48}$Sc cross-sections based on reported TTY measurements (the latter is not medically relevant, but $^{48}$Sc production is important as it is a radioactive impurity). A similar attempt has already been proposed in [21] for the study of $^{34m}$Cl production. In this work, we verify this approach for above-mentioned reactions while employing a different, straight-forward numerical algorithm (our Python code is submitted in the Supplementary Materials to this paper).

## 2. Materials and Methods

In our recent work [20], we reported TTY for $^{43}$Ca(p,n)$^{43}$Sc on $^{43}$CaCO$_3$ (90% $^{43}$Ca) targets, $^{44}$Ca(p,n)$^{44g}$Sc and $^{44}$Ca(p,n)$^{44m}$Sc on $^{44}$CaCO$_3$ (94.8% $^{44}$Ca) as well as $^{48}$Ca(p,2n)$^{47}$Sc and $^{48}$Ca(p,n)$^{48}$Sc on $^{48}$CaCO$_3$ (97.1% $^{48}$Ca). Those TTY values are directly related to cross-sections by the following formula [22,23]:

$$TTY(E) = \frac{H\,N_A}{Z\,e\,m\,\tau} \int_{E_{min}}^{E_{max}} \frac{\sigma(E)}{dE/dx(E)} dE$$

where $H$ is target enrichment, $N_A$ is Avogadro's number, $\tau$ is the mean lifetime of a radioisotope, $Z$ is the ionization number of the projectile, $e$ is the elementary charge, $m$ is the atomic mass of the target, $E_{max}$ and $E_{min}$ are the maximal and minimal energy of the projectile penetrating the target (in case of TTY, $E_{min}$ <= reaction threshold), respectively, $\sigma$ is the cross-section for the nuclear reaction, and $dE/dx$ is the stopping-power of the projectile according to the aerial density of the target. Here, we describe the attempt to obtain the energy dependence of the cross-section (the excitation function) based on the experimental $TTY_{exp}(E)$ [MBq/μAh] values for different projectile energies $E$. These data are supplemented by an assumption $TTY_{exp}(E_{thr}) = 0$, where $E_{thr}$ denotes the energy threshold for this reaction.

The crucial factor is the choice of the function used to describe the TTY energy dependence. The number of parameters of the function used to fit the data should be restricted, as the number of the experimental data points is usually limited. Therefore, we propose a simple shape,

$$TTY_{fit}(E) = d + \frac{ac}{2}\left( \sqrt{\pi}\,(b - E_{thr})\,erf\left\{\frac{E-b}{a}\right\} - a\,exp\left\{\frac{-(E-b)^2}{a^2}\right\} \right)$$

which fulfils several important criteria. This function is monotonically increasing, as $TTY(E)$ should be. Most importantly, its derivative is a modified q-Weibull distribution [24],

$$\frac{dTTY_{fit}}{dE}\left[\frac{MBq}{\mu Ah}\right] = MAX\left[0;\ c\,(E - E_{thr})\,exp\left\{\frac{-(E-b)^2}{a^2}\right\}\right]$$

which reflects the global shape of the (p,n) and (p,2n) excitation functions, commonly used in the field of the production of medical radioisotopes. The request $TTY_{exp}(E_{thr}) = 0$ provides the condition

$$d = \frac{a^2 c}{2} \, exp\left\{ \frac{-(b - E_{thr})^2}{a^2} \right\}$$

and limits the number of $TTY_{fit}$ parameters to 3: $a$, $b$ and $c$. Once those parameters are obtained, the cross-section values can be estimated as

$$\sigma(E)[mb] = \frac{\tau[h]Ze[C]m[u]}{N_A H} \cdot \frac{\mathrm{d}TTY_{fit}}{\mathrm{d}E}\left[\frac{MBq}{\mu Ah}\right] \cdot \frac{\mathrm{d}E}{\mathrm{d}x}\left[\frac{MeV}{mg/cm^2}\right] \cdot 10^{42}$$

In our case, TTY measurements were obtained on $CaCO_3$ targets instead of metallic Ca. Therefore, we used $dE/dx(E)$ values corresponding to the energy loss in calcium carbonate (provided by SRIM software [25]), $m = 100$ u to address the mass of $CaCO_3$, and $H$ as the level of enrichment of employed material. We have also calculated the 95% confidence band for $TTY_{fit}(E)$ fit and reconstructed the cross-section. Details of our calculations are shown and explained in the Python code attached to this paper.

Alternatively, in [21], the cross-section was reconstructed after fitting the TTY curve by calculating target yields (TY) for thicknesses corresponding to 0.1 MeV projectile energy loss each 1 MeV and multiplied by projectile range. This method assumes the constant stopping-power in each layer. In our approach, this simplification was not necessary.

## 3. Results and Discussion

In Figures 1–5, we show the TTY data and the reconstructed cross-sections for $^{43}Ca(p,n)^{43}Sc$, $^{44}Ca(p,n)^{44g}Sc$, $^{44}Ca(p,n)^{44m}Sc$, $^{48}Ca(p,2n)^{47}Sc$ and $^{48}Ca(p,n)^{48}Sc$ reactions (the fit parameters are shown in Table 2 while the reconstructed cross-section values are listed in Table 3). We compare them with the experimental cross-section in [26–35], with the recommended values from [36], with the predictions of the EMPIRE [37] evaporation code (version 3.2.2 Malta) and with the TENDL-2017 cross-section library [38]. All reconstructions exhibit a similar shape to the model predictions and measured cross-section values, indicating the validity of modified q-Weibull distribution in estimating the global shape of the (p,n) and (p,2n) excitation functions.

We have also checked our reconstruction method by implementing the approach in [20]. We obtained similar values (marked on the plots) with a visible correction near the threshold in the $^{44m}Sc$ case (Figure 3) but also with the discontinuity fragments due to the numerical approach. Since the mentioned paper does not provide the recommended $TTY_{fit}$ function, we adopted ours.

**Table 2.** Parameters of the $TTY_{fit}$ (for Figures 1–5) obtained with least square method for different nuclear reactions and the $\chi 2/dof$ values for each fit. Parameter $d$ is calculated from $a$, $b$, $c$ and $E_{thr}$.

| Parameter | $^{43}Ca(p,n)^{43}Sc$ | $^{44}Ca(p,n)^{44g}Sc$ | $^{44}Ca(p,n)^{44m}Sc$ | $^{48}Ca(p,2n)^{47}Sc$ | $^{48}Ca(p,n)^{48}Sc$ |
|---|---|---|---|---|---|
| $E_{thr}$ [MeV] | 3.07 | 4.54 | 4.81 | 8.93 | theory: 0.51 adopted: 3.0 |
| $a$ [MeV] | 10(2) | 8.8(6) | 13.7(7) | 14.7(9) | 8.0(8) |
| $b$ [MeV] | 4.5(6) | 4.8(1.0) | 7.0(1) | 9.05(14) | 4.2(7) |
| $c$ [MBq(μAh)$^{-1}$ (MeV)$^{-2}$] | 7.1(9) | 24.5(1.2) | 0.075(3) | 1.57(6) | 2.5(2) |
| $d$ [MBq/μAh] | 348 | 952 | 6.82 | 169 | 79.4 |
| $\chi 2/dof$ | 1.30 | 0.57 | 6.11 | 1.05 | 1.79 |

In the case of $^{43}Sc$ data (Figure 1), the recent experimental results [34] are significantly lower than other measurements (by a factor of 2 around 10 MeV proton energy). The experimental results for TTY are quite linear in the measured proton energy range and do not reach the expected saturation, so the resulting excitation function is relatively flat and does not reproduce any of the previous measurements.

This reaction might require further validation, as with the extension of TTY measurements up to 30 MeV proton energy.

A general agreement is observed for $^{44g}$Sc (Figure 2), both with the theoretical models and experimental results, although again the data by [34] are lower than the measurements. More discrepancies are observed in the case of $^{44m}$Sc (Figure 3). The excitation function obtained from TTY measurements does not show the peak seen in the experiments and in model calculations and overestimates the values near the reaction threshold. We suspect that the problem with this reconstruction might be related to the offset of TTY data, as only in case of $^{44m}$Sc are the TTY values below model predictions at low energies and above them at higher energies, which causes the reconstructed excitation function to be flatter.

For $^{47}$Sc (Figure 4), the shape of the reconstruction reflects the shape predicted by both model calculations. While our results provide about 10% lower values compared to the models, recent measurements [35] indicate similar values at low energies but about 20% higher values at maximum.

Finally, we decided to adopt the arbitrary value of $E_{thr} = 3.0$ MeV as a parameter for $^{48}$Sc fit (Figure 5) to satisfy the visible and significant TTY build-up at this energy rather than the actual threshold (0.51 MeV). This might be explained by the fact that the shape of the function used for the fit does not adequately describe the behavior of the cross-section at energies much below the Coulomb barrier. Since the cross-section values far below the Coulomb barrier are very small, they do not contribute significantly to the TTY values. The extracted cross-section values are in line with the data in [30] at lower energies and in [35] at higher energies.

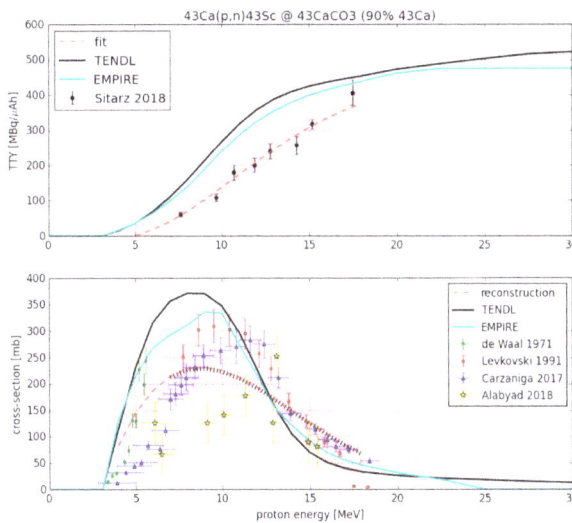

**Figure 1.** Reconstruction of $^{43}$Ca(p,n)$^{43}$Sc cross-section (bottom) based on the fit to TTY data on $^{43}$CaCO$_3$ enriched in 90% $^{43}$Ca (top). The cross-section data points are taken from [26,31,33,34].

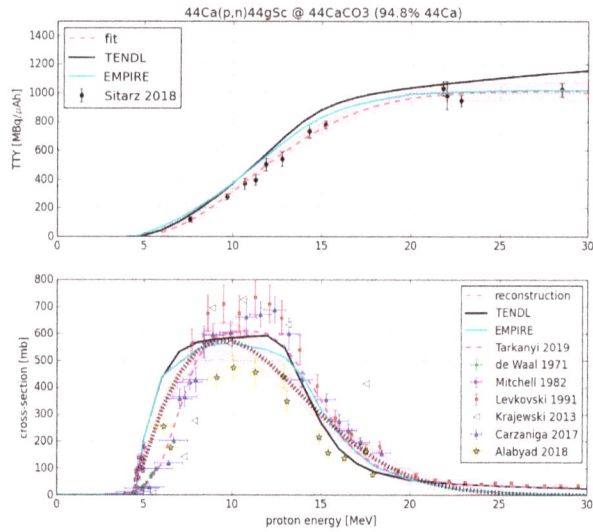

**Figure 2.** Reconstruction of $^{44}$Ca(p,n)$^{44g}$Sc cross-section (bottom) based on the fit to TTY data on $^{44}$CaCO$_3$ enriched in 94.8% $^{44}$Ca (top). The cross-section data points are taken from [26,29,31–34,36].

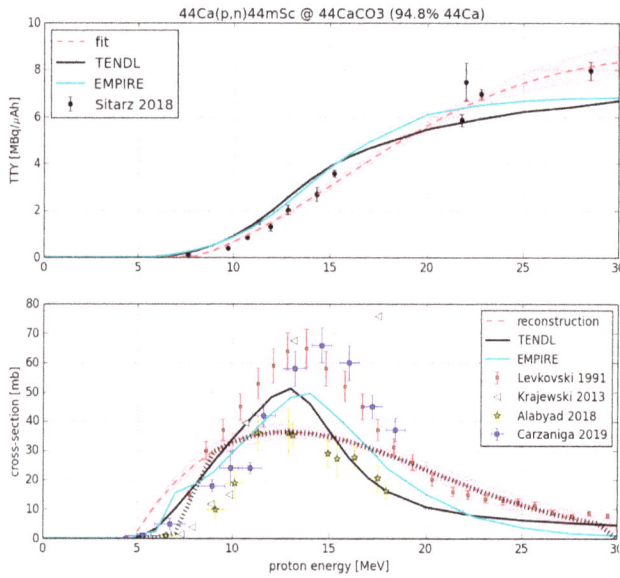

**Figure 3.** Reconstruction of $^{44}$Ca(p,n)$^{44m}$Sc cross-section (bottom) based on the fit to TTY data on $^{44}$CaCO$_3$ enriched in 94.8% $^{44}$Ca (top). The cross-section data points are taken from [31,32,34,35].

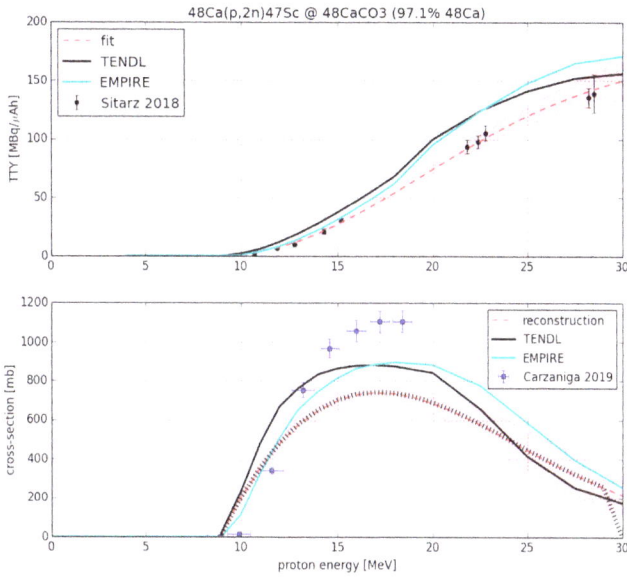

**Figure 4.** Reconstruction of $^{48}$Ca(p,2n)$^{47}$Sc cross-section (bottom) based on the fit to TTY data on $^{48}$CaCO$_3$ enriched in 97.1% $^{48}$Ca (top). The cross-section data points are taken from [35].

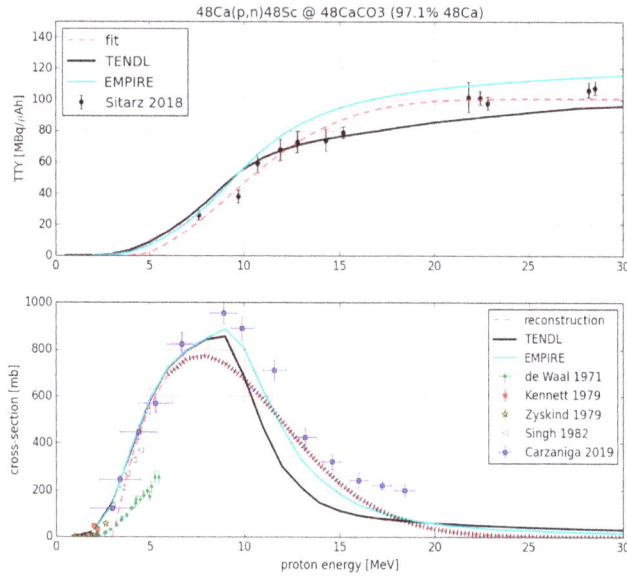

**Figure 5.** Reconstruction of $^{48}$Ca(p,n)$^{48}$Sc cross-section (bottom) based on the fit to TTY data on $^{48}$CaCO$_3$ enriched in 97.1% $^{48}$Ca (top). Here, we adopted an arbitrary threshold of 3 MeV. The cross-section data points are taken from [26–28,30,35]. The results from [27,28,30] are averaged.

**Table 3.** Cross-section values for different nuclear reactions deduced from the thick target yield data from [20].

| E [MeV] | σ [mb] | | | | |
|---|---|---|---|---|---|
| | $^{43}$Ca(p,n)$^{43}$Sc | $^{44}$Ca(p,n)$^{44g}$Sc | $^{44}$Ca(p,n)$^{44m}$Sc | $^{48}$Ca(p,2n)$^{47}$Sc | $^{48}$Ca(p,n)$^{48}$Sc |
| 5 | 145(20) | 115(6) | 2.11(8) | 0 | 552(48) |
| 6 | 187(26) | 314(13) | 11.7(5) | 0 | 692(66) |
| 7 | 214(29) | 450(20) | 19.2(8) | 0 | 761(77) |
| 8 | 228(28) | 534(26) | 25.1(1.1) | 0 | 744(78) |
| 9 | 231(24) | 572(31) | 29.6(1.3) | 14.2(6) | 740(71) |
| 10 | 225(18) | 572(32) | 32.9(1.4) | 199(7) | 674(58) |
| 11 | 211(12) | 543(30) | 35.0(1.4) | 353(14) | 586(43) |
| 12 | 193(10) | 493(25) | 36.2(1.3) | 478(18) | 490(29) |
| 13 | 172(15) | 431(19) | 36.4(1.2) | 576(20) | 393(20) |
| 14 | 148(21) | 363(15) | 35.9(1.0) | 647(20) | 304(18) |
| 15 | 125(26) | 297(14) | 34.7(8) | 698(20) | 228(21) |
| 16 | 104(29) | 235(15) | 33.1(7) | 729(19) | 164(22) |
| 17 | 84(30) | 180(16) | 31.0(7) | 739(18) | 115(21) |
| 18 | 66(30) | 134(16) | 28.7(8) | 734(18) | 78(19) |
| 19 | | 97(16) | 26.1(1.0) | 716(21) | 51(15) |
| 20 | | 68(14) | 23.4(1.1) | 684(26) | 32(12) |
| 21 | | 47(12) | 20.8(1.3) | 647(31) | 20(9) |
| 22 | | 31(10) | 18.2(1.4) | 600(35) | 12(6) |
| 23 | | 20(7) | 15.7(1.4) | 550(40) | 7(4) |
| 24 | | 13(6) | 13.5(1.4) | 499(43) | 4(3) |
| 25 | | 8(4) | 11.3(1.4) | 445(46) | 2.0(1.6) |
| 26 | | 5(3) | 9.5(1.4) | 393(47) | 1.0(9) |
| 27 | | 2.7(1.8) | 7.8(1.3) | 343(48) | 0.5(5) |
| 28 | | 1.5(1.1) | 6.4(1.2) | 297(47) | 0.3(3) |
| 29 | | 0.8(7) | 5.1(1.1) | 253(45) | 0.12(16) |
| 30 | | 0.4(4) | 4.0(9) | 213(43) | 0.05(8) |

## 4. Conclusions and Summary

We have presented an attempted numerical method for cross-section evaluation based on the thick target yield (TTY) measurements obtained from the irradiation of thick targets (in which the energy of a projectile is reduced to the reaction threshold). This method is based on fitting a function with three free parameters to TTY data points and using its analytical derivative to obtain the cross-section. The fitting requires the knowledge of the reaction threshold and a sufficient number of experimental points to represent the shape of the TTY curve, including the saturation region.

Using this approach, we were able to obtain a useful estimation of cross-sections for the production of medically important $^{43}$Sc, $^{44g}$Sc, $^{44m}$Sc, $^{47}$Sc, and $^{48}$Sc radioisotopes via (p,n) and (p,2n) reactions on Ca. The results were compared to the already measured cross-sections and to the model predictions. General agreement is observed; however, not all experimental results confirm our reconstructions, particularly those near the reaction threshold. In conclusion, our algorithm can provide good insights for the (p,xn) excitation function, but improvements are necessary.

**Supplementary Materials:** The following are available online at http://www.mdpi.com/2410-390X/3/2/29/s1.

**Author Contributions:** M.S.: formal analysis, software, methodology, writing—original draft, writing—review & editing; J.J.: conceptualization; F.H.: supervision, writing—review & editing; T.M.: conceptualization, methodology, supervision, writing—review & editing; K.S.: supervision, writing—review & editing; W.Z.: conceptualization, supervision, writing—review & editing.

**Funding:** Part of this work was performed within the framework of the EU Horizon 2020 project RIA-ENSAR2 (654 002). This research was also partly supported by the Polish Funding Agency NCBiR grant No. DZP/PBS3/2319/2014 and by a grant from the French National Agency for Research called "Investissements d'Avenir", Equipex Arronax-Plus noANR-11-EQPX-0004, Labex IRON noANR-11-LABX-18-01, and ISITE NEXT no. ANR-16-IDEX-0007.

**Acknowledgments:** The authors sincerely thank Anna Stolarz and Agnieszka Trzcińska (from Heavy Ion Laboratory, Warsaw) as well as Aleksander Bilewicz (from Institute of Nuclear Chemistry and Technology, Warsaw) for valuable discussions and constructive suggestions regarding this work. The PhD cotutelle scholarship from the French Government for Mateusz Sitarz is acknowledged.

**Conflicts of Interest:** The authors declare no conflict of interest.

## References

1. Koumarianou, E.; Pawlak, D.; Korsak, A.; Mikolajczak, R. Comparison of receptor affinity of $^{nat}$Sc-DOTA-TATE versus $^{nat}$Ga-DOTA-TATE. *Nucl. Med. Rev.* **2011**, *14*, 85–89. [CrossRef]
2. Walczak, R.; Krajewski, S.; Szkliniarz, K.; Sitarz, M.; Abbas, K.; Choiński, J.; Jakubowski, A.; Jastrzębski, J.; Majkowska, A.; Simonelli, F.; et al. Cyclotron production of $^{43}$Sc for PET imaging. *EJNMMI Phys.* **2015**. [CrossRef] [PubMed]
3. Bunka, M.; Müller, C.; Vermeulen, C.; Haller, S.; Türler, A.; Schibli, R.; van der Meulen, N.P. Imaging quality of $^{44}$Sc in comparison with five other PET radionuclides using Derenzo phantoms and preclinical PET. *Appl. Radiat. Isot.* **2016**, *110*, 129–133. [CrossRef] [PubMed]
4. Domnanich, K.A.; Müller, C.; Farkas, R.; Schmid, R.M.; Ponsard, B.; Schibli, R.; Türler, A.; van der Meulen, N.P. $^{44}$Sc for labeling of DOTA- and NODAGA-functionalized peptides: Preclinical in vitro and in vivo investigations. *EJNMMI Radiopharm. Chem.* **2016**. [CrossRef]
5. Domnanich, K.A.; Eichler, R.; Müller, C.; Jordi, S.; Yakusheva, V.; Braccini, S.; Behe, M.; Schibli, R.; Türler, A.; van der Meulen, N.P. Production and separation of $^{43}$Sc for radiopharmaceutical purposes. *EJNMMI Radiopharm. Chem.* **2017**. [CrossRef] [PubMed]
6. Grignon, C.; Barbet, J.; Bardiès, M.; Carlier, T.; Chatal, J.F.; Couturier, O.; Cussonneau, J.P.; Faivre, A.; Ferrer, L.; Girault, S.; et al. Nuclear medical imaging using β+ γ coincidence from $^{44}$Sc radio-nuclide with liquid xenon as detection medium. *Nucl. Instrum. Meth. A* **2017**, *571*, 142–145. [CrossRef]
7. Lang, C.; Habs, D.; Parodi, K.; Thirolf, P.G. Sub-millimeter nuclear medical imaging with reduced dose application in positron emission tomography using β+ γ coincidences. *JINST* **2013**, *9*, P01008. [CrossRef]
8. Thirolf, P.G.; Lang, C.; Parodi, K. Perspectives for Highly-Sensitive PET-Based Medical Imaging Using β+γ Coincidences. *Acta Phys. Pol. A* **2015**, *127*, 1441–1444. [CrossRef]
9. Huclier-Markai, S.; Kerdjoudj, R.; Alliot, C.; Bonraisin, A.C.; Michel, N.; Haddad, F.; Barbet, J. Optimization of reaction conditions for the radiolabeling of DOTA and DOTA-peptide with $^{44m/44}$Sc and experimental evidence of the feasibility of an in vivo PET generator. *Nucl. Med. Biol.* **2014**, *41*, e36–e43. [CrossRef] [PubMed]
10. Alliot, C.; Audouin, N.; Barbet, J.; Bonraisin, A.C.; Bossé, V.; Bourdeau, C.; Bourgeois, M.; Duchemin, C.; Guertin, A.; Haddad, F.; et al. Is there an interest to use deuteron beams to produce non-conventional radionuclides? *Front. Med.* **2015**. [CrossRef]
11. Alliot, C.; Kerdjoudj, R.; Michel, N.; Haddad, F.; Huclier-Markai, S. Cyclotron production of high purity $^{44m,44}$Sc with deuterons from $^{44}$CaCO$_3$ targets. *Nucl. Med. Biol.* **2015**, *42*, 524–529. [CrossRef]
12. Duchemin, C.; Guertin, A.; Haddad, F.; Michel, N.; Métivier, V. Production of scandium-44m and scandium-44g with deuterons on calcium-44: Cross-section measurements and production yield calculations. *Phys. Med. Biol.* **2015**, *60*, 17. [CrossRef]
13. Domnanich, K.A.; Müller, C.; Benešová, M.; Dressler, R.; Haller, S.; Köster, U.; Ponsard, B.; Schibli, R.; Türler, A.; van der Meulen, N.P. $^{47}$Sc as useful β- emitter for the radiotheragnostic paradigm: A comparative study of feasible production routes. *EJNMMI Radiopharm. Chem.* **2017**. [CrossRef]
14. International Atomic Energy Agency. Call for Coordinated Research Project "Therapeutic Radiopharmaceuticals Labelled with New Emerging Radionuclides ($^{67}$Cu, $^{186}$Re, $^{47}$Sc)". 2015. Available online: https://www.iaea.org/projects/crp/f22053 (accessed on 13 May 2019).
15. International Atomic Energy Agency. 2015. Available online: http://cra.iaea.org/cra/stories/2015-09-30-F22053-New_Emerging_Radionuclides.html (accessed on 13 May 2019).
16. Müller, C.; Bunka, M.; Haller, S.; Köster, U.; Groehn, V.; Bernhardt, P.; van der Meulen, N.P.; Türler, A.; Schibli, R. Promising Prospects for $^{44}$Sc-/$^{47}$Sc-Based Theragnostics: Application of $^{47}$Sc for Radionuclide Tumor Therapy in Mice. *J. Nucl. Med.* **2014**, *55*, 1658–1664. [CrossRef]

17. Müller, C.; Domnanich, K.A.; Umbricht, C.A.; van der Meulen, N.P. Scandium and terbium radionuclides for radiotheranostics: Current state of development towards clinical application. *Br. J. Radiol.* **2018**. [CrossRef]
18. International Atomic Energy Agency. Live Chart of Nuclides. Available online: https://www-nds.iaea.org/relnsd/vcharthtml/VChartHTML.html (accessed on 12 May 2019).
19. Szkliniarz, K.; Sitarz, M.; Walczak, R.; Jastrzębski, J.; Bilewicz, A.; Choiński, J.; Jakubowski, A.; Majkowska, A.; Stolarz, A.; Trzcińska, A.; Zipper, W. Production of medical Sc radioisotopes with an alpha particle beam. *Appl. Radiat. Isot.* **2016**, *118*, 182–189. [CrossRef]
20. Sitarz, M.; Szkliniarz, K.; Jastrzębski, J.; Choiński, J.; Guertin, A.; Haddad, F.; Jakubowski, A.; Kapinos, K.; Kisieliński, M.; Majkowska, A.; et al. Production of Sc medical radioisotopes with proton and deuteron beams. *Appl. Radiat. Isot.* **2018**, *142*, 104–112. [CrossRef]
21. Nagatsu, K.; Fukumura, T.; Takei, M.; Szelecsényi, F.; Kovács, Z.; Suzuki, K. Measurement of thick target yields of the $^{nat}S(\alpha,x)^{34m}Cl$ nuclear reaction and estimation of its excitation function up to 70 MeV. *Nucl. Inst. Meth. Phys. Res. B* **2008**, *266*, 709–713. [CrossRef]
22. Phelps, M.E. *PET: Molecular Imaging and Its Biological Applications*; Springer: New York, NY, USA, 2004.
23. Pedroso de Lima, J.J. *Nuclear Medicine Physics Series in Medical Physics and Biomedical Engineering, Cyclotron and Radionuclide Production, Series in Medical and Biomedical Engineering*; CRC Press: Boca Raton, FL, USA, 2010.
24. Jose, K.K.; Naik, S.R. On the q-Weibull Distribution and Its Applications. *Com. Stat. Th. Meth.* **2009**, *38*, 912–926. [CrossRef]
25. Ziegler, J.F.; Ziegler, M.D.; Biersack, J.P. SRIM Code, Version 2008.04. Available online: http://www.srim.org/ (accessed on 12 May 2019).
26. de Waal, T.J.; Peisach, M.; Pretorius, R. Activation cross sections for proton-induced reactions on calcium isotopes up to 5.6 MeV. *J. Inorg. Nucl. Chem.* **1971**, *33*, 2783–2789. [CrossRef]
27. Kennett, S.R.; Switkowski, Z.E.; Paine, B.M.; Sargood, D.G. Yield Measurements in the Reactions $^{48}Ca(p,\gamma)^{49}Sc$ and $^{48}Ca(p,n)^{48}Sc$. *J. Phys. G* **1979**, *5*, 399. [CrossRef]
28. Zyskind, J.L.; Davidson, J.M.; Esat, M.T.; Spear, R.H.; Shapiro, M.H.; Fowler, W.A.; Barnes, C.A. Cross Section Measurements and Thermonuclear Reaction Rates for $^{48}Ca(p,\gamma)^{49}Sc$ and $^{48}Ca(p,n)^{48}Sc$. *Nucl. Phys. A* **1979**, *315*, 430–444. [CrossRef]
29. Mitchell, L.W.; Anderson, M.R.; Kennett, S.R.; Sargood, D.G. Cross-sections and thermonuclear reaction rates for $^{42}Ca(p,\gamma)^{43}Sc$, $^{44}Ca(p,\gamma)^{45}Sc$, $^{44}Ca(p,n)^{44}Sc$ and $^{45}Sc(p,n)^{45}Ti$. *J. Nucl. Phys. A* **1982**, *380*, 318–334. [CrossRef]
30. Singh, G.; Kailas, S.; Saini, S.; Chatterjee, A.; Balakrishnan, M.; Mehta, M.K. Reaction $^{48}Ca(p,n)^{48}Sc$ from E(p) = 1.885 to 5.1 MeV. *Pramana* **1982**, *19*, 565–577. [CrossRef]
31. Levkovskij, V.N. *Cross-section of medium mass nuclide activation (A=40-100) by medium energy protons and alpha-particles (E=10-50 MeV)*; Inter-Vesi: Moscow, Russia, 1991.
32. Krajewski, S.; Cydzik, I.; Abbas, K.; Bulgheroni, A.; Simonelli, F.; Holzwarth, U.; Bilewicz, A. Cyclotron production of $^{44}Sc$ for clinical application. *J. Radiochim. Acta* **2013**. [CrossRef]
33. Carzaniga, T.S.; Auger, M.; Braccini, S.; Bunka, M.; Ereditato, A.; Nesteruk, K.P.; Scampoli, P.; Türler, A.; van der Meulen, N. Measurement of $^{43}Sc$ and $^{44}Sc$ production cross-section with an 18 MeV medical PET cyclotron. *Appl. Radiat. Isot.* **2017**, *129*, 96–102. [CrossRef] [PubMed]
34. Alabyad, M.; Mohamed, G.V.; Hassan, H.E.; Takács, S.; Ditrói, F. Experimental measurement and theoretical calculations for proton, deuteron and α-particle induced nuclear reactions on calcium: Special relevance to the production of $^{43,44}Sc$. *J. Radioan. Nucl. Chem.* **2018**, *316*, 119–128. [CrossRef]
35. Carzaniga, T.S.; Braccini, S. Cross-section measurement of $^{44m}Sc$, $^{47}Sc$, $^{48}Sc$ and $^{47}Ca$ for an optimized $^{47}Sc$ production with an 18 MeV medical PET cyclotron. *Appl. Radiat. Isot.* **2019**, *143*, 18–23. [CrossRef]
36. Tárkányi, F.T.; Ignatyuk, A.V.; Hermanne, A.; Capote, R.; Carlson, B.V.; Engle, J.W.; Kellett, M.A.; Kibédi, T.; Kim, G.N.; Kondev, F.G.; et al. Recommended nuclear data for medical radioisotope production: diagnostic positron emitters. *J. Radioan. Nucl. Chem* **2019**, *319*, 533–666. [CrossRef]

37. Herman, M.; Capote, R.; Carlson, B.V.; Obložinský, P.; Sin, M.; Trkov, A.; Wienke, H.; Zerkin, V. EMPIRE: Nuclear Reaction Model Code System for Data Evaluation. *Nucl. Data Sheets* **2007**, *108*, 2655–2715. [CrossRef]
38. Koning, A.J.; Rochman, D. Modern Nuclear Data Evaluation with The TALYS Code System. *Nucl. Data Sheets* **2012**, *113*, 2841–2934. [CrossRef]

![instruments logo] *instruments*

*Article*

# Design of a Thorium Metal Target for $^{225}$Ac Production at TRIUMF

Andrew K.H. Robertson [1,2], Andrew Lobbezoo [1], Louis Moskven [1], Paul Schaffer [1,3] and Cornelia Hoehr [1,4,*]

[1]  Life Sciences Division, TRIUMF, Vancouver, BC V6T 2A3, Canada; arobertson@triumf.ca (A.K.H.R.); lobbezaj@mcmaster.ca (A.L.); lmoskven@triumf.ca (L.M.); pschaffer@triumf.ca (P.S.)
[2]  Department of Physics and Astronomy, University of British Columbia, Vancouver, BC V6T 1Z1, Canada
[3]  Department of Radiology, University of British Columbia, Vancouver, BC V5Z 1M9, Canada
[4]  Department of Physics and Astronomy, University of Victoria, Victoria, BC V8W 2Y2, Canada
*   Correspondence: choehr@triumf.ca; Tel.: +1-604-222-1047

Received: 26 December 2018; Accepted: 11 February 2019; Published: 15 February 2019

**Abstract:** With recent impressive clinical results of targeted alpha therapy using $^{225}$Ac, significant effort has been directed towards providing a reliable and sufficient supply of $^{225}$Ac to enable widespread using of $^{225}$Ac-radiopharmaceuticals. TRIUMF has begun production of $^{225}$Ac via spallation of thorium metal with 480 MeV protons. As part of this program, a new $^{225}$Ac-production target system capable of withstanding the power deposited by the proton beam was designed and its performance simulated over a range of potential operating parameters. Special attention was given to heat transfer and stresses within the target components. The target was successfully tested in two irradiations with a 72–73 µA proton beam for a duration of 36.5 h. The decay corrected activity at end of irradiation (average $\pm$ standard deviation) was (524 $\pm$ 21) MBq (14.2 mCi) and (86 $\pm$ 13) MBq (2.3 mCi) for $^{225}$Ac and $^{225}$Ra, respectively. These correspond to saturation yields of 72.5 MBq/µA for $^{225}$Ac and 17.6 MBq/µA for $^{225}$Ra. Longer irradiations and production scale-up are planned in the future.

**Keywords:** actinium-225; thorium; spallation; proton target; ANSYS; targeted alpha therapy

## 1. Introduction

Targeted alpha therapy (TAT) using radiopharmaceuticals that combine suitable alpha-emitting radionuclides with cancer-targeting biomolecules has demonstrated an ability to treat late stage cancers by harnessing the cytotoxic high linear energy transfer (LET) of alpha radiation [1–7]. When combined with a disease-specific targeting biomolecule, the short-range of alpha radiation also limits the radiation dose delivered to surrounding healthy tissues. These properties make TAT especially promising for treatment of small and radio-resistant tumours and microscopic malignancies where dose escalation is not possible with conventional radiotherapy. Actinium-225 ($^{225}$Ac) is an alpha-emitter of particular ability for TAT due to its favourable half-life ($t_{1/2}$ = 9.9 d), chemical properties, and the multiple (4) alpha emissions in its decay chain [8].

While clinical results have demonstrated the potential of $^{225}$Ac-radiopharmaceuticals [9–13], the development of these drugs is slowed by the limited supply of the radionuclide. While the majority of the approximately 63 GBq global annual $^{225}$Ac supply is derived from the decay of $^{229}$Th ($t_{1/2}$ = 7600 y) through the $^{225}$Ac parent isotope $^{225}$Ra ($t_{1/2}$ = 14.9 d) [14–19], these $^{229}$Th sources remain fixed. The resulting low annual $^{225}$Ac availability has spurred many recent efforts to increase $^{225}$Ac production via particle accelerators—most notably those of the US Department of Energy's Isotope Program [16,20–26].

Of potential alternative accelerator-based [225]Ac-production methods, the proton-induced spallation of thorium at TRIUMF's 500 MeV Isotope Production Facility has potential to produce significant quantities of [225]Ac [14]. The co-production of [225]Ra during the spallation process is also of interest, as [225]Ra can serve as a generator of additional [225]Ac.

To make [225]Ac via thorium spallation at TRIUMF, thorium targets are bombarded with 480 MeV protons at a beam current of up to 100 µA. During the irradiation the thorium is hermetically sealed within a target capsule. For safe and reliable operation it is crucial that this hermetic seal around the thorium is at all times able to withstand any thermally induced mechanical stresses resulting from the large power deposition into the target from the proton beam, since such a "target failure" could enable the release of gaseous and volatile radionuclides co-produced within the thorium during the spallation process. Careful modelling and design of new targets must therefore be done to ensure safe and successful irradiations.

Herein, we introduce the design of a new target system for [225]Ac production through proton-induced thorium spallation at the TRIUMF 500 MeV Isotope Production Facility (IPF). The input and assumption for the modeling are described and discussed, and modeling results and initial operational experiences are presented.

## 2. Materials and Methods

### 2.1. TRIUMF's 500 MeV Isotope Production Facility

TRIUMF's 500 MeV Isotope Production Facility (IPF) was first conceived in 1978, and modelled after a similar facility at Brookhaven National Laboratory [27]. Historically, IPF has been primarily used for the irradiation of molybdenum targets for the production [82]Sr/[82]Rb generators), as well as occasional CsCl and KCl targets. However, the facility has received little use in recent years despite routinely receiving proton beam. IPF is located near the end of beamline 1A (BL1A), the main beamline of TRIUMF's 500 MeV cyclotron [28]. With the exception of TRIUMF's Thermal Neutron Facility [29]—which is downstream of the BL1A beam dump—the majority of BL1A users are located upstream from IPF and are not affected by its operations. Typical beam parameters for IPF are described in Section 2.2.

IPF consists of a 30 cm diameter, 8 m tall column of cooling water. The target station is located in the bottom of the water column at beam level. Above the water column, a shielded transfer hot cell is used to bring targets in or out of the facility and to move them in or out of the beam. Within the transfer cell, targets are inserted into one of six cassettes which are lowered into the cooling water column and down to target station by a chain drive. A total of 12 targets can be simultaneously irradiated at IPF—each cassette can hold a pair of up to 8 mm thick targets, separated by a 2.5 mm gap (see Figure 1b). When isotope production targets are not in use, helium gas targets are inserted to displace cooling water from the beam path, minimizing heat loads on the IPF cooling water heat exchanger. During irradiation, targets are submerged in the water column and housed in one of six cassettes through which recirculating cooling water is pumped (see Figure 1a,b). The cooling water flow of 114 L/min is routed through the station's 8 water circuits, six of which each feed individual cassettes. The two remaining water circuits are used for radial cooling of the target station's entrance and exit windows. Flow rates through individual cassettes are estimated to be between 7.2 and 23.4 L/min, depending on the location of the cassette. Cooling water temperature typically measures 25–28 °C at the target.

A typical IPF target is shown in Figure 1c: two Inconel® 600 windows (0.127 mm thickness) are welded to either side of a ring-shaped stainless steel target frame (10.1 cm diameter, 8.4 mm thickness), sealing within the 8.2 mm thick, 7.6 cm diameter puck of target material (ex. a puck of molybdenum dioxide or potassium chloride salt) [27]). As shown in Figure 1, the targets interface with other IPF components in three ways: alignment grooves on the bottom of the frame fix the orientation of the target within the cassette; a thermocouple hole at the bottom of the target has a thermocouple inserted in it during irradiation to monitor target temperature; and a tapped hole in the side of the frame allows targets to be securely picked up and manipulated by a threaded rod. More details regarding the IPF facility and its targets are provided by Burgerjon et al. [27].

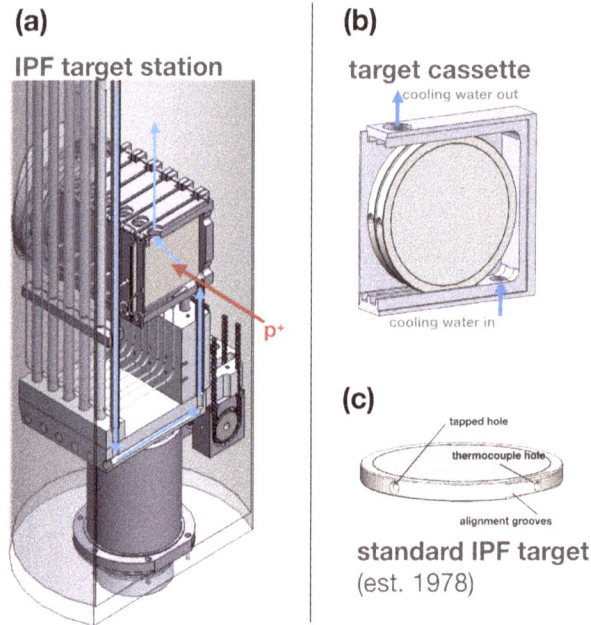

**Figure 1.** (**a**) IPF target station at beam level, with six cassettes inserted into the beam. The arrows indicate cooling water flow direction; (**b**) An IPF cassette holding two targets; (**c**) A typical IPF target.

*2.2. Proton Beam Parameters*

For 6–7 months per year, BL1A typically operates with 100–110 μA of 480 MeV protons at extraction from the TRIUMF main cyclotron, with the ability to increase this up to 170 μA. Before reaching IPF, beam losses occur due to two beryllium muon production targets (T1 and T2). Depending on the thickness of these targets, the proton beam reaching IPF can have an energy between 451 and 472 MeV and a current that is reduced by 15–40% to 60–94 μA under typical operating conditions. The beam profile at IPF is measured by a multi-wire scanner located 2 m before IPF. The Gaussian beam typically has a $2\sigma$ width between 25 and 35 mm in the horizontal and vertical directions, depending on the beam tune and thickness of the T1 and T2 targets.

Due to the variability in the beam conditions at IPF and the desire to design an $^{225}$Ac production target that is compatible with all BL1A operating conditions, multiple operational cases are considered when modelling irradiations of the thorium targets. A symmetrical Gaussian beam shape is assumed, with $2\sigma$ beam widths between 15 and 40 mm. Currents between 60 and 120 μA are also considered. Beam energy is fixed at 454 MeV (the lowest estimated beam energy at IPF), since at 454–470 MeV a <4% change in beam energy negligibly affects isotope production rates and heat loads on the target as both cross-sections and stopping powers are relatively flat in this energy range [30,31].

## 2.3. Thorium Target

### 2.3.1. Design Considerations

Since safe and reliable $^{225}$Ac production requires the hermetic seal of the target to withstand any thermally induced stresses caused by the proton beam, the target design must also maximize heat transfer from the thorium to the cooling water through the hermetic seal.

Post-irradiation processing of the target also requires that the thorium be dissolved. While a previous IPF thorium target prototype used ThO$_2$ [32], this is not suitable for routine $^{225}$Ac production, due to the high insolubility of ThO$_2$ [33]. Thorium metal was chosen as the target material as it can be readily dissolved post-irradiation. The metal also has other advantageous properties such as a high melting point and high density.

While the high density of thorium metal provides greater scaleability for maximizing $^{225}$Ac production if simultaneous irradiation of larger thorium quantities are desired, current needs only require the irradiation of thin thorium foils <1 mm in thickness. Therefore, the 8 mm wide hermetic target casing used for decades at IPF had to be modified in this target design.

### 2.3.2. Mechanical Assembly

The thorium target consists of an stainless steel 316 (SS316) outer frame that interfaces with IPF cassettes, and the thorium target material that is sealed within the frame by two Inconel® 718 windows (0.127 mm thickness). Currently, the thorium metal target material consists of a 60 mm diameter, 0.25 mm thick foil, purchased from IBI Labs (International Bio-Analytical Industries, Inc., Boca Raton, FL, USA).

Sealing of the thorium within the target is done in two stages. First, the thorium is sealed within a target sub-assembly that consists of a thin inner welding ring (1 mm thick, 76 mm diameter), to which both windows are electron-beam (EB) welded (Figure 2a). The sub-assembly is then EB welded to the target frame (Figure 2b). A photo of the finished target is shown in Figure 2c.

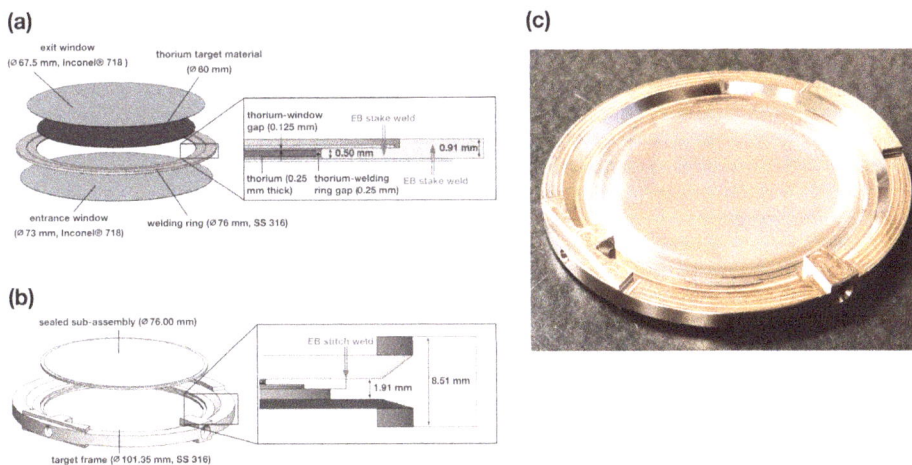

**Figure 2.** (**a**) Target sub-assembly, including the thorium foil, welding ring, and entrance and exit windows. Note the location of EB welds that bind the windows to the welding ring, sealing the thorium within; (**b**) Welding of the sub-assembly to the target frame; (**c**) Photo of the thorium target. Note the inward deflection of the Inconel® 718 window.

Relevant material properties for the thorium, SS316, and Inconel® 718 components of the target assembly are shown in Table 1. Yield and ultimate strength values for Inconel® 718 were obtained from the materials certification provided by the manufacturer (American Special Metals, Coral Springs, FL, USA). Material composition by mass for each target component is provided in Table 2.

All material properties used for modelling of the target were obtained either directly from the manufacturer, or the Knovel Engineering Data and Technical Reference Database [34]. Thorium material properties were also obtained from the ASM International databases [35].

**Table 1.** Base properties at room temperature for target components.

| Property | Inconel® 718 | Thorium | SS 316 |
|---|---|---|---|
| Density (g/cm$^3$) | 8.19 | 11.72 | 8 |
| Young's Modulus (GPa) | 199 | 72.4 | 193 |
| Thermal expansion coefficient (μm/mK) | 13 | 11.1 | 16.3 |
| Poisson's ratio | 0.3 | 0.27 | 0.28 |
| Yield strength (MPa) | 460 | 144 | 290 |
| Ultimate strength (MPa) | 895 | 219 | 580 |
| Melting Point (K) | 1533 | 2028 | 1673 |
| Thermal conductivity (W/m$^2$K) | 11.1 | 13.86 | 14.6 |

**Table 2.** Elemental composition of alloys used in L124 IPF targets.

| Material | Composition (% Mass) | | | | | | | | | |
|---|---|---|---|---|---|---|---|---|---|---|
| | Cu | Cr | Fe | Mn | Si | C | S | Mo | Ni | P |
| Inconel® 718 | 0.3 | 17 | 23.6 | 0.35 | 0.35 | 0.08 | 0.015 | 3.3 | 55 | 0.015 |
| SS 316 | - | 16 | 70 | 0.5 | 0.25 | 0.083 | 0.083 | 2 | 11 | 0.083 |
| thorium | >99.5% purity according to manufacturer and confirmed by ICP-MS analysis | | | | | | | | | |

### 2.3.3. Areal Contact at Thorium-Window Interface

Contact between the thorium foil and the target assembly window is an important factor when considering how effectively the target will be cooled during irradiation, as the majority of power deposited in the target assembly is expected to be removed to the cooling water via heat transfer over this barrier.

As shown in Figure 2a, before the thorium is sealed within the sub-assembly, a 0.125 mm gap exists between the thorium and each window. Sealing of the foil under the vacuum provided by the EB welding chamber results in a first mode deflection of the target windows: atmospheric pressure acts on the sealed, evacuated sub-assembly, forcing the windows to bend inwards (visible in Figure 2c). This ensures contact between the thorium and windows at the center of the target where heat loads from the proton beam are highest.

In order to measure the area for which the thorium and windows are in contact, the thickness of a manufactured sub-assembly was measured using a profilometer (Nanovea ST400). Shown in Figure 3, this indicates that the inner 49.8 mm of the thorium foil contacts the windows, as defined by the points where the windows stop deflecting inwards. These 49.8 mm represent 83.0% of the 60 mm thorium foil diameter, or 68.9% of the surface area. A second profile (not shown), made perpendicular to the one in Figure 3, also indicated 48.2 mm of contact (80.4% of the diameter).

For simulations of thorium target temperatures during irradiation, a conservative 45 mm wide contact region (75% of the diameter, and 56.3% of the surface area) is assumed. However, it should be noted that the amount of contact during the irradiation is expected to increase from measurements shown in Figure 3: while the profile measurement was done at atmosphere, additional pressure on the target windows will be present during irradiation due to the depth of the target in the cooling water column (8 m below the water surface).

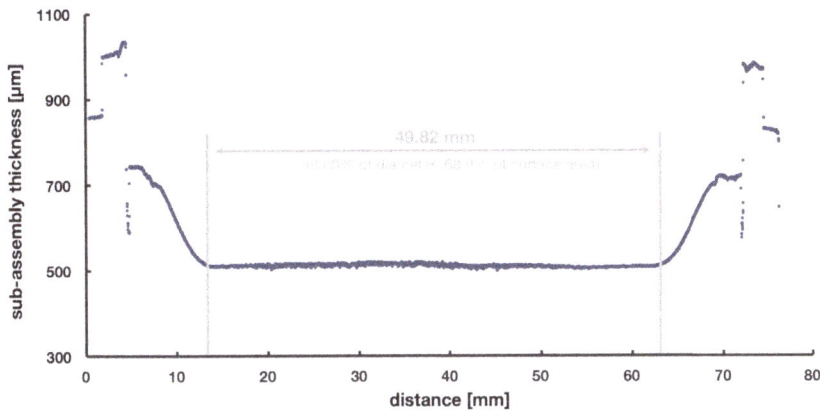

**Figure 3.** Plot of the thorium target sub-assembly thickness. Measurement of the inward deflection of the window, as also seen in Figure 2c, indicates the region for which the thorium and windows are in contact. The expected thickness in the centre of the sub-assembly (assuming contact of the thorium with both windows) is 504 µm.

## 2.3.4. Thermal Contact Resistance at Thorium-Window Interface

In addition to the contact area between the thorium and windows, the thermal contact resistance at this interface is also an important factor when considering how effectively the target will be cooled during irradiation. The greatest predictors of contact resistance are contact pressure, surface roughness, hardness, and yield strength. Thorium is soft and has a polished finish, so the contact resistance would likely be small. While limited data exists for thorium contact resistance measurements, comparisons can be made to similar materials. Several copper and aluminum alloys have similar properties to that of thorium; however, since aluminum values for contact resistance are higher it is more conservative to compare the thorium to aluminum (6000 series). The aluminum alloy with the closest material properties to thorium is aluminum 6061-T4 (Table 3). Table 4 also shows approximated contact resistances for several comparable metals. The aluminum-aluminum contact resistance is the highest of the comparable metals at 0.0005 m$^2$K/W. Other sources show a more detailed estimation of Aluminum 6061-T4 stress for the pressure range (0.1–0.2 MPa) the target will operate in [34]. Since these sources show a resistivity of below 0.001 m$^2$K/W the simulations will make the conservative approximation of 0.001 m$^2$K/W as the target contact resistance.

**Table 3.** Comparison of thorium to select aluminum and copper alloys in terms of material properties relevant to thermal contact resistance.

| Material | Yield Strength (MPa) | Brinell Hardness |
|---|---|---|
| thorium | 144 | 60–90 |
| aluminum 6061-T4 | 146 | 65–89 |
| aluminum 2024-T4 | 395 | 120–150 |
| aluminum 7075-T6 | 503 | 150–191 |
| copper 1010 | 305 | 105–123 |

**Table 4.** Thermal contact resistances for select materials under vacuum conditions.

| Interface | Thermal Contact Resistance (m$^2$K/W) |
| --- | --- |
| iron-aluminum | 0.00002 |
| copper-copper | 0.0001 |
| aluminum-aluminum | 0.00045 |
| stainless-stainless | 0.005 |
| ceramic-ceramic | 0.002 |

## 2.4. Thermomechanical Modelling

### 2.4.1. Power Deposition

Energy deposited by 454 MeV protons within each target component were simulated using SRIM [30] and results for each target component in MeV/proton are shown in Table 5. These values are then used to create power distribution profiles for each component that incorporate the beam's current and Gaussian profile. A MATLAB script is used to generate these profiles and output them as contours that can be imported directly into ANSYS CFX (version 19.0). ANSYS CFX then interpolates between these contours to determine the 3-dimensional power deposition distribution for each target component. Please note that the beam is assumed to have a Gaussian shape symmetrical in the x- and y-directions, with beam width specified by the 2σ-value of the Gaussian. Since beam current and width may change between individual IPF irradiations (see Section 2.2), multiple beam width and current values are considered.

**Table 5.** Energy deposited in thorium target components by 454 MeV protons.

| Component | Material | Thickness (mm) | Energy Deposited (MeV/Proton) |
| --- | --- | --- | --- |
| thorium foil | thorium | 0.25 | 0.423 |
| entrance window | Inconel® 718 | 0.127 | 0.21 |
| exit window | Inconel® 718 | 0.127 | 0.21 |
| target frame | SS 316 | 8.51 | 8.306 |
| welding ring | SS 316 | 0.91 | 1.466 |

### 2.4.2. Thermal Modelling

ANSYS CFX simulations were used to model heat transport within the target during irradiation. Power deposition distributions and cooling water flow simulations were combined to model the thermal response of an irradiated and cooled target. Radiative heat transfer was excluded from the thermal simulations and only conductive heat flow was considered.

ANSYS CFX simulations used a $\kappa - \epsilon$ turbulence model to represent the mixed laminar and turbulent flow conditions present inside the target cassette. This model does not deal with issues of flow recovery, unconfined flows, flow separation or flow reattachment. While flow recovery, flow separation, and flow reattachment were observed in simulations, the flow rate is low enough for these effects to be negligible. Therefore, the $\kappa - \epsilon$ turbulence model is a robust, stable, and conservative model to use [36].

Properties of inlet cooling water flow across the targets were set to 28 °C , and a total flow rate of 0.120 L/s per cassette was used (corresponding to the lowest calculated value for mass flow rate through a given cassette, as detailed in Section 2.1). Due to the absence of flow measurements for individual cassettes, flow sensitivity checks were conducted to determine the effects of reduced flow on the simulation results.

The numerical accuracy of the solutions was determined based on the approach found in the *Journal of Fluid Engineering* [37]. The technique implements the iterative process of decreasing mesh size and solving the simulation until temperature, heat transfer and flow changes are negligible between mesh changes. The Grid Converged Index is then used to estimate the rate of convergence of the solution and truncation errors. Based on this standard check, the maximum error in the simulated temperature results is 9 K for the target windows.

### 2.4.3. Mechanical Modelling

To evaluate potential damage to the target during irradiation, stresses for the thorium foil and target windows are analyzed using a linear stress-strain model. Greater importance is placed on the integrity of the windows that provide a hermetic seal around the target during irradiation—damage to the thorium foil alone is not considered a target failure. Use of a linear model reduces computation time but limits the accuracy of results to cases where simulated stresses are below the yield stress of the materials. Therefore, when assessing the potential for target failure, the maximum stress in the target windows is, conservatively, compared to the yield stress of Inconel® 718.

To further reduce computation time, simulations also assumed temperature independence of material properties within the target assembly. This introduces an additional limitation: results will only be accurate within temperature ranges for which critical material properties are constant. For the Inconel® windows—the components critical when evaluating potential for target failure—this can be assumed between 0 and 600 °C . The most critical Inconel® 718 properties—the ultimate and yield strengths—drastically decrease above 600 °C . Please note that other Inconel® 718 properties such as Young's modulus, Poisson's ratio, and thermal conductivity are also temperature dependent, but are mostly constant over the same 0 to 600 °C range. Due to its rare and radioactive nature, similar material property temperature-dependence data is challenging to find for thorium.

In addition to thermally induced stresses in the target caused by the proton beam, the target also experiences stresses caused by atmospheric and hydrostatic pressures. Since the thorium is sealed within the sub-assembly under vacuum conditions, 0.1 MPa of atmospheric pressure is exerted on the Inconel® windows after the target is removed from the EB welding chamber. The windows experience an additional 0.1 MPa (0.2 MPa total) when submerged under 8 m of water in the IPF target station during irradiation.

In the centre of the target where the windows press against the thorium, this 0.2 MPa of stress is negligible compared to the stress caused by heat during irradiation (on order of $10^2$ MPa). Higher stresses are experienced near the edge of the window in the region where it deflects across the edge of the welding ring. Since accurate modelling of stresses at such boundaries is challenging, tests of the target's ability to withstand these pressures were conducted: after the target was placed in a helium pressure chamber at 0.5 MPa for 2 h, no damage was observed. Based on this result, atmospheric and hydrostatic pressures were neglected when modelling the stresses in the thorium target during irradiation.

Mechanical simulation boundary conditions were selected on the perimeter of the thermal loading region to determine conservative worst case stresses. Similar to methods described in Section 2.4.2, various iterations of mesh size and type were completed to determine the grid independence of the mechanical solution before final simulations were completed.

### 2.5. Yield Measurements

After test irradiations, the thorium foil was dissolved in 10 M $HNO_3$ + 12.5 mM HF and evaporated to a thorium nitrate salt before re-dissolution in 80.0 mL of 1 M $HNO_3$, in preparation for the radiochemical separation of $^{225}$Ac from the target (described in an upcoming publication). A small portion (<100 μL of the redissolved target was removed and analyzed by gamma ray spectroscopy with a N-type Co-axial HPGe gamma spectrometer from Canberra fitted with a 0.5 mm beryllium window. The detector was calibrated (energy and efficiency) with a 20 ml $^{133}$Ba and $^{152}$Eu source. The dead time was less than 2%. The amount of $^{225}$Ra produced was quantified using the 40 keV gamma line of $^{225}$Ra, while the amount of $^{225}$Ac produced was quantified using the 218 keV gamma line of $^{221}$Fr (secular equilibrium between $^{225}$Ac and daughter $^{221}$Fr can be assumed at the time of measurement).

## 3. Results and Discussion

### 3.1. Modelling and Sensitivity Analysis

Table 6 shows the power deposited by the proton beam in each target component and the total sum, accounting for different beam shapes and beam currents. The beam current was varied from 60 μA to 120 μA and the beam width from 15 mm to 40 mm to cover all realistic cases. While the total deposited power scales with the beam current as expected, variation due to the beam width are 23% for all beam currents. It should be noted that as the beam width increases, the total power deposited in the thorium foil decreases since some of the proton beam is, undesirably, no longer hitting the thorium but hits the welding ring and frame instead. As an example, the resulting temperature and thermally induced stress for a 100 μA, 20 mm beam are shown in Figure 4, along with the cooling water flow profiles for a 0.12 L/s cassette flow. The maximum temperature and stress in the window follow roughly the beam shape.

**Table 6.** Energy deposited in thorium target components by 454 MeV protons.

| Current (μA) | Width (mm) | Power Deposition in Target (W) | | | | | |
|---|---|---|---|---|---|---|---|
| | | Thorium | ent. Window | Exit Window | Welding Ring | Frame | Total |
| 60 | 15 | 25.42 | 12.63 | 12.63 | 0 | 0 | 50.68 |
| | 20 | 25.28 | 12.63 | 12.63 | 0.01 | 0.05 | 50.61 |
| | 25 | 24.59 | 12.63 | 12.63 | 0.01 | 0.88 | 50.75 |
| | 30 | 23.16 | 12.63 | 12.63 | 0.17 | 4.06 | 52.64 |
| | 35 | 21.21 | 12.63 | 12.63 | 0.27 | 10.04 | 56.78 |
| | 40 | 19.08 | 12.61 | 12.62 | 0.35 | 17.63 | 62.29 |
| 80 | 15 | 33.89 | 16.84 | 16.84 | 0 | 0 | 67.57 |
| | 20 | 33.71 | 16.84 | 16.84 | 0.02 | 0.07 | 67.48 |
| | 25 | 32.79 | 16.84 | 16.84 | 0.02 | 1.18 | 67.67 |
| | 30 | 30.88 | 16.84 | 16.84 | 0.22 | 5.41 | 70.19 |
| | 35 | 28.28 | 16.84 | 16.84 | 0.36 | 13.39 | 75.7 |
| | 40 | 25.44 | 16.82 | 16.83 | 0.46 | 23.5 | 83.05 |
| 100 | 15 | 42.36 | 21.05 | 21.05 | 0 | 0 | 84.46 |
| | 20 | 42.14 | 21.05 | 21.05 | 0.02 | 0.09 | 84.35 |
| | 25 | 40.99 | 21.05 | 21.05 | 0.02 | 1.47 | 84.58 |
| | 30 | 38.6 | 21.05 | 21.05 | 0.28 | 6.77 | 87.74 |
| | 35 | 35.36 | 21.05 | 21.05 | 0.45 | 16.74 | 94.63 |
| | 40 | 31.8 | 21.02 | 21.04 | 0.58 | 29.38 | 103.82 |
| 120 | 15 | 50.83 | 25.26 | 25.26 | 0 | 0 | 101.35 |
| | 20 | 50.57 | 25.26 | 25.26 | 0.03 | 0.11 | 101.22 |
| | 25 | 49.19 | 25.26 | 25.26 | 0.03 | 1.77 | 101.5 |
| | 30 | 46.32 | 25.26 | 25.26 | 0.33 | 8.12 | 105.29 |
| | 35 | 42.43 | 25.25 | 25.25 | 0.53 | 20.08 | 113.55 |
| | 40 | 38.16 | 25.22 | 25.25 | 0.69 | 35.26 | 124.58 |

To summarize, thermal results determined by ANSYS CFX for all beam parameters considered, are shown in Figure 5. Similarly, the resulting maximal stresses on the thorium foil and Inconel® windows are shown in Figure 6. As expected, increases in beam current and decreases in beam width result in higher maximum temperatures and stresses on the target windows. However, using safety factors defined relative to each material's yield strength, it can be seen that all beams of width >20 mm and <100 μA result in safety factors on the target window >1. Window temperatures also remain within the region of accuracy (<600 °C) as defined in Section 2.4.3. These limits are well within the range of typical beam parameters at IPF (Section 2.2), meaning that [225]Ac production at IPF can occur downstream of experimental users without any alterations to the beam or impact to other users of the beamline.

Some of the assumptions and input parameters to the thermomechanical simulations were simplified in order to reduce computation time. The temperature-independence of material properties, the linear stress-strain curves used for each material, and the mechanical simulation boundary conditions on the windows are examples that limit the accuracy of the simulations. However, these are balanced by conservative assumptions such as thermal contact resistance value used for the thorium-inconel interface. Combined with conservative stress limits on the hermetic seal of the target (relative to the yield stress of Inconel® 718 as opposed to the much higher ultimate stress), these assumptions provide confidence that these targets can be irradiated safely under the simulated conditions.

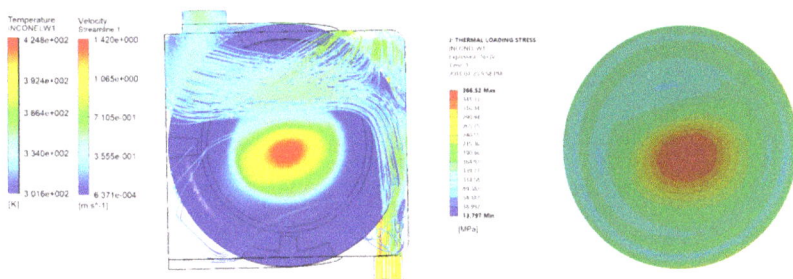

**Figure 4.** (**a**) Example of simulated flow streamlines and temperatures; and (**b**) stresses in the target entrance window for a 100 µA, 20 mm beam.

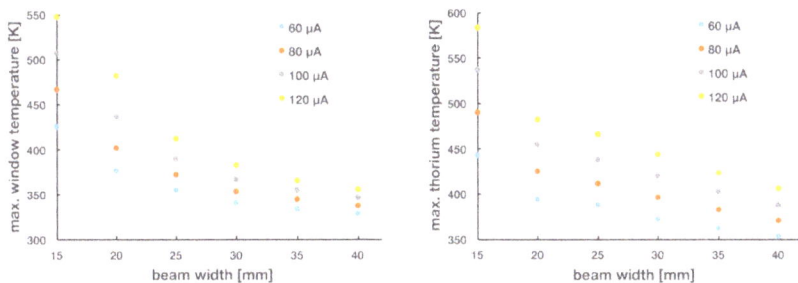

**Figure 5.** Maximum simulated temperatures within target window (**left**) and thorium foil (**right**) components during irradiations of varying beam current and width.

To test the sensitivity of the window and thorium foil temperature on the achieved thermal contact resistance between the thorium foil and the window, simulations were carried out varying the resistance from 0.0005 m$^2$K/W to 0.003 m$^2$K/W. The window temperature, cooled by the cooling water, is almost unaffected even for a heat transfer resistance underestimated by a factor of three. The thorium foil temperature, however, raises from 442 K to 506 K, potentially causing damage to the thorium foil.

Similarly, we varied the water flow in the cassette holding the targets from the nominal 0.12 kg/s down to 0.006 kg/s. Although the temperature of both the window and the thorium foil increase, even at a 95% reduction in water flow, the temperature only increases by less than 15%. Please note that a zero flow situation is interlocked by the target thermocouples, which would trip off the proton beam in the absence of cooling water flow.

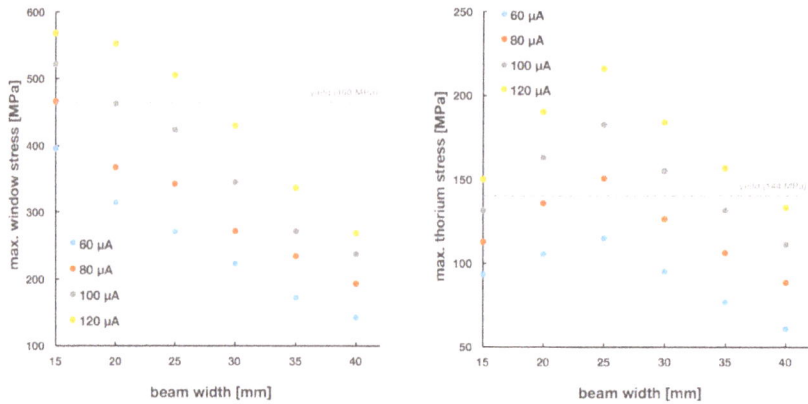

**Figure 6.** Maximum simulated stresses resulting from temperatures within window (**left**) and thorium (**right**) target components during irradiations. Please note that increase in thorium foil stresses as beam width increases (from 15 to 25 mm) is due to deposition of heat in regions of the thorium that are not in contact with the window.

*3.2. Test Irradiations*

Two irradiations of this target design have been completed to date. Both targets were irradiated for 36.5 h with an average beam current of 72.5 μA. Beam width was between 28–32 mm in the vertical and 30–35 mm in the horizontal directions.

After allowing targets to decay for 7 days, the thorium foil was removed from the target. In both cases, no damage, warping, or significant discoloration of the foil was observed.

The amount of $^{225}$Ac and $^{225}$Ra produced were measured via gamma spectroscopy. The decay corrected activity at EOB (average $\pm$ standard deviation) was (524 $\pm$ 21) MBq (14.2 mCi) and (86 $\pm$ 13) MBq (2.3 mCi) for $^{225}$Ac and $^{225}$Ra, respectively. These correspond to saturation yields of 72.5 MBq/μA for $^{225}$Ac and 17.6 MBq/μA for $^{225}$Ra.

While the $^{225}$Ac quantities produced so far may be modest, significant scaleability exists for $^{225}$Ac production at TRIUMF's IPF. An increase in irradiation time to a full $^{225}$Ac half-life (9.9 days), would increase production by a factor of at least 8. The simultaneous irradiation of 12 targets is also possible. Combined, these two factors alone suggest $^{225}$Ac production at IPF could be readily scaled to at least 42 GBq (1.1 Ci) every 10 days, without requiring changes to the beam parameters or target design. Further increases in $^{225}$Ac and $^{225}$Ra yields could theoretically be achieved by increasing the proton beam current or the thickness of the 0.25 mm thick thorium foil (IPF targets can accommodate targets up to 8 mm thick); however, these would require a reassessment of the safety of such an irradiation—using the methodology presented herein—and a potential redesign of the target to prevent target failure.

The saturation yields for a 12-target irradiation would increase to 870 MBq/μA and 211 MBq/μA for $^{225}$Ac and $^{225}$Ra, respectively. Previous studies by others for the production of $^{225}$Ac via $^{232}$Th spallation have reported $^{225}$Ac saturation yields of 444 MBq/μA at 90 MeV and 1140 MBq/μA at 192 MeV [24,25]. However, these studies do not report yields for $^{225}$Ra production, which is known to be smaller at the reported energies of irradiation [31,38,39]. The $^{225}$Ra produced has the potential to not only provide approximately 0.8 MBq of $^{225}$Ac per 1 MBq of $^{225}$Ra [14], but also to provide generator-produced $^{225}$Ac with significantly reduced $^{227}$Ac impurities, a long-lived ($t_{1/2}$ = 21.8 y) alpha-emitting radioisotope with low regulatory restrictions on waste disposal and accidental intake [32].

## 4. Conclusions

A target system for accelerator-based production of $^{225}$Ac from thorium metal at TRIUMF was presented and discussed. The thermomechanical response of the target to irradiation by the proton beam was thoroughly evaluated against conservative thresholds for target safety and the target was found to be compatible with the existing beam at typically received at the irradiation facility.

The target system was tested in two irradiations with 72 μA of 450 MeV protons. After removal from the beam neither the target windows nor the thorium foil showed signs of warping, discoloration or any other signs of excessive heat or mechanical stress. With this system we were able to produce (524 ± 21) MBq of $^{225}$Ac (14.2 mCi) and (86 ± 13) MBq of $^{225}$Ra (2.3 mCi) within a 36 h irradiation, corresponding to saturation yields of 72.5 MBq/μA for $^{225}$Ac and 17.6 MBq/μA for Raa. Longer irradiations to produce higher quantities are planned for the near future. It is estimated that the simultaneous irradiation of 12 targets over a 240 h period could produce 42 GBq of $^{225}$Ac and 7 GBq of $^{225}$Ra without requiring any changes to the target or beam parameters—further scaleability exists if these parameters would are also be changed.

**Author Contributions:** Conceptualization, P.S. and A.K.H.R.; methodology, L.M. and A.K.H.R.; formal analysis, A.L. and A.K.H.R.; data curation, A.K.H.R. and A.L.; writing—original draft preparation, A.K.H.R. and C.H.; writing—review and editing, C.H., P.S., and A.L.; supervision, C.H.; project administration, A.K.H.R.; funding acquisition, P.S.

**Funding:** TRIUMF receives funding via a contribution agreement with the National Research Council of Canada.

**Acknowledgments:** The authors would like to thank the following individuals for their valuable input into the design of the thorium target presented in this work: Ewart Blackmore, Gabriel Cojocaru, Tim Goodsell, Yetvart Hosepyan, Juergen Kaefer, Max Kinakin, Anders Mjos, Sam Varah, and Bob Welbourn.

**Conflicts of Interest:** The authors declare no conflict of interest.

## References

1. McDevitt, M.R.; Sgouros, G.; Finn, R.D.; Humm, J.L.; Jurcic, J.G.; Larson, S.M.; Scheinberg, D.A. Radioimmunotherapy with alpha-emitting nuclides. *Eur. J. Nucl. Med.* **1998**, *25*, 1341–1351. [CrossRef] [PubMed]
2. Couturier, O.; Supiot, S.; Degraef-Mougin, M.; Faivre-Chauvet, A.; Carlier, T.; Chatal, J.F.; Davodeau, F.; Cherel, M. Cancer radioimmunotherapy with alpha-emitting nuclides. *Eur. J. Nucl. Med. Mol. Imaging* **2005**, *32*, 601–614. [CrossRef] [PubMed]
3. Mulford, D.A.; Scheinberg, D.A.; Jurcic, J.G. The promise of targeted {alpha}-particle therapy. *J. Nucl. Med.* **2005**, *46* (Suppl. 1), 199S–204S. [PubMed]
4. Brechbiel, M.W. Targeted alpha-therapy: Past, present, future? *Dalton Trans.* **2007**, 4918–4928. [CrossRef] [PubMed]
5. Kim, Y.S.; Brechbiel, M.W. An overview of targeted alpha therapy. *Tumor Biol.* **2012**, *33*, 573–590. [CrossRef] [PubMed]
6. Baidoo, K.E.; Yong, K.; Brechbiel, M.W. Molecular pathways: Targeted alpha-Particle radiation therapy. *Clin. Cancer Res.* **2013**, *19*, 530–537. [CrossRef] [PubMed]
7. Elgqvist, J.; Frost, S.; Pouget, J.P.; Albertsson, P. The potential and hurdles of targeted alpha therapy—Clinical trials and beyond. *Front. Oncol.* **2014**, *3*, 324. [CrossRef] [PubMed]
8. Miederer, M.; Scheinberg, D.A.; McDevitt, M.R. Realizing the potential of the Actinium-225 radionuclide generator in targeted alpha particle therapy applications. *Adv. Drug Deliv. Rev.* **2008**, *60*, 1371–1382. [CrossRef] [PubMed]
9. Kratochwil, C.; Bruchertseifer, F.; Giesel, F.L.; Weis, M.; Verburg, F.A.; Mottaghy, F.; Kopka, K.; Apostolidis, C.; Haberkorn, U.; Morgenstern, A. 225Ac-PSMA-617 for PSMA targeting alpha-radiation therapy of patients with metastatic castration-resistant prostate cancer. *J. Nucl. Med.* **2016**, *57*, 1941–1944. [CrossRef] [PubMed]
10. Jurcic, J.G.; Rosenblat, T.L. Targeted alpha-particle immunotherapy for acute myeloid leukemia. In *2014 American Society of Clinical Oncology Educational Book*; American Society of Clinical Oncology: Alexandria, VA, USA, 2014; pp. e126–e131. [CrossRef]

11. Kratochwil, C.; Giesel, F.L.; Bruchertseifer, F.; Mier, W.; Apostolidis, C.; Boll, R.; Murphy, K.; Haberkorn, U.; Morgenstern, A. 213Bi-DOTATOC receptor-targeted alpha-radionuclide therapy induces remission in neuroendocrine tumours refractory to beta radiation: A first-in-human experience. *Eur. J. Nucl. Med. Mol. Imaging* **2014**, *41*, 2106–2119. [CrossRef]

12. Allen, B.; Singla, A.; Rizvi, S.; Graham, P.; Bruchertseifer, F.; Apostolidis, C.; Morgenstern, A. Analysis of patient survival in a Phase I trial of systemic targeted α-therapy for metastatic melanoma. *Immunotherapy* **2011**, *3*, 1041–1050. [CrossRef] [PubMed]

13. Kratochwil, C.; Bruchertseifer, F.; Rathke, H.; Hohenfellner, M.; Giesel, F.L.; Haberkorn, U.; Morgenstern, A. Targeted α-Therapy of Metastatic Castration-Resistant Prostate Cancer with 225 Ac-PSMA-617: Swimmer-Plot Analysis Suggests Efficacy Regarding Duration of Tumor Control. *J. Nucl. Med.* **2018**, *59*, 795–802. [CrossRef] [PubMed]

14. Robertson, A.K.H.; Ramogida, C.F.; Schaffer, P.; Radchenko, V. Development of 225Ac Radiopharmaceuticals: TRIUMF Perspectives and Experiences. *Curr. Radiopharm.* **2018**, *11*. [CrossRef] [PubMed]

15. Zhuikov, B.L. Successes and problems in the development of medical radioisotope production in Russia. *Phys. Uspekhi* **2016**, *59*, 481–486. [CrossRef]

16. Radchenko, V.; Engle, J.W.; Wilson, J.J.; Maassen, J.R.; Nortier, F.M.; Taylor, W.A.; Birnbaum, E.R.; Hudston, L.A.; John, K.D.; Fassbender, M.E. Application of ion exchange and extraction chromatography to the separation of actinium from proton-irradiated thorium metal for analytical purposes. *J. Chromatogr. A* **2015**, *1380*, 55–63. [CrossRef]

17. Aliev, R.A.; Ermolaev, S.V.; Vasiliev, A.N.; Ostapenko, V.S.; Lapshina, E.V.; Zhuikov, B.L.; Zakharov, N.V.; Pozdeev, V.V.; Kokhanyuk, V.M.; Myasoedov, B.F.; et al. Isolation of Medicine-Applicable Actinium-225 from Thorium Targets Irradiated by Medium-Energy Protons. *Sol. Extr. Ion Exch.* **2014**, *32*, 468–477. [CrossRef]

18. Morgenstern, A. Bismuth-213 and actinium-225—Generator performance and evolving therapeutic applications of two generator-derived alpha-emitting radioisotopes. *Curr. Radiopharm.* **2012**, *5*, 221–227. [CrossRef]

19. International Atomic Energy Agency. *Technicial Meeting on Alpha eMitting Radionuclides and Radiopharmaceuticals for Therapy*; Technical Report; International Atomic Energy Agency: Vienna, Austria, 2013.

20. Weidner, J.W.; Mashnik, S.G.; John, K.D.; Ballard, B.; Birnbaum, E.R.; Bitteker, L.J.; Couture, A.; Fassbender, M.E.; Goff, G.S.; Gritzo, R.; et al. 225Ac and 223Ra production via 800 MeV proton irradiation of natural thorium targets. *Appl. Radiat. Isot.* **2012**, *70*, 2590–2595. [CrossRef]

21. Weidner, J.W.; Mashnik, S.G.; John, K.D.; Hemez, F.; Ballard, B.; Bach, H.; Birnbaum, E.R.; Bitteker, L.J.; Couture, A.; Dry, D.; et al. Proton-induced cross sections relevant to production of 225Ac and 223Ra in natural thorium targets below 200 MeV. *Appl. Radiat. Isot.* **2012**, *70*, 2602–2607. [CrossRef]

22. Engle, J.W.; Mashnik, S.G.; Weidner, J.W.; Wolfsberg, L.E.; Fassbender, M.E.; Jackman, K.; Couture, A.; Bitteker, L.J.; Ullmann, J.L.; Gulley, M.S.; et al. Cross sections from proton irradiation of thorium at 800 MeV. *Phys. Rev. C Nucl. Phys.* **2013**, *88*. [CrossRef]

23. Cutler, C.; Mausner, L. Energetic protons boost BNL isotope production TRIUMF targets alpha therapy. *Cern Cour.* **2016**, *56*, 32–35.

24. Griswold, J.; Medvedev, D.; Engle, J.; Copping, R.; Fitzsimmons, J.; Radchenko, V.; Cooley, J.; Fassbender, M.; Denton, D.; Murphy, K.; et al. Large scale accelerator production of 225Ac: Effective cross sections for 78–192 MeV protons incident on 232Th targets. *Appl. Radiat. Isot.* **2016**, *118*, 366–374. [CrossRef] [PubMed]

25. Griswold, J.R. Actinium-225 Production via Proton Irradiation of Thorium-232. Ph.D. Thesis, Univeristy of Tennessee, Knoxville, TN, USA, 2016.

26. NorthStar Medical Radioisotopes. *Production of Actinium-225 via High Energy Proton Induced Spallation of Thorium-232*; Technical Report; NorthStar Medical Radioisotopes: Madison, WI, USA, 2011.

27. Burgerjon, J.J.; Pate, B.D.; Blaby, R.E.; Page, E.G.; Lenz, J.; Trevitt, B.T. The TRIUMF 500 MeV, 100 uA Isotope Production Facility. In Proceedings of the 27th Conference on Remote System Technology, San Francisco, CA, USA, 12–16 November 1979.

28. Bylinskii, I.; Craddock, M.K. The TRIUMF 500 MeV cyclotron: The driver accelerator. *Hyperfine Interact.* **2013**, *225*, 9–16. [CrossRef]

29. Blackmore, E.W.; Dodd, P.E.; Shaneyfelt, M.R. Improved capabilities for proton and neutron irradiations at TRIUMF. In Proceedings of the 2003 IEEE Radiation Effects Data Workshop, Monterey, CA, USA, 25 July 2003. [CrossRef]

30. Ziegler, J.F.; Ziegler, M.; Biersack, J. SRIM—The stopping and range of ions in matter (2010). *Nucl. Instrum. Methods Phys. Res. Sect. B Beam Interact. Mater. Atoms.* **2010**, *268*, 1818–1823. [CrossRef]
31. Chadwick, M.; Herman, M.; Obložinský, P.; Dunn, M.; Danon, Y.; Kahler, A.; Smith, D.; Pritychenko, B.; Arbanas, G.; Arcilla, R.; et al. ENDF/B-VII.1 Nuclear Data for Science and Technology: Cross Sections, Covariances, Fission Product Yields and Decay Data. *Nucl. Data Sheets* **2011**, *112*, 2887–2996. [CrossRef]
32. Robertson, A.K.H.; Ladouceur, K.; Nozar, M.; Moskven, L.; Ramogida, C.F.; D'Auria, J.; Sossi, V.; Schaffer, P. Design and Simulation of Thorium Target for Ac-225 Production. *AIP Conf. Proc.* **2017**, *1845*, 020019–1–020019–5. [CrossRef]
33. Hyde, E.K. *The Radiochemistry of Thorium*; Vol. NAS-NS 300, Subcommittee on Radiochemistry, National Academy of Sciences–National Research Council: Berkeley, CA, USA, 1960. [CrossRef]
34. Knovel Engineering Technical Reference Information Database. Available online: https://app.knovel.com/web/ (accessed on 13 February 2019).
35. International, A. *Properties and Selection: Nonferrous Alloys and Special-Purpose Materials*; ASM International: Almere, The Netherlands, 1991; Volume 2, p. 1300. [CrossRef]
36. Mohammadi, B.; Pironneau, O. *Analysis of the K-Epsilon Turbulence Model*; Wiley: Hoboken, NJ, USA, 1994.
37. Journal of Fluids Engineering Editorial Policy Statement on the Control of Numerical Accuracy. Technical Report. Available online: https://www.asme.org/wwwasmeorg/media/ResourceFiles/Shop/Journals/JFENumAccuracy.pdf (accessed on 13 February 2019).
38. Experimental Nuclear Reaction Data (EXFOR). Available online: http://www.nndc.bnl.gov/exfor/exfor.htm (accessed on 13 February 2019).
39. ENDF: Evaluated Nuclear Data File. Available online: https://www-nds.iaea.org/exfor/endf.htm (accessed on 13 February 2019).

*instruments*

MDPI

*Technical Note*

# Introduction of the New Center for Radiopharmaceutical Cancer Research at Helmholtz-Zentrum Dresden-Rossendorf

Martin Kreller [1,*], Hans Jürgen Pietzsch [1], Martin Walther [1], Henrik Tietze [2], Peter Kaever [2], Torsten Knieß [1], Frank Füchtner [1], Jörg Steinbach [1] and Stephan Preusche [1]

[1]  Helmholtz-Zentrum Dresden-Rossendorf, Institute of Radiopharmaceutical Cancer Research, Bautzner Landstraße 400, 01328 Dresden, Germany; h.j.pietzsch@hzdr.de (H.J.P.); m.walther@hzdr.de (M.W.); t.kniess@hzdr.de (T.K.); f.fuechtner@hzdr.de (F.F.); j.steinbach@hzdr.de (J.S.); s.preusche@hzdr.de (S.P.)
[2]  Helmholtz-Zentrum Dresden-Rossendorf, Department of Research Technology, Bautzner Landstraße 400, 01328 Dresden, Germany; h.tietze@hzdr.de (H.T.); p.kaever@hzdr.de (P.K.)
*   Correspondence: m.kreller@hzdr.de; Tel.: +49-351-260-4029

Received: 20 December 2018; Accepted: 24 January 2019; Published: 30 January 2019

**Abstract:** A new Center for Radiopharmaceutical Cancer Research was established at the Helmholtz-Zentrum Dresden-Rossendorf in order to centralize radionuclide production, radiopharmaceutical production and the chemical and biochemical research facilities. The newly installed cyclotron is equipped with two beamlines, two target selectors and several liquid, gas and solid target systems. The cyclotron including the target systems and first results of beam characterization measurements as well as results of the radionuclide production are presented. The produced radionuclides are automatically distributed from the targets to the destination hot cells. This process is supervised and controlled by an in-house developed system.

**Keywords:** cyclotrons; radionuclide production; solid, liquid and gas targets

## 1. Introduction

Radiopharmaceutical research and the production of radiopharmaceuticals have a long history at the Research Center in Rossendorf. The production of radiopharmaceuticals started in 1958 with a nuclear research reactor (10 MW) and the Cyclotron U-120 (Leningrad). A broad scale of radiolabeled products based on $^{14}C$, $^{131}I$, $^{123}I$, $^{32}P$, $^{75}Se$, $^{67}Ga$, $^{85}Sr$, $^{111}In$, $^{211}At$ and fission radionuclides such as $^{90}Sr/^{90}Y$, $^{99}Mo$ were provided. Furthermore, the Research Center was the second producer of fission $^{99}Mo/^{99m}Tc$-generators.

The year 1997 marked the official opening of Rossendorf PET-Center for research and application including the manufacturing authorization for PET drugs. The marketing authorization includes [$^{18}F$]FDG (GlucoRos), [$^{18}F$]Fluoride (NaFRos) and [$^{18}F$]FDOPA (DOPARos). Furthermore, there are 15 different radiopharmaceuticals available on demand. In the past, radionuclide production, pharmacological research and the pharmaceutical production were located at different places at Helmholtz-Zentrum Dresden-Rossendorf (HZDR) [1].

The new Center for Radiopharmaceutical Cancer Research (ZRT), as it is shown in Figure 1, was established to centralize the main units: a high current proton cyclotron, a radiopharmaceutical production—GMP (Good manufacturing practice) unit including the quality control, laboratories for PET-radiochemistry, chemical laboratories, laboratories for biochemical investigations inside and outside the controlled area, laboratories for small animal imaging (small animal PET/CT, PET/MR, SPECT, OI, MR) and a laboratory animal facility (mice and rats).

**Figure 1.** Impression of the Center for Radiopharmaceutical Tumor Research. The vault of the cyclotron is shown in the front. Picture: HZDR/Frank Bierstedt.

The ZRT completes the research infrastructure of the Institute of Radiopharmaceutical Cancer Research. In close cooperation, the six departments at the institute are developing and testing radioactive drugs for cancer diagnostics and therapy. The corresponding processes are accompanied from the idea (clinical need) to the introduction of novel drugs into clinical practice. The radiopharmaceuticals are produced for research and for hospital use.

## 2. The Production of Radiopharmaceuticals

Due to the short half-lives of $^{18}$F and $^{11}$C, the radionuclides and the corresponding PET radiopharmaceuticals are produced in one site, which means the cyclotron, GMP production area and quality control are located in one building. If the medical application is not situated in the campus area, the transport time should not exceed one half-life of the radionuclide. As a rule of thumb, the time lag between radionuclide production and patient investigation should not exceed three half-life periods, consequently the production of PET radiopharmaceuticals is faced with a number of challenges.

As is well known, a number of prerequisites or work steps must be fulfilled for the production of a radiopharmaceutical: The radiolabelling reaction, i.e., the "introduction" of the PET radionuclide into the biomolecule, is performed by synthesis modules in lead-shielded hot cells to minimize the radiation exposure to the staff. All production steps have to be performed under clean room conditions and according to GMP guidelines. For that purpose a clean room area of about 200 m$^2$ was established within the ZRT. It is equipped with 14 hot cells under clean room conditions B and C (see Figure 2).

The majority of the routinely produced radiopharmaceuticals ([$^{18}$F]FDG ([$^{18}$F]fluoro-2-desoxy-D-glucose), [$^{18}$F]FDOPA (3,4-dihydroxy-6-[$^{18}$F]-fluoro-L-phenylalanine), Sodium [$^{18}$F]fluoride, [$^{11}$C]Methionin, [$^{18}$F]FMISO ([$^{18}$F]Fluoromisonidazole)) are applied for cancer diagnostics. [$^{18}$F]FDOPA and [$^{18}$F]Flutemetamol (Vizamyl©) are produced for diagnostics of neurodegenerative disorders.

**Figure 2.** GMP area including the hot cells for the radiopharmaceutical production.

The ZRT has also created optimal conditions for the quality control of radiopharmaceuticals. The low level of non-radiolabeled compound present in the radiopharmaceutical precludes the application of common analytical techniques like NMR (Nuclear magnetic resonance) and mass spectroscopy. Identity, purity and radiochemical purity of the pharmaceuticals have to be assessed with (radio)-HPLC (High-performance liquid chromatography) and thin layer chromatography. Potential contamination is determined by gas chromatography. Further physicochemical tests include the purity of the radionuclide, pH-value and osmolarity. Additionally, the sterility and absence of pyrogens of the radiopharmaceutical is assessed. It is noteworthy that this entire quality control process has to be performed for every new batch and should be completed within the shortest time possible, ordinarily within 30 min, to ensure adequate time for patient application.

## 3. The TR-Flex Cyclotron

The former cyclotron of the HZDR, an IBA Cyclone 18/9, was put into operation in autumn 1996. After 18 years of routine operation, comprehensive upgrades would have to be necessary to fulfill the new demands in the second decade of the 21st century. On the other hand HZDR could not forego the production of radionuclides with the Cyclone 18/9 during the ZRT building phase. Thus, HZDR decided to install a new cyclotron with higher ion energy and higher ion bean current in ZRT building and not to move the Cyclone 18/9.

The new TR-Flex cyclotron, shown in Figure 3, from Advanced Cyclotron Systems Inc. (ACSI, Richmond, BC, Canada) [2] was put into operation in 2017. The cyclotron is equipped with two extraction ports. Both extraction foils are radially movable to adjust the energy of the extracted proton beam in the range of 18 MeV up to 30 MeV. Two beamlines are connected behind a combo magnet on the extraction port 1. Two 4 port target selectors are installed at one beamline and the second extraction port. The cyclotron and the targetry is characterized by the following key parameters:

- Acceleration of $H^-$ and extraction of $H^+$ ions
- External multi-cusp ion source, ion current up to 300 μA
- Adjustable energy in the range of 18 MeV (14 MeV) up to 30 MeV
- dual beam operation with split ratio 1:100 to 50:50
- Two $[^{18}F]F^-$ water targets and one $[^{18}F]F_2$ gas target
- One $[^{11}C]CH_4$ gas target and one $[^{11}C]CO_2$ gas target
- One 30° and one 90° solid state target

**Figure 3.** The TR-Flex cyclotron at the HZDR. The beamline 1B with a 4-port target selector is shown in the foreground. The second 4-port target selector is at the opposite side of the Cyclotron. Picture: HZDR/Frank Bierstedt.

The TR-Flex is in stable and reliable operation now. Although, the cyclotron is designed to extract ions in the range of 18 MeV up to 30 MeV it is of real interest to extract ions at lower ion energies. The reaction cross section for a lot of radionuclides are higher and the impurities are lower for lower ion energies. For example $^{64}Ni(p,n)^{64}Cu$ production should be done below 15 MeV. Hence, experiments were done to determine the lowest possible ion extraction energy. It was possible to extract ions at energies as low as 14 MeV at the beamline extraction port. Autoradiography measurements at ion beam energies of 14 MeV and 30 MeV were executed to determine the profile of the proton beam hitting the solid target. The Autoradiographic measurement of a 30 µA beam current with an energy of 30 MeV is shown in Figure 4.

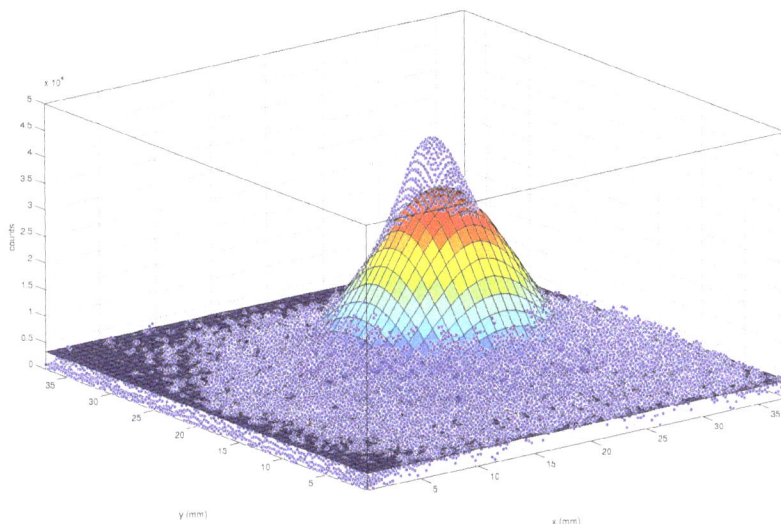

**Figure 4.** Autoradiographic measurement of an irradiated gold disk at the 90°-solid state target, beam energy 30 MeV. Blue dots: measured values; colored surface: fitted curve.

A two dimensional gaussian function was fitted to the measured profile to determine the beam size in x- and y- direction.

$$I(x,y) = I_0 \cdot e^{-\left(\frac{(x-\mu_x)^2}{2 \cdot \sigma_x^2} + \frac{(y-\mu_y)^2}{2 \cdot \sigma_y^2}\right)} \qquad (1)$$

We measured a pretty well shaped beam profile for lower and higher energies at the target selector at the end of the beamline as it is written in Table 1. The maxium beam current is limited to 50 μA for low energies, because of a higher beam loss in the beamline below an energy of 18 MeV.

**Table 1.** Determined Beam size in x- and y- direction.

| $E$ (MeV) | $FWHM_x$ (mm) | $FWHM_y$ (mm) |
|-----------|---------------|---------------|
| 14        | 13.6(2)       | 13.9(2)       |
| 30        | 11.8(2)       | 12.0(2)       |

The first production runs for $^{64}$Cu using 14 MeV protons for the $^{64}$Ni(p,n)$^{64}$Cu reaction [3,4] were carried out and evaluated. Typical irradiation parameters for the copper production are an ion current of 50 μA and an irradiation time of 90 min. The molar activity of the $^{64}$Cu is about 1 TBq/μmol. We achieved an activity of 15 GBq that is corresponding to a saturation yield of 3.8 GBq/μA.

The following radionuclides have been produced reliably with the TR-Flex since the beginning of 2018. Typical production parameters of the new TR-Flex and the achieved activities are presented in Table 2.

**Table 2.** Typical production parameters of the TR-Flex at the HZDR.

| Isotope | Chem. Form | (Typ. Current) | Irr. Time | Acitvity | Sat. Yield |
|---------|-----------|----------------|-----------|----------|------------|
| $^{18}$F | F$^-$ | 80 μA | 20 min | 95 GBq | 10.5 GBq/μA |
| $^{18}$F | F$^-$ | 105 μA | 70 min | 355 GBq | 10.0 GBq/μA |
| $^{18}$F | F$_2$ | 30 μA | 60 min | 20 GBq | 2.2 GBq/μA |
| $^{11}$C | CO$_2$ | 40 μA | 35 min | 155 GBq | 5.5 GBq/μA |
| $^{11}$C | CH$_4$ | 30 μA | 40 min | 55 GBq | 2.1 GBq/μA |
| $^{64}$Cu | Cu | 50 μA | 90 min | 15 GBq | 3.8 GBq/μA |

## 4. The Radionuclide Distribution System

A new Radionuclide Distribution System was developed and installed by the Department of Research Technology at HZDR. The liquid and gas targets are unloaded through capillaries to a central hot cell. Henceforward the radionuclides can be distributed automatically to the GMP unit and the research hot cells.

The system controls the target unload and the transport to the hot cells within the whole building. The gas is transported by stainless steel capillaries with an inner diameter of 1.4 mm and the liquid is transported by PTFE capillaries with an inner diameter of 0.8 mm. The transport distances can reach up to 100 m. The supervision of the relevant parameters and interlock system for the radiation protection (shielding of the hot cells, correct transportation path, correct ventilation system) and generation of the target unload clearance signal sent to the cyclotron is done automatically by the Radionuclide Distribution System.

The cyclotron targets are unloaded to the central hot cell "0". Several multi position valves in this hot cell allow the activity dosing and distribution to 25 hot cells in the whole Center for Radiopharmaceutical Cancer Research. Normally this is an automatic transport but also an manual operation and a so-called emergency mode, that allows to abrogate the interlock system, is possible. A schematic view of the distribution is shown in Figure 5.

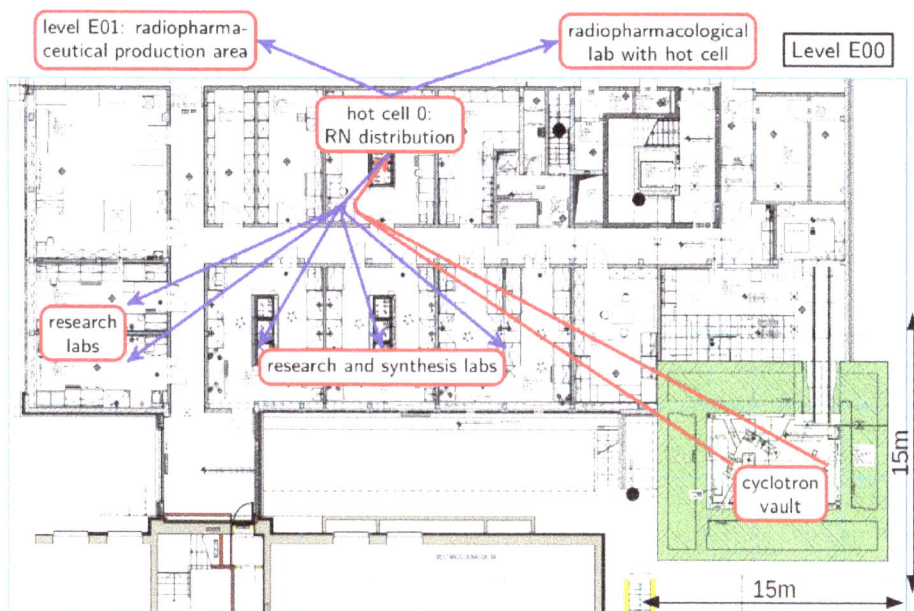

**Figure 5.** Schematic view of the radionuclide distribution in the building. Red lines: unload the targets to the hot cell "0"; blue lines: radionuclide distribution within the ZRT building.

Solid targets are unloaded to a transport container on a hand cart. An unload clearance signal is generated when the hand cart is docked at the solid target system and the cooling water blow out as well as the unload process is done by the control system of the cyclotron.

## 5. Conclusions

In our contribution we presented the new Center for Radiopharmaceutical Cancer Research including the new production and research infrastructure. Furthermore, we introduced the new cyclotron TR-Flex including first results of the radionuclide production and beam characterization measurements. The Institute of Radiopharmaceutical Cancer Research is now concentrated within two interconnected buildings at HZDR. The new research and production units are fully operational. The new research complex remarkably expands the capabilities for high-end research. The parameters of the new cyclotron open new opportunities with regards to the yields of produced radionuclides and the usable nuclear reactions. This expands the range of producible radionuclides.

Some improvements to the solid target system will be done in the near future. The first point is to modify the solid target system to use targets thicker than 2 mm. Furthermore, an energy degrader will be designed and installed to reduce the energy at the beamline target to below 14 MeV. The production of further radionuclides will be started step by step.

In comparison to similar medical institutions, the TR-Flex cyclotron gives the opportunity to produce radionuclides also by the (p,2n) and (p,3n) reaction. The close cooperation with the National Center for Tumor Diseases Dresden and the National Center for Radiation Research in Oncology Dresden offers unique possibilities for research and application of radiopharmaceuticals.

**Author Contributions:** Conceptualization, F.F. and J.S. and S.P. and M.K.; software, H.T.; formal analysis, M.W. and M.K.; investigation, M.W. and S.P. and M.K.; writing—original draft preparation, M.K. and T.K. and J.S.; writing—review and editing, S.P. and T.K. and M.W. and M.K.; visualization, M.K.; supervision, J.S. and P.K. and F.F. and H.-J.P; project administration, F.F. and J.S.

**Funding:** This research received no external funding.

**Conflicts of Interest:** The authors declare no conflict of interest.

## References

1.  Spies, H.; Steinbach, J. Fifty years of radiopharmacy at Rossendorf. *J. Label. Compd. Radiopharm.* **2007**, *50*, 895–902. [CrossRef]
2.  Watt, R.; Gyles, W.; Zyuzin, A. Building on TR-24 success: Advanced Cyclotron Systems Inc. launches a new cyclotron model. *J. Radioanal. Nucl. Chem.* **2015**, *305*, 93–98. [CrossRef]
3.  McCarthy, D.W.; Shefer, R.E.; Klinkowstein, R.E.; Bass, L.A.; Margeneau, W.H.; Cutler, C.S.; Anderson, C.J.; Welch, M.J. Efficient production of high specific activity $^{64}$Cu using a biomedical cyclotron. *Nucl. Med. Biol.* **1997**, *24*, 35–43. [CrossRef]
4.  Thieme, S.; Walther, M.; Pietzsch, H.-J.; Henniger, J.; Preusche, S.; Maeding, P.; Steinbach, J. Module-assisted preparation of $^{64}$Cu with high specific activity. *Appl. Radiat. Isot.* **2012**, *70*, 602–608. [CrossRef] [PubMed]

*instruments*

MDPI

*Technical Note*

# A Compact Quick-Release Solid Target System for the TRIUMF TR13 Cyclotron

Stefan Zeisler, Benjamin Clarke, Joel Kumlin, Brian Hook, Samuel Varah and Cornelia Hoehr *

Life Sciences Division, TRIUMF, Vancouver, BC V6T 2A3, Canada; zeisler@triumf.ca (S.Z.);
clarke25@uwindsor.ca (B.C.); kumlin@artms.ca (J.K.); bhook@triumf.ca (B.H.); svarah@triumf.ca (S.V.)
* Correspondence: choehr@triumf.ca; Tel.: +1-604-222-1047

Received: 29 December 2018; Accepted: 7 February 2019; Published: 12 February 2019

**Abstract:** A new solid target system for the TRIUMF TR13 cyclotron that can accommodate target discs with a 1-2-mm thickness and a 28-mm diameter has been developed. The target system design is based on a modified clamping mechanism of a KF-40 vacuum connector, and comprises an easy and quick ejection mechanism for the target plate. The new quick-release target system decreases the retrieval time of the irradiated target to less than 1 minute and is expected to reduce the radiation burden to operating staff by a factor of ~10.

**Keywords:** PET; medical isotopes; solid target system

---

## 1. Introduction

The TRIUMF Life Sciences Division produces positron-emitting radiopharmaceuticals for clinical applications as well as a variety of isotopes for research. The clinical program uses exclusively $^{11}$C- and $^{18}$F-labeled tracers; however, radiometals such as $^{68}$Ga, $^{64}$Cu, $^{44}$Sc and $^{89}$Zr have recently become increasingly important for the development of new chelators and radiolabeled biomolecules [1].

All isotopes for the Life Sciences program are produced on TRIUMF's TR13 cyclotron [2,3], a self-shielded 13 MeV negative ion accelerator, which is capable of extracting two proton beams of up to 50 µA simultaneously. Irradiations are usually performed at beam currents of up to 25 µA; the produced radioisotopes are used in house or delivered to collaborators in the Vancouver area. The TR13 has two target selectors (one on either beam port), which can accommodate four targets each (Figure 1a). Space on and around the selectors is rather limited, requiring all targets to be of compact design.

Radiometals are commonly produced in solid targets (see for example [4–8]), which typically consist of a backing plate and the actual target material. The target material may be electroplated, sputter-coated or fused onto the backing. The target is then placed into a holder on the beam port of the cyclotron. During irradiation, the target needs to be water-cooled to avoid overheating. Some designs include a thin metal foil that separates the target system from the cyclotron vacuum and prevents the sputtered target material from contaminating the beam port. In order to withstand intense beams, this foil requires cooling, commonly by a helium jet.

Figure 1. (**a**) TR13 four position target selector. (**b**) Standard TR13 solid target holder.

For several years, the Life Sciences program has relied on a rather basic solid target system, which consists of three components as illustrated in Figure 1b: (1) a nose cone, compatible with the ports on the TR13 target selector; (2) a helium-cooling assembly, which directs helium jets onto both the entrance foil separating the helium from the cyclotron vacuum and the target plate; and (3) the target holder with water-cooling channels. This system accepts discs with a 35-mm diameter and a 1-mm thickness. The entire assembly is held together concentrically with four screws, which apply sufficient force onto the O-rings located in each component to seal the water and the helium circuits. In order to retrieve the target disc after bombardment, the system needs to be manually removed from the target selector, which requires disconnecting the water- and helium-cooling lines. The target is then transported to a fume hood, where it is completely dismantled, and the irradiated disc is removed. The entire retrieval process takes 8–10 minutes on average, with operators having to spend approximately 5 minutes in close proximity to the highly radioactive targets on the selector.

While this simple system was deemed adequate for the occasional irradiation of a solid target at moderate beam currents, it is not suitable for frequent (daily) use due to the resulting unacceptably high radiation dose to technical staff. TRIUMF's guidelines limit the maximum daily exposure of Nuclear Energy Workers to 500 μSv and mandate that any radiation burden must be kept as low as reasonably achievable (ALARA principle). Annual whole-body exposure must not exceed 10 mSv.

The issue of handling radioactive target plates could be overcome by installing a remote-controlled target assembly that releases the target into a dedicated transfer system, thus avoiding handling the irradiated target altogether. However, due to space constraints around the TR13 target selector and the lack of penetrations through the cyclotron, shielding the design of an automated system was found to be virtually impossible.

Most of the operator radiation exposure occurs during the disconnecting of the tube fittings on the two water and two helium lines and during the removal of the target assembly from the selector. As such, we developed a new target system that permits the rapid extraction of the irradiated disc in situ while leaving the cooling lines in place.

## 2. Design

### 2.1. Functional Description and Specifications

The quick-release solid target system incorporates previously engineered and tested technologies with carefully adapted additions to fulfill the following functional specifications:

- Water-cooling on the back of the target plate to sufficiently cool aluminum, tantalum, silver or rhodium discs at 12.8 MeV, 30 μA proton current;

- Helium-cooling assembly to cool the front entrance foil and the front of the target plate;
- Eliminate the need to disconnect any cooling lines for target retrieval;
- Controlled, reliable unloading of target;
- Assembly must incorporate a current pickup to measure the beam current on target during irradiation;
- System must be able to accommodate 28-mm diameter, 1–2-mm thick discs;
- Design must be compatible with current nose cone, cooling lines, and electrical components.

Figure 2 shows three-dimensional renderings of the new system assembly (a) and a longitudinal cross-section (b). The new system is based on a slightly modified KF-40 flange functioning as the quick-release and sealing mechanism. A standard KF clamp is used to connect the preassembled nose cone/helium window component and the water-cooling port, which holds the target disc.

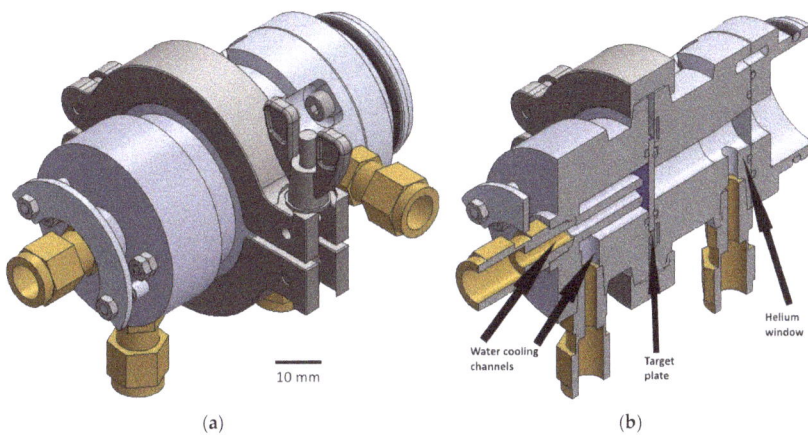

(a)  (b)

**Figure 2.** (a) New target system assembly in closed position. The water-cooling inlet line is attached to the axially centered fitting on the left; the water outlet is at the bottom left; the beam enters from the top right. (b) Cross-section showing the water-cooling channels, the helium-cooling window, and the target plate.

The helium window and water port were designed to maintain similar features and functions as the previous system. The cooling water flows directly onto the center of the back of the target plate, and exits the component parallel to the target plate as before. This allows for the most accessible configuration when the target is mounted onto the target selector. The helium window is compatible with the original nose cone. The helium continues to provide cooling to the front of the target plate, as well as the front entrance foil. Both components use standard tube fittings to maintain compatibility with the existing cooling lines. The water port was designed to incorporate a novel holding and release mechanism that ensures reliable handling of the target plate during removal.

The principle of the operation of the target system is as follows: After the target has been irradiated, the operator enters the target area and releases the KF clamp without removing any cooling lines. At this point, the target plate is still held on the water port by a spring-loaded target holding ring. The target holder is placed over a shielding container; pushing the release mechanism opens a gap between the target holding ring and the target, which allows the plate to drop into the container.

## 2.2. Detailed Description of Components

### 2.2.1. Water Port

The water port (Figure 3) was designed to perform several important functions in the system. It contains the target holder, the target plate ejection mechanism and the water-cooling channels for the target plate. Four holes lay on a 33-mm diameter reference circle, concentric with the center bore. Each hole is 4.25 mm in diameter and penetrates through the entire body. The holes are used as guides for the release the mechanism's components. Anchor studs and extension springs are inserted into two of the holes to hold the target plate in place, and release pins are inserted into the other two holes. A 018 O-ring groove surrounds the water channel in the center (018 size O-ring as per AS568 standard: inner diameter 18.77 mm, outer diameter 22.33 mm, cross-section 1.78 mm). This O-ring size was chosen for its inner diameter, which is larger than the given water channel bore but smaller than the given target plate diameter. This difference in size leaves enough room on either side to provide a reliable water seal on the plate face. Raised circular segments with an inner radius of 14.5 mm function as target plate-locating features for consistent centering without hindering plate ejection. These dimensions let the 28-mm diameter target plate sit relatively loose for easy release. The water-cooling line is designed to create an impinging jet that is directed to the center of the target plate. The water disperses radially and exits the component parallel to the target plate.

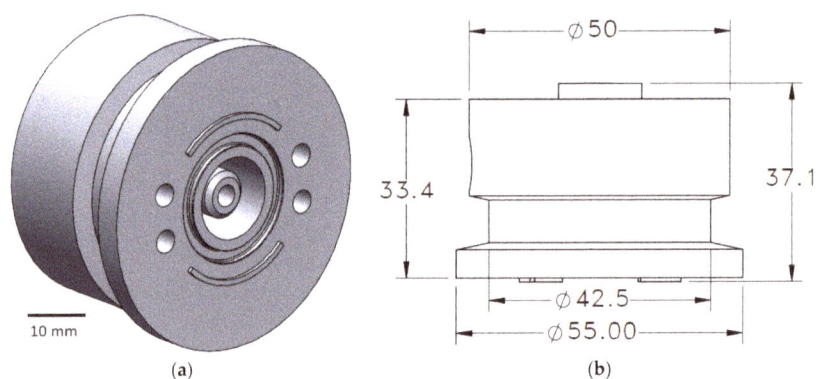

**Figure 3.** (a) Water port (inner face), showing the water-cooling channel in the center, circular segments for target plate centering and the four holes housing the anchor studs with springs and release pins. (b) Water port (side view), dimensions in mm.

The various diameters of the component are shown in Figure 3b. The flange diameters of 42.5 and 55 mm are the same as the standard KF-40 flange. The larger diameter fits and is in direct contact with the KF-40 clamp, while the smaller diameter provides enough space for the clamp to surround and compress the flange without obstruction. The 50-mm diameter on the back was chosen to match the previous system size and was deemed appropriate for comfortable manual handling.

The thickness of the component was dimensioned to fit the length of the extension spring and anchor stud of the release mechanism. To assemble the system with these internal components, the water port thickness must be within the range of approximately 30–33.5 mm. An extrusion where the water inlet port is placed extends the nozzle length. Both inlet and outlet use 1/8" NPT (national pipe thread) ports to be compatible with current water-cooling lines.

### 2.2.2. Target Holding Ring and Release Mechanism

The target holding ring, shown in Figure 4, is part of the release mechanism and in direct contact with the target plate to hold it in place. An inner diameter of 26 mm gives 1 mm of material around the ring to hold the outer surface of the target plate. This contact is only designed for positioning and does not need to be helium- or water-tight. Two anchor studs to connect to the springs and two countersunk holes lay on a 33-mm diameter reference circle. This diameter aligns the ring with the water port to integrate the release mechanism. The countersunk holes are made to fit M2 flathead screws. The screw size and style were chosen to maintain a flush surface since the ring, when fully assembled, sits within the KF-40 clamp. The outer diameter of the ring acts as a centering guide for assembling the system; it aligns with the centering grooves of the helium window, as discussed below. The ring is 1 mm thick, leaving additional space between the face of the ring and the surface of the helium window centering groove. This space prevents the contact of the ring with the surface of the helium window, which would obstruct compression of the O-rings by the KF-40 clamp.

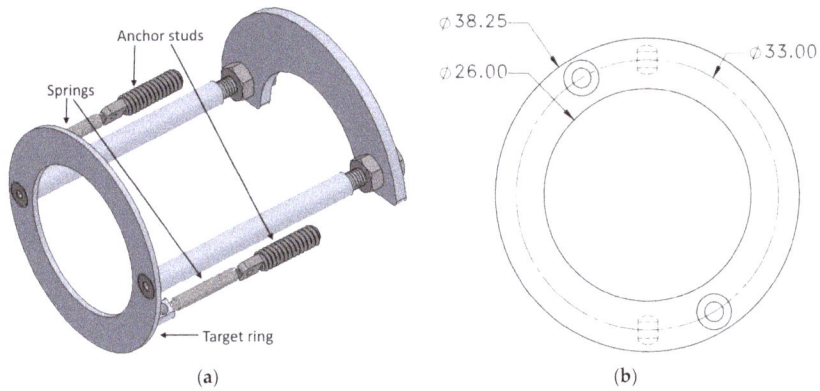

(a)                                                                 (b)

Figure 4. (a) Partially assembled release mechanism showing the target ring on the left, the release pins connected to the release plate on the right and the springs and anchor studs. (b) Target ring (top view), dimensions in mm.

The anchor studs were designed in a similar fashion to commercial extension spring anchors. The studs have a hole to latch onto the hook of an extension spring and are thin enough to fit inside the holes in the water port.

The release mechanism comprises the target holding ring, two shafts, and a release plate, facilitating one-handed operation. Figure 4a shows a partially assembled release mechanism and illustrates how the target ring is fastened to both shafts by M2 flathead screws. On the other end of each pin is a 10-mm long M3 outer thread. The release plate is held between two hex nuts. The release plate is shaped as a half circle to clear the water-cooling fitting connected to the back of the water port.

Figure 5a,b are different representations of the release mechanism integrated within the water port. To eject the target, the operator presses on the release plate, forcing the shafts forward. This pushes the target holding ring to the open position, which allows the plate to drop out of the holder. When the system is in its equilibrium position, the extension springs contract, forcing the target holding ring to sit flush with the face of the target plate or, when not loaded, the water port. The purpose of the release mechanism is not to provide a compressive force on the target face to seal against the O-ring. It is designed strictly to hold the target in place during loading and irradiation, and to minimize mishandling during unloading.

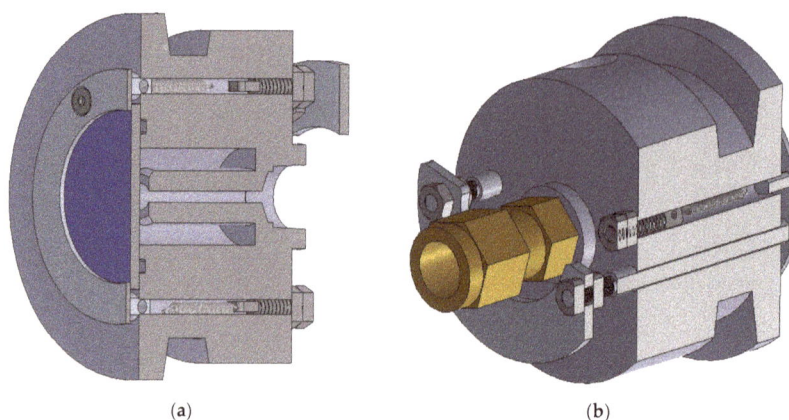

(a)                                                    (b)

**Figure 5.** (**a**) Release mechanism (cross-section). (**b**) Release mechanism (side view).

2.2.3. Helium Window Assembly

The helium window acts as a separator between the cyclotron and the target plate. The window is fitted with a thin aluminum foil that isolates the internal target assembly components from the cyclotron vacuum. The helium path, as seen in Figure 6a, splits into two directions as it enters the component. One jet is used to cool the aluminum foil; the second jet impinges on the front face of the target plate.

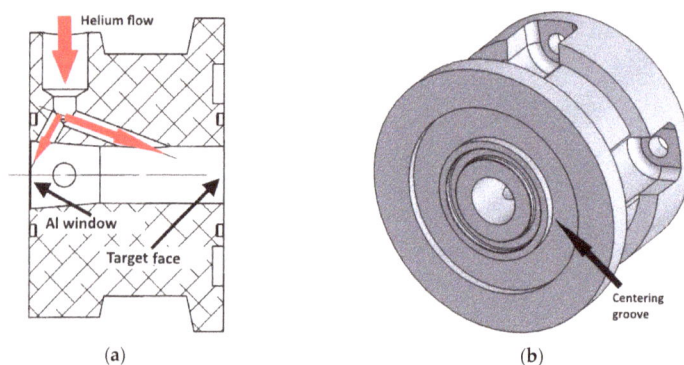

(a)                                                    (b)

**Figure 6.** (**a**) Helium window showing the two channels that guide the helium flow into the center of the aluminum foil on the left and the target plate on the right (cross-section). (**b**) Helium window inner face.

Figure 6b shows the inner face of the helium window. The centering groove is dimensioned to fit the target holding ring in the assembly. To seal the helium window and the water port, both components are pressed together by the KF clamp. The flange of the helium window is dimensioned identically to that of the water port.

The helium window is fastened to the standard TR13 nose cone with four M4 screws in slots. As such, the front part of the assembly does not need to be removed for target retrieval. It is only disassembled for maintenance, i.e., for the replacement of the aluminum foil or the O-rings.

### 2.2.4. Sealing Mechanism

The sealing mechanism uses a KF-40 clamp for primary sealing and quick-release functions. Since a commercial clamp is used for a slightly different function, the adjusted flange dimensions are critical. A standard KF-40 clamp, when assembled with a T-ring and O-ring, has a thickness of 11.75 mm. However, our quick-release system requires a range of 1–2 mm in thickness to accommodate different target plate thicknesses.

Figure 7 shows a rendering of the modified quick-release flange. The thickness of the flange varies from 11.5 mm to 12.5 mm, depending on the thickness of the target plate. This range is asymmetrical about the benchmark of 11.75 mm because the standard clamp is designed with more room on the higher tolerance. Therefore, when using a 1-mm thick target plate, the minimum flange thickness is sufficient to prevent the clamp from bottoming out on the flanges. A 2-mm thick target backing also allows the clamp to close around the flanges.

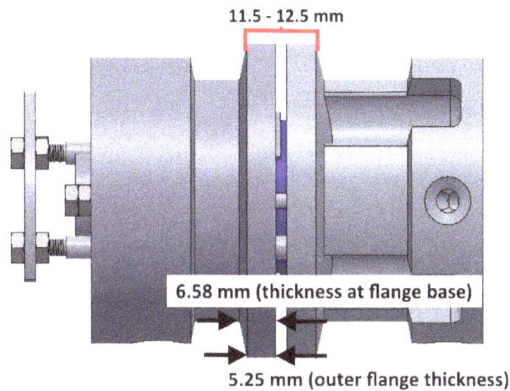

**Figure 7.** Quick-release flange system (assembled).

Each flange of our quick-release system extends outwards from the outermost diameter at a 12-degree angle, which is consistent with the standard KF-40 flange. To achieve the correct dimensions, both the helium window and the water port have outer flange thicknesses of 5.25 mm. This thickness increases to 6.58 mm at the base of each flange.

### 3. Results

The new quick-release target system was manufactured in house and assembled. The cooling systems were found to be leak-tight. The release mechanism was thoroughly bench-tested in terms of functionality and mechanical robustness and was proven to be very reliable for both 1-mm and 2-mm thick target plates. The target plate can be removed from the target holder in less than one minute.

At the time of this report, the new system was installed on the target selector of the TR13 cyclotron. A series of experiments with varying beam currents will be conducted in the near future to explore its behavior under irradiation conditions.

### 4. Conclusions

We have developed and manufactured a novel and compact quick-release solid target system based on a modified KF flange assembly. The target system can hold circular plates with thicknesses varying between 1 mm and 2 mm, and provides helium-cooling on the target plate front and water-cooling on the target back. The retrieval of the target plate does not require the removal of the cooling lines. Compared to the currently used basic target holder, the new quick-release system reduces the time required to extract a target plate, from up to 10 minutes to less than 1 minute. It is expected that the radiation exposure to operators will be reduced by a factor of ~10.

**Author Contributions:** Conceptualization, S.Z. and C.H.; methodology, S.Z., J.K., B.H. and B.C.; formal analysis, B.C. and C.H.; investigation, B.C., S.V., J.K. and S.Z.; resources, S.Z. and C.H.; writing—original draft preparation, B.C., S.Z.; writing, S.Z., B.C., J.K., J.S. and C.H.; visualization, B.C. and S.Z.; supervision, S.Z. and C.H.; project administration, C.H.; funding acquisition, C.H.

**Funding:** TRIUMF receives federal funding via a contribution agreement with the National Research Council of Canada. This work was supported by the Natural Sciences and Engineering Research Council (NSERC) via the Discovery Grant program (RGPIN-2016-03972).

**Acknowledgments:** The conceptual design of our system was inspired by a conversation with the cyclotron staff of the University of Wisconsin (Madison), Department of Medical Physics.

**Conflicts of Interest:** The authors declare no conflict of interest.

## References

1.  Hoehr, C.; Bénard, F.; Buckley, K.; Crawford, J.; Gottberg, A.; Hanemaayer, V.; Kunz, P.; Ladouceur, K.; Radchenko, V.; Ramogida, C.; et al. Medical isotope production at TRIUMF–from imaging to treatment. *Phys. Procedia* **2017**, *90*, 200–208. [CrossRef]
2.  Buckley, K.R.; Huser, J.; Jivan, S.; Chun, S.K.; Ruth, T.J. $^{11}$C-methane production in small volume, high pressure gas targets. *Radiochim. Acta* **2000**, *88*, 201–205. [CrossRef]
3.  Laxdal, R.E.; Altman, A.; Kuo, T. Beam measurements on a small commercial cyclotron. In Proceedings of the EPAC 94, London, UK, 27 June–1 July 1995; pp. 545–547.
4.  Hanemaayer, V.; Benard, F.; Buckley, K.R.; Klug, J.; Kovacs, M.; Leon, C.; Ruth, T.J.; Schaffer, P.; Zeisler, S.K. Solid targets for $^{99m}$Tc production on medical cyclotrons. *J. Radioanal. Nucl. Chem.* **2013**, *299*, 1007–1011. [CrossRef]
5.  Lin, M.; Waligorski, G.J.; Lepera, C.G. Production of curie quantities of $^{68}$Ga with a medical cyclotron via the $^{68}$Zn(p,n)$^{68}$Ga reaction. *Appl. Radiat. Isot.* **2018**, *133*, 1. [CrossRef] [PubMed]
6.  Zeisler, S.; Limoges, A.; Kumlin, J.; Siikanen, J.; Hoehr, C. Fused zinc target for the production of gallium radioisotopes. *Instruments* **2019**, *3*, 10. [CrossRef]
7.  Carzaniga, T.S.; Auger, M.; Braccini, S.; Bunka, M.; Ereditato, A.; Nesteruk, K.P.; Scampoli, P.; Tuerler, A.; van der Meulen, N. Measurement of $^{43}$Sc and $^{44}$Sc production cross-section with an 18 MeV medical PET cyclotron. *Appl. Radiat. Isot.* **2017**, *129*, 96–102. [CrossRef] [PubMed]
8.  Valdovinos, H.F.; Hernandez, R.; Graves, S.; Ellison, P.A.; Barnhart, T.E.; Theuer, C.P.; Engle, J.W.; Cai, W.; Nickles, R.J. Cyclotron production and radiochemical separation of $^{55}$Co and $^{58m}$Co from $^{54}$Fe, $^{58}$Ni and $^{57}$Fe targets. *Appl. Radiat. Isot.* **2017**, *130*, 90–101. [CrossRef] [PubMed]

*instruments*

MDPI

*Communication*

# Solid Target System with In-Situ Target Dissolution

**William Z. Gelbart [1,*] and Richard R. Johnson [2]**

[1]    Advanced Systems Designs Inc., Garden Bay, BC V0N 1S1, Canada
[2]    Deptment of Physics and Astronomy, University of British Columbia, Vancouver, BC V6T 1Z4, Canada;
       richardrjohnson81738@gmail.com
*    Correspondence: gelbart@advancesystems.ca or williamgelbart@gmail.com

Received: 27 December 2018; Accepted: 5 February 2019; Published: 11 February 2019

**Abstract:** A significant number of medical radioisotopes use solid, often metallic, parent materials. These materials are deposited on a substrate to facilitate the cooling and handling of the target during placing, irradiation, and processing. The processing requires the transfer of the target to a processing area outside the irradiation area. In this new approach the target is processed at the irradiation site for liquid only transport of the irradiated target material to the processing area. The design features common to higher energy production target systems are included in the target station. The target is inclined at 14 degrees to the beam direction. The system has been designed to accept an incident beam of 15 to 16 mm diameter and a beam power between 2 and 5 kW. Thermal modeling is presented for targets of metals and compounds. A cassette of five or 10 prepared targets is housed at the target station as well as a target dissolution assembly. Only the dissolved target material is transported to the chemistry laboratory so that the design does not require additional irradiation area penetrations. This work presents the design, construction, and modeling details of the assembly. A full performance characterization will be reported after the unit is moved to a cyclotron facility for beam related measurements.

**Keywords:** radioisotopes; medical radioisotopes; radioisotope targetry; solid radioisotope targets; radioisotope target processing; medical cyclotrons

---

## 1. Introduction

In solid target irradiation systems the solid target must be transferred to the irradiation position for irradiation and then removed for processing.

There are commercial solid target systems available. For example IBA offers a system called NIRTA [1]. Comecer offers the ALCEO system [2] and ARTMS offers a solid target system [3]. Users have adapted these systems to their needs. For example Carzaniga et al. at the University of Bern reported their developments based on a NIRTA system at the conference in reference [4]. The general features of these systems are that they use disk target bodies with normal beam entrance and transport the irradiated metallic target to the radiochemistry area for subsequent processing.

In the simplest form each target is transferred manually between the hot-cell and the accelerator. More sophisticated systems use some form of mechanical or pneumatic transfer. They are not only more complex and expensive, but require large passages to the accelerator vault or target cave, passages that are difficult to add to existing installations and that can create radiation "leaks".

The system reported here uses a target inclined at 14 degrees to the beam and the target is processed at the irradiation site for liquid transport of the irradiated target material to the radiochemistry area. The new design that is reported here is an evolution of high current target designs adapted to lower current accelerator facilities that may not have the required target or cyclotron vault penetrations.

This self-contained unit automatically places the targets in the irradiation position and at the end of irradiation transfers them to an integral dissolution cell located at the target station itself. Fresh targets—up to 10 in the regular configuration—are contained in a detachable cassette that can be easily and quickly replaced once all the targets are irradiated and processed. In most cyclotron installations the cassette change can coincide with the regular maintenance shutdown.

The advantage of this approach is that only the dissolution liquid and the liquid dissolved material are transferred between the target system and the processing hot-cell. The liquids are moved through a small diameter tube that is easily routed through the existing cable ducts. There is no need for large, dedicated passages for the pneumatic transfer pipes (in the case of pneumatically transferred targets) and no need for a large hot-cell to house the target receiver nor any manual manipulation of the target in the hot-cell; the dissolution liquid is sent directly to the chemistry module.

## 2. Materials and Methods

A small aluminum cage contains all the system's components. Three pneumatic actuators remove a fresh target from the cassette, place it in the irradiation chamber, and insert it in the dissolution vessel at the end of the irradiation. The vessel is equipped with heaters and the temperature of the liquid controlled. The system, indicating the main components, is shown in Figure 1.

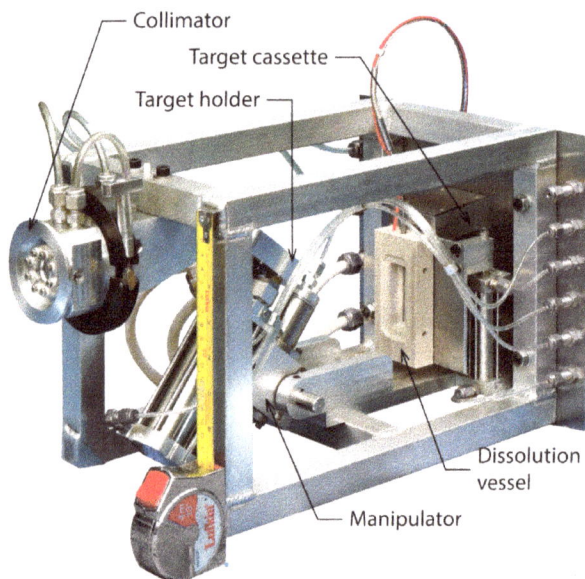

**Figure 1.** Target system assembly.

The solid target substrate is usually made from copper or silver. The face of the target is clad with the irradiated material and cooling channels on the back are designed to dissipate the heat load while maintaining the target face temperature at the desired level. The target is placed at 14 degrees to the beam and the irradiated area forms an ellipse of approximately 13 mm × 52 mm.

The system is designed for a 15 mm–16 mm diameter, roughly circular beam. If no beam focusing elements are present in the beam-line the target location is set to a distance at which the beam diverges to this size. In most small cyclotrons this distance is in the order of less than one meter. To obtain the correct beam coverage on the target material and to get closer to the optimal "top hat" beam cross section, the beam is collimated to 13 mm diameter by a conical, water cooled collimator attached to the front of the system's irradiation chamber; 10% to 15% of the beam power is dissipated in the collimator. Both the collimator beam current and the target beam current are monitored. To provide accurate target current reading the target and the irradiation chamber form an electrically insulated Faraday cage.

An optional four sector insulated mask (shown in Figure 1) can be installed on the front surface of the collimator. This allows accurate beam positioning, essential in beam-lines with steering/focusing elements and helpful in all instances when beam drifts shift the impact spot. With a well centered on target beam the mask segments will not read any current, and current reading on the sectors will indicate a drift in that direction. The sectors are made out of pure silver and the sector currents are fed-through to the outside via an all-metal seal. Though silver does activate with a 6 h half-life positron from Cd107, and Cd109 auger low energy X-rays with 463 days half-life, the former decays quickly and the latter are very low energy X-rays. Silver has a high heat conductivity and mechanical strength and is a desirable material for collimators that are in the beam periphery.

The angle of the target spreads the beam over a larger area and allows thinner target material cladding (about 25% of the thickness compared to a 90 degrees target). This not only helps with the heat transfer, but allows cladding with materials that cannot easily be deposited in thicker layers. Figure 2 shows the target and the irradiated area dimensions.

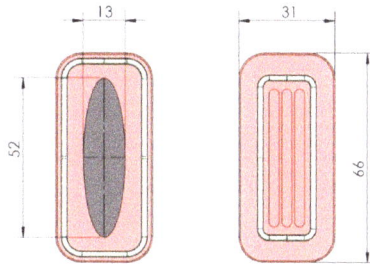

**Figure 2.** Solid target, front and back. The **left** figure shows the front irradiated area in grey, the O ring in yellow and the target substrate in orange. The **right** figure shows the cooling channels milled into the substrate and the O ring in yellow. The dimensions are in mm.

The sequence of operation of the system is shown in Figures 3–7.

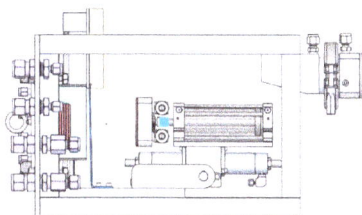

**Figure 3.** Ready for target placing.

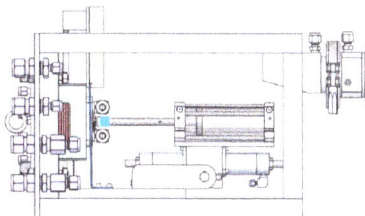

**Figure 4.** Grabbing the target from the cassette.

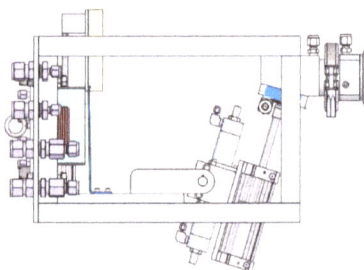

**Figure 5.** The target in irradiation position.

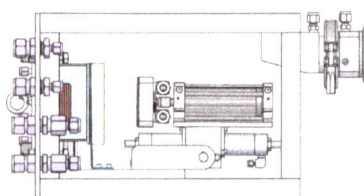

**Figure 6.** The irradiated target ready for dissolution.

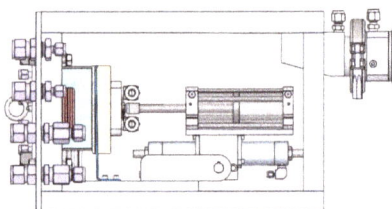

**Figure 7.** The target placed for dissolution.

Fresh targets (up to 10 usually, with a practical limit of 20) are pre-loaded in the detachable target cassette. The cassette snaps in place and can be easily changed in under a minute using a long-reach tool.

Used irradiated targets are released at the end of dissolution and dropped into a shielded container placed under the system and can be left there until the radioactivity levels drop.

In most cases the targets are not reused and the used targets can be stored on the container until disposal time. Figure 8 shows the empty cassette.

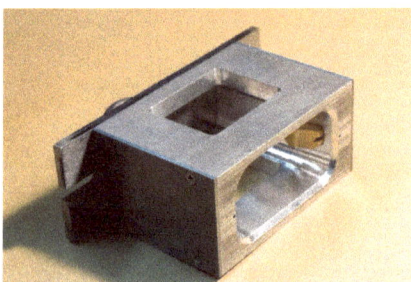

**Figure 8.** The target cassette.

## 3. Results

This solid target irradiation station design is the result of close to forty years of experience building and operating solid targets for radioisotope production and is yet another refinement of a similar system designed and built previously [5–11]. Construction methods and materials were carefully chosen to ensure low activation and to minimize radiation damage to the components. Whenever possible, aluminum, ceramics, and polyimides were used. This design criterion was followed for parts that were manufactured and commercial components such as connectors and fittings only available in other materials were kept in those materials. The only elastomer seals are on the target. These are, of course, used only for one irradiation run. The seals in all the air cylinders are100% graphite thus eliminating the dominant weakness of air actuators in radiation areas since the commercial elastomer seals are exposed to high radiation fields. Water and air tubing are polyurethane. They can serve for a number of years under normal operating conditions.

The target substrate can be supplied in different configurations of the number and size of cooling channels to optimize the cost versus the power requirement. A simplified, cost effective design (Figure 2) is capable of dissipating 2 KW of beam power on target (representing a typical 15 MeV cyclotron delivering 150 μA total beam) with only moderate surface temperature increase. As an example, Figure 9 shows the result of a simulation of the copper target with 60 micron molybdenum cladding and 6 L/min cooling water flow. 150 μA, a 2.3 KW total beam collimated to 2 KW beam on target.

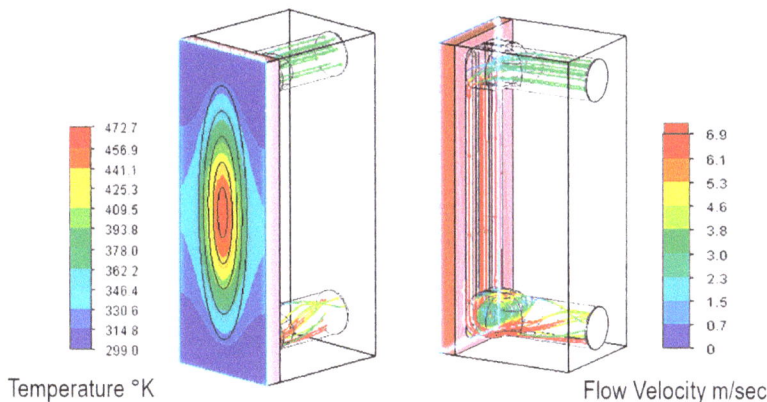

**Figure 9.** Flow and thermal simulation showing the water trajectories and velocity in m/sec (on the **right**) and surface temperature contours in degrees K (on the **left**).

Different cooling channel configurations and higher water flow can increase the power capability up to 5 KW (nine 0.8 mm wide × 1.7 mm deep water channels) with a cost increase of the target fabrication of about 50%. All the water channel designs use turbulent water flow with no cavitation along the channels.

As is the case for all solid targets the "power dissipation capability" depends not only on the supplied water flow and beam distribution on the target face, but on the maximum allowable target surface temperature. This temperature is a function of the target material cladding and the ability of the cladding to withstand this temperature. An excessive temperature manifests itself as mechanical stress flaking, and actual cladding melting. (Note that sputtering of the target material is not possible since there are no energetically ions or electrons to induce the process.) Reducing the target temperature is the important reason for the target inclination to the beam since the beam power is then distributed over a larger area. The thinner cladding also reduces mechanical stress at the boundary of the cladding and substrate. Careful uniform preparation of the cladding reduces mechanical stress as well.

The remaining variable is the power delivered to the cladding so different materials will have different maximum beam currents. In the above examples and ratings a full contact between the cladding and the substrate is assumed. The surface temperature must, of course, be below the melting point of the target material and for low melting point materials the power should be adjusted accordingly by the accelerator operator limiting the beam current.

Special face-grooved solid targets that can support non-metallic target materials were built and tested (Figure 10a). This type of target can be used with materials that can be melted into the grooves (melting point lower than the substrate).

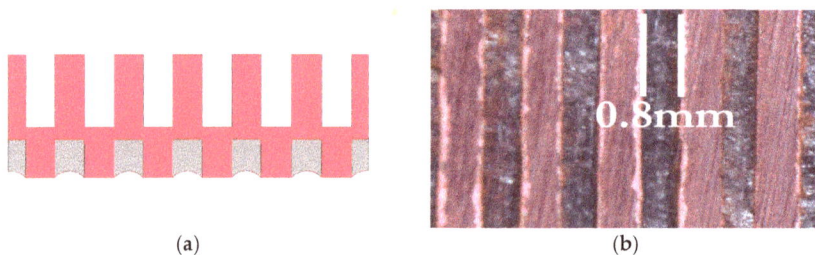

(a)                                                                 (b)

**Figure 10.** (**a**) Grooved target cross-section; the target material is grey while red is the copper substrate. (**b**) Face close-up with fused target material (grey color).

Figure 10b is a close-up of a typical grooved target face with rubidium chloride (MP 718 °C) fused between the grooves for higher energy irradiations and Figure 11 is a representative example of thermal modeling of the central segment of the target (highest heat flux region) for a copper substrate with 0.8 mm wide × 1 mm deep face grooves and nine 0.8 mm wide × 1.7 mm deep water channels. The beam is a 4 KW beam on target and the cooling water flow 10 L/min.

**Figure 11.** Grooved target thermal modeling.

The dissolution cell can be configured for a different target material. It has a heater provision for up to 100 °C heating, and agitation, if needed, is accomplished by inert gas bubbling. A stream selector manifold for solvents and wash liquids is located in the radiochemistry laboratory and delivered to the cell through capillary tubing.

The whole system is small with a footprint of only 350 mm × 200 mm and 240 mm vertical clearance. The system requires a supply of cooling water and compressed air. All target manipulation and processing is done remotely and automatically with interlocks ensuring proper sequences. Control of the process is done by a small, dedicated Programmable Logic Controller (PLC) and can be integrated with the accelerator controls.

## 4. Discussion and Summary

The design of a solid target station that uses targets tilted at 14 degrees to distribute the heat over a larger area, together with at target station target material dissolution after irradiation with subsequent liquid transfer to the radiochemistry laboratory was presented. Heat distribution modeling was conducted for several target substrate geometries. Power limits varied between 2 and 5 kW. The different cooling channel configurations add more cost to the target substrates of up to an additional 50%. The target station has been fabricated and a series of bench tests without radioactivity have begun. The full description of the system and performance measurements with the beam will be reported after the assembly has been moved to a cyclotron facility.

The bench tests will establish the operation limits and confirm the automated procedures. In addition, the bench tests will extend the operational parameters to their limits to study failure mechanisms. Since this system is in a high radiation area and repairs are difficult, a very high meantime between failure is required. For example, the dissolution cell will be tested with non-radioactive materials to confirm the chemical processes for the target. The first two targets to be tested are zinc and molybdenum. Note that the dissolution cell is vertical so that after processing a target the cell must be thoroughly cleaned and dried before the spent target is removed so that no liquids or radioactive materials are released into the area. This is performed automatically as a part of the process cycle.

The target station will finally be moved to a cyclotron facility for hot testing. That location has not been finalized.

## 5. Patents

A United States patent, covering some aspects of the system has been submitted.

**Author Contributions:** Conceptualization, Validation, Analysis, Resources, Writing, Review and editing, and Funding were the responsibility of W.Z.G. and R.R.J.

**Funding:** This research received no external funding.

**Conflicts of Interest:** The authors declare no conflict of interest.

## References

1. Nirta®Target Technology. Available online: www.iba-radiopharmasolutions.com/products/target-technology (accessed on 4 February 2019).
2. ALCEO Solid Target Processing System. Available online: www.comecer.com/alceo-solid-target-processing-system (accessed on 4 February 2019).
3. Quantm Irradiation System™. Available online: https://artmsproducts.com (accessed on 4 February 2019).
4. Carzaniga, T.S.; Haffner, P.; Turler, A.; Braccini, S. Solid Target development at BERN. In Proceedings of the 17th International Workshop on Targetry and Target Chemistry (WTCC17), Coimbra, Portugal, 27–31 August 2018.
5. Gelbart, W.Z.; Stevenson, N.R.; Johnson, R.R.; Nortier, F.M.; Orzechowski, J.; Cifarelli, F. High Current Radioisotope Production with Solid Target System. In Proceedings of the International Conference on Particle Accelerators, Washington, DC, USA, 17–20 May 1993; Volume 4, pp. 3099–3101. [CrossRef]
6. Ho, W.; Bakhtiari, S.; Gelbart, W.Z.; Stevenson, N.R. High Current Encapsulated Target and Target System for Radioisotope Production. In Proceedings of the 1997 Particle Accelerator Conference, Vancouver, BC, Canada, 16 May 1997; Volume 3, pp. 3842–3844. [CrossRef]
7. Gelbart, W.Z.; Stevenson, N.R. Solid Targetry Systems: A Brief History. In Proceedings of the 15th International Conference on Cyclotrons and their Applications, Caen, France, 14–19 June 1998; pp. 90–93.
8. Gelbart, W.Z.; Johnson, R.R.; Abeysekera, B. Solid target irradiation and transfer system. *AIP Conf. Proc.* **2012**, *1509*, 141–145.
9. Gelbart, W.Z.; Johnson, R.R. Irradiation Targets for Accelerator Production of 99mTc. In Proceedings of the 14th International Workshop on Targetry and Target Chemistry (WTCC14), Playa del Carmen, Mexico, 26–29 August 2012.

10. Gelbart, W.Z. Solid Target System. In Proceedings of the Society of Nuclear Medicine and Molecular Imaging, Vancouver, BC, Canada, 8–12 June 2013.
11. Gelbart, W.Z.; Johnson, R.R. 5th Generation High Current Solid Target Irradiation System. In Proceedings of the 15th International Workshop on Targetry and Target Chemistry (WTTC15), Prague, Czech Republic, 14 August 2014.

*instruments*
MDPI

*Article*

# Enhancement and Validation of a 3D-Printed Solid Target Holder at a Cyclotron Facility in Perth, Australia

**Sun Chan [1,*], David Cryer [1] and Roger I. Price [1,2]**

[1]   Sir Charles Gairdner Hospital, Department of Medical Technology and Physics, Nedlands 6009, Australia;
      david.cryer@health.wa.gov.au (D.C.); roger.price@health.wa.gov.au (R.I.P.)
[2]   Department of Physics, University of Western Australia, Nedlands 6009, Australia
*    Correspondence: sun.chan@health.wa.gov.au

Received: 31 December 2018; Accepted: 30 January 2019; Published: 2 February 2019

**Abstract:** A 3D-printed metal solid target using additive manufacturing process is a cost-effective production solution to complex and intricate target design. The initial proof-of-concept prototype solid target holder was 3D-printed in cast alloy, Al–7Si–0.6Mg (A357). However, given the relatively low thermal conductivity for A357 ($\kappa_{max}$, 160 W/m·K), replication of the solid target holder in sterling silver (SS925) with higher thermal conductivity ($\kappa_{max}$, 361 W/m·K) was investigated. The SS925 target holder enhances the cooling efficiency of the target design, thus achieving higher target current during irradiation. A validation production of $^{64}$Cu using the 3D-printed SS925 target holder indicated no loss of enriched $^{64}$Ni from proton bombardment above 80 μA, at 11.5 MeV.

**Keywords:** cyclotron; targetry; solid target; metal 3D-printing; target temperature; radiometals; radionuclides

---

## 1. Introduction

We investigated enhancement of the design and material used in the construction of an existing 3D-printed cast alloy (A357) solid target holder for improved thermal management and shuttle movement. The enhanced solid target, constructed in sterling silver, enables the target to withstand the maximum extracted target current from an IBA Cyclone® 18/18 cyclotron at a degraded energy of 11.5 MeV. The improved thermal management of solid target material significantly improved the target yield and/or irradiation time, in order to achieve equivalent isotopic activity. The 'non-slanted' target design ensures compactness and a minimal amount of expensive enriched target material needed for each irradiation. The 3D-printed sterling silver solid target incorporates the experimentally derived thermocouple measurement configuration [1] for continuous temperature monitoring during an irradiation. The cost of the 3D-printed sterling silver target holder, at $650 USD, offers an economical alternative to the traditional subtractive manufacturing technique.

## 2. Materials and Methods

In order to satisfy the complex and intricate design of the cooling mechanism, 3D-printing was used over traditional manufacturing techniques. The sterling silver (SS925) solid target holder was ordered online via the Shapeways 3D-printing manufacturing portal. Due to the inability to directly translate the design for different 3D-printed metals, minor changes were made to the cooling cavity to comply with the material specifications and manufacturing requirements.

The previous 3D-printed target [1] was fabricated from industrial cast alloy (Al–7Si–0.6Mg, A357) on the ReaLizer® SLM (selective laser melting) platform. The target body was annealed at 500 °C for 4 hours, and then furnace-cooled to room temperature to purge the material of excess Si. The theoretical maximum thermal conductivity for cast alloy using this method is 160 W/m·K. Thermal conductivity of sterling silver (SS925) is approximately 2.2 times higher than A357, with a theoretical maximum at 361 W/m·K. The composition of SS925 is predominately silver and copper, 92.5% and 7.5%, respectively. Commercial 3D-printing for sterling silver utilizes a lost-wax process, where a plaster mold of the 3D-printed design is created before molten silver is added into the cast. The 3D-printed holders had minimal post-fabrication treatment with light polishing of accessible surfaces only. In-house machining of the 3D-printed part included O-ring grooves, thermocouple hole, and polishing the interface where the target material is located.

Figures 1 and 2 show the 3D-printed target holder inside the complete solid targetry assembly. The mounting flange for the solid target incorporates a graphite collimator and graphite degrader (dia. 10 mm, thickness 1.0 mm).

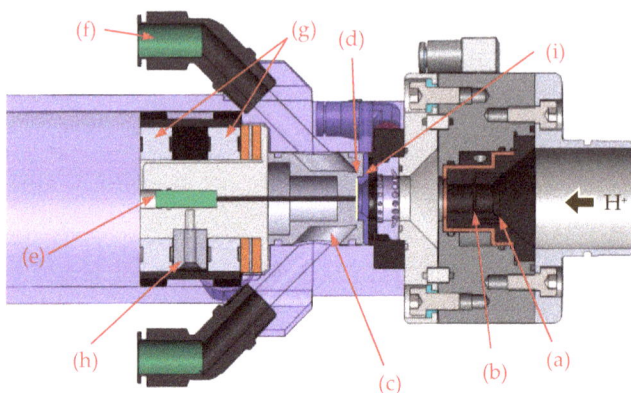

**Figure 1.** Schematic of cast alloy solid target: (a) 10 mm graphite collimator with removable crown; (b) graphite degrader; (c) 3D-printed A357 target holder; (d) target material; (e) type-K thermocouple; (f) 8 mm ID water line; (g) ceramic bearings; (h) lead ballast; (i) magnetically coupled Nb front cover.

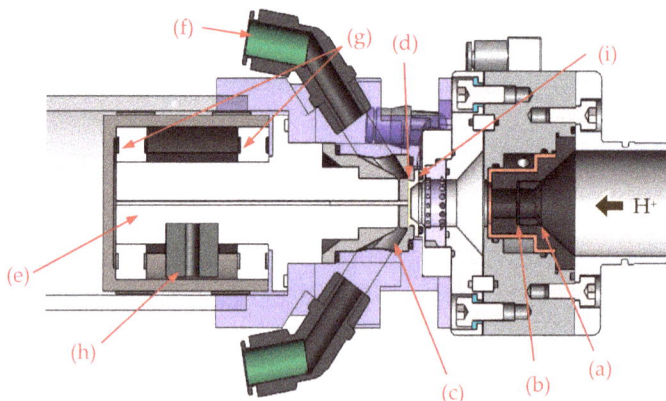

**Figure 2.** Schematic of 3D-printed sterling silver solid target: (a) 10 mm graphite collimator with removable crown; (b) graphite degrader; (c) 3D-printed SS925 target holder; (d) target material; (e) type-K thermocouple; (f) 8 mm ID water line; (g) ceramic bearings; (h) lead ballast; (i) magnetically coupled Nb front cover.

The graphite degrader and collimator are water-cooled to dissipate the heat generated from degrading the cyclotron primary beam energy from 18 to 11.5 MeV. The degrader is interchangeable for specific energy requirements, and is easily accessible from the front by removing the graphite collimator crown.

The new SS925 target system has the same floating design as the previous A357 solid target shuttle [1]. The target body is encapsulated in two ceramic bearings, and free to rotate in the center. A lead ballast is used to self-align the water openings (in the horizontal plane) during the loading process. The complete solution allows the target shuttle to be loaded and unloaded pneumatically from outside the cyclotron bunker. Further enhancement was made to the location of the coolant O-ring seals by placing them on the vertical plane, to assist with the loading and unloading of the target shuttle without resistance, with details shown in Figure 3. The complete target shuttle is pressurized from behind to maintain a tight seal for both water and helium cooling cavities.

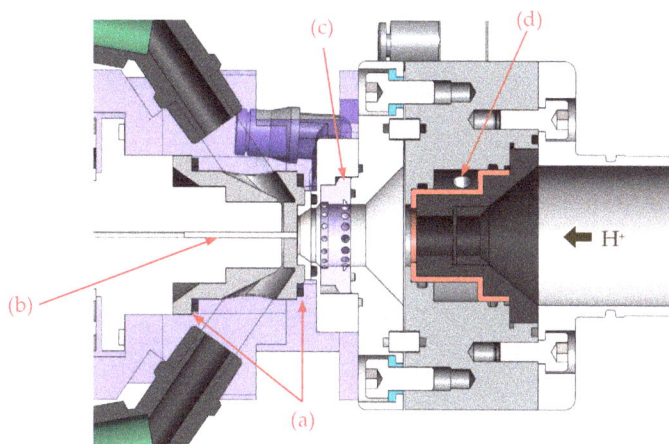

**Figure 3.** Details of 3D-printed sterling silver target (SS925) showing cooling and O-ring seals: (a) O-ring seals; (b) type-K thermocouple; (c) He cooling for target material and vacuum window; (d) water jacket for graphite collimator and degrader.

Simulations were conducted using SolidWorks 2017/18 CAD package with flow simulation CFD (computational fluid dynamics). The proton beam was modeled as a surface heat source on the target material and assumes that the total power is absorbed on the material surface [1]. The simulation does not account for helium flow on the surface of the target material, since helium cooling is primarily used for the vacuum window only. The CFD package calculated the temperatures for the target material with water applied into the solid target body at an adjusted flow rate of 20 L/min.

The simulated target reflects the same physical dimensions of an electroplated target for the production of $^{64}$Cu. The model target is composed of a Ni layer (100 µm, dia. 7.0 mm) on top of an Au foil (125 µm, dia. 15 mm). The theoretical maximum temperature was calculated for target currents from 40 to 120 µA (at 11.5 MeV) in 10 µA increments. Results were compared for both cast alloy (A357) and sterling silver (SS925) solid target designs.

The solid target shuttle has an embedded Type-K thermocouple (dia. 1.5 mm), located in the center and mounted laterally to the beam path (refer to Figures 2 and 3). The thermocouple is in direct contact with the target material for improved sensitivity and to minimize the surface area exposed to the proton beam. The integration of a thermocouple inside the target holder provides continuous temperature monitoring of the target material during irradiation. Temperature measurements for the 3D-printed target were conducted using a single 125 µm pure Au foil irradiated at 40 µA up to 80 µA, in increments of 10 µA. The sterling silver target and irradiated Au foil were visually inspected for thermal damage for target currents >80 µA.

The 3D-printed sterling silver solid target was validated with the production of radio-copper using enriched $^{64}$Ni electroplated on 125 μm Au foil. This was irradiated at 40 μA to 80 μA for 5 minutes at increments of 10 μA. The maximum temperature was recorded for each increment and compared to the theoretical results from the simulation.

## 3. Results

The advantage of the lost-wax casting process over additive metal 3D-printing techniques is the low porosity of the final metallic form, with a relative density is equivalent to the silver granules used in the casting process, 10.37 g/cm$^3$. As a result, the thermal conductivity is closer to the physical properties of pure silver, 10.49 g/cm$^3$. SLM and other additive 3D-printing processes have moderate porosity with a difficult to determine final relative density. The microstructure from SLM 3D-printing is highly dependent on the scanning speed, hatch spacing, laser power, layer thickness, and the properties of the powder. The relative density for SLM 3D-printed cast alloy can vary from 77% to 97.5% [2], thereby influencing the thermal capability of the target holder. The main disadvantage of the lost-wax casting process is the lower resolution, with wider minimum gap size and greater consideration for escapement holes for the casting material. The surface finish for SS925 is superior to A357, as shown in Figure 4a,b. The internal structures are well defined for SS925, with low obstruction to water flow compared to A357. SLM and other additive manufacturing processes require additional post-fabrication machining and polishing. Since the internal cooling structure is inaccessible once the 3D production is complete, the lost-wax casting process is superior for this design and application.

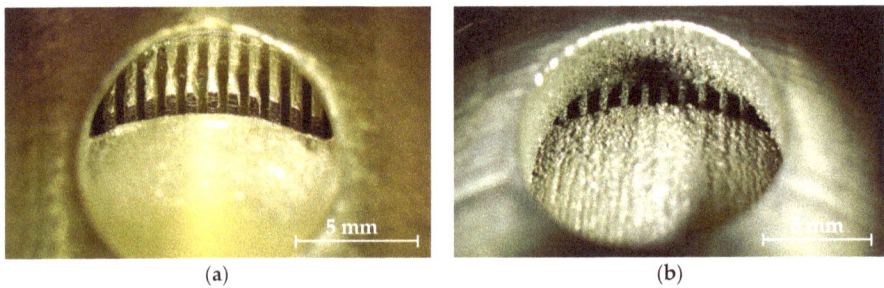

**Figure 4.** Internal structure of the water cooling cavity. (**a**) SS925 solid target holder; (**b**) A357 solid target holder.

The maximum theoretical temperatures calculated for the A357 and SS925 solid target stations are shown in Table 1 below. As expected, given the higher thermal conductivity of sterling silver compared to cast alloy (A357), the calculated temperature for SS925 is significantly lower at all target currents.

**Table 1.** Maximum calculated temperatures for A357 and SS925 solid target stations.

| Target Current (μA) | Power (W) | Cast Alloy (A357) (°C) | Sterling Silver (SS925) (°C) |
|---|---|---|---|
| 40 | 460 | 121 | 94 |
| 50 | 575 | 147 | 112 |
| 60 | 690 | 172 | 130 |
| 70 | 805 | 197 | 147 |
| 80 | 920 | 221 | 164 |
| 90 | 1035 | 246 | 182 |
| 100 | 1150 | 270 | 198 |
| 110 | 1265 | 294 | 214 |
| 120 | 1380 | 318 | 231 |

The maximum temperature calculated by the CFD package is observed between the interface of the target material and the surface heat source. In CFD, the thermal advantage of a 3D-printed SS925 target is ~30% compared to A357, and ~60% better than the existing compact solid target system [1]. The simulated temperatures are below the melting points of all relevant target materials, such as Ni ($T_m$ = 1455 °C), Au ($T_m$ = 1064 °C), Y ($T_m$ = 1526 °C), and sterling silver ($T_m$ = 893 °C).

The SS925 solid target system was manually loaded for testing without a PLC-controlled pneumatic loading system. Figure 5 shows the SS925 solid target with the pneumatic shuttle.

(a)

**Figure 5.** Solid target assembly with target shuttle and SS925 target holder (a).

Table 2 shows the maximum temperature recorded during the irradiation of the Au foil (without electrodeposited layer) on the SS925 solid target station at various target currents.

**Table 2.** Maximum recorded temperature and visual inspections at 80 μA.

| Target Current (μA) | Solid Target Max. Temp. (°C) | Visual Inspection, Damage (Yes/No) |
|:---:|:---:|:---:|
| 0 | 16 | - |
| 40 | 73 | - |
| 50 | 86 | - |
| 60 | 99 | - |
| 70 | 113 | - |
| 80 | 122 | No |

Visual inspection of the Au foil post-irradiation showed no visible signs of damage, Figure 6b. Beam losses due to divergence are reduced by placing the target material closer to the exit port of the cyclotron. The average beam loss for this design is <25%, compared to >50% for the existing compact solid target on the end of a 30 cm beamline.

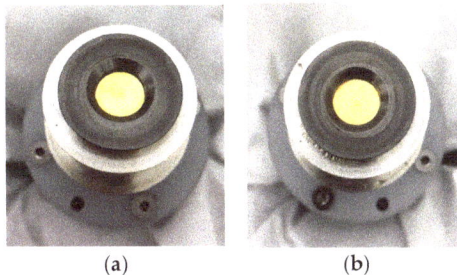

(a)                    (b)

**Figure 6.** (a) Au foil before irradiation on the 3D-printed SS925 solid target holder. (b) Au foil after being subjected to experimental protocol as described above.

A small lump was observed at the center of the Au foil, which indicates the protrusion of the thermocouple from the SS925 body. This reduces the surface contact between the Au foil and the target body, therefore affecting the cooling efficiency.

The maximum temperatures observed during the irradiation of $^{64}$Ni plated on Au for target SS925 is shown in Table 3 below. Once this maximum had been established and the target current stabilized, the temperature remained at a plateau with little variation. The calculated peak temperature is shown in column 3 of Table 3, for comparison with these experimental results. By contrast, from previous experimental results with the A357 target holder [1], the temperature of the assembly would fluctuate erratically if the limit of the material physical property was reached.

**Table 3.** Maximum recorded temperature for $^{64}$Ni on Au using the new solid target system.

| Target Current (µA) | Max. Recorded Solid Target Temp. (°C) | Theoretical Temp. (°C) [1] |
|---|---|---|
| 0 | 16 | 18 |
| 40 | 64 | 87 |
| 50 | 74 | 103 |
| 60 | 86 | 119 |
| 70 | 102 | 135 |
| 80 | 118 | 151 |

[1] A thickness of 100 µm of Ni material is used in the theoretical model.

The total mass of the target (enriched $^{64}$Ni and Au) was 478.3 mg with 16 mg of plated enriched $^{64}$Ni. The equivalent thickness of plated $^{64}$Ni was approximately 46.7 µm. The total mass of the target, post-irradiations, was 478.2 mg, which indicates no loss of enriched material from proton bombardment at 80 µA. Figure 7 shows the electroplated enriched $^{64}$Ni on Au pre- and post-irradiation, mounted on the SS925 target holder.

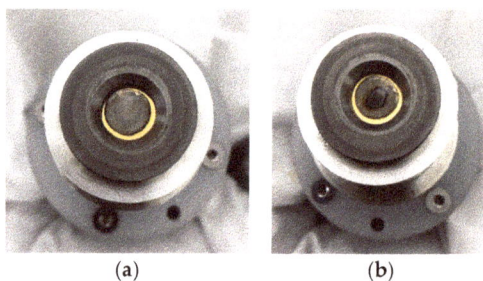

(a)      (b)

**Figure 7.** (a) $^{64}$Ni on Au before irradiation of the 3D-printed SS925 solid target holder; (b) $^{64}$Ni on Au after all irradiations as described above, and shown in Table 3.

As determined from SRIM [3], ~1.12 MeV of the degraded primary beam is deposited in the Ni layer, ~5.15 MeV in the Au backing, and the residual 5.23 MeV beam is deposited in the target holder. A thickness of ~100 µm of Ag and stainless steel will effectively stop the residual beam completely. Proton interaction with the thermocouple junction is negligible, since the thermocouple material is encased in a stainless steel sheath with thickness >100 µm. Radionuclide production of $^{106m}$Ag ($T_{1/2}$ = 8.28 d) and $^{107}$Cd ($T_{1/2}$ = 6.5 h), from the 5.23 MeV residual beam, has low excitation functions of <5 mb and <30 mb, respectively [4]. Proton activation of the Ag target holder can be eliminated if the thickness of target material or the gold backing is increased by ~100 µm.

We observed no thermal damage of the Au foil at 80 µA, with no loss of electroplated $^{64}$Ni material. This indicates an improved thermal management of the target current due to the higher thermal conductivity of the SS925 material. The irradiated $^{64}$Ni material appeared tarnished from proton bombardment in Figure 7b, and is visually similar to a routine 40 µA $^{64}$Cu production on the existing solid target station. The imprint of the beam is clearly visible on the target surface, indicating a sharp approximately Gaussian profile.

Given the assumptions used in the theoretical model for proton bombardment as a homogeneous surface heat source, and the difference in the thickness of the actual enriched $^{64}$Ni layer used experimentally, some differences between theoretical temperatures and measured temperatures for the new solid target system are to be expected. An adjusted simulation model to reflect the true target thickness (46.7 μm) shows that the theoretical model overestimates the temperature by ~35%. The margin of error between true and calculated temperature is indicative of the poor contact between the target material and the SS925 target holder. In addition, the thermocouple transverses the water cavity with active cooling of stainless steel sheath, which may affect the final temperature reading. Given the temperature measurement is used as a guide for target integrity only, the ability to continuously monitor the temperature during an irradiation is invaluable, despite the error. Further refinement to the design, with insulation of the thermocouple from the water channel, is currently in progress. Calibration with destructive testing and thicker Au backing will likely increase the accuracy of the measured temperature and validate the minimal direct influence of the proton beam on thermocouple measurement.

## 4. Conclusions

Theoretically and experimentally, the new solid target design shows significant improvement in temperature management compared to the existing compact solid target system. The new SS925 solid target system offers a more robust and long-term solution with the convenience of loading and unloading the target material outside of the cyclotron bunker. It also moves the target material closer to the cyclotron exit port, thus reducing the beam losses due to divergence along an external beam line. The system is capable of higher target currents without the need to slant the target material to the incident proton beam. This ensures a moderate starting material cost, given the comparatively small amount [5] of enriched $^{64}$Ni material needed for electroplating, compared with traditional slanted targets. Other advantages are the minimal engineering impact to peripheral systems used in the pre- and post-irradiated target preparation and separation processes.

Mounting the thermocouple in the lateral position protects the tip from the incident proton beam and provides direct contact with the target backing. The thermocouple is an effective tool for both beam alignment and live feedback used to monitor the target material during an irradiation. Minor adjustment to the length is needed to minimize its protrusion from the flat surface. This will ensure better target material contact with the SS925 target holder.

Further investigation is needed to validate the temperature observed during a standard irradiation (80 μAh) and to complete the shuttle loading and unloading mechanism. Once the operating temperature profile of the new design is confirmed, it can be adapted to the existing solid target system for greater thermal management at higher target currents.

The SS925 target holder is comparatively economical and efficient, once the design has satisfied all 3D-printing fabrication requirements. Future work will aim to confirm that the 3D-printed SS925 target holder has relatively low activation from proton bombardment at beam energies <11.5 MeV [4], and to compare the yields produced for both $^{64}$Cu and $^{89}$Zr, between our existing compact solid target and the new SS925 solid target system.

**Author Contributions:** Conceptualization, S.C. and D.C.; Methodology, S.C.; Software simulations, S.C.; Fabrication, D.C.; Validation, S.C. and D.C.; Investigation, S.C. and D.C.; Resources, R.I.P.; Writing—original draft preparation, S.C.; Writing—review and editing, S.C. and R.I.P.; Supervision, R.I.P.

**Acknowledgments:** The technical and scientific support of the Radiopharmaceutical Production and Development Laboratory (RAPID, MTP) staff is gratefully acknowledged.

**Conflicts of Interest:** Authors declare no conflict of interest.

## References

1. Chan, S.; Cryer, D.J.; Price, R.I. Development of a New Solid Target Station at Cyclotron Facility in Perth, Australia. In Proceedings of the 16th International Workshop on Targetry and Target Chemistry, Santa Fe, NM, USA, 29 August–1 September 2016.
2. Aboulkhair, N.T.; Everitt, N.M.; Ashcroft, I.; Tuck, C. Reducing porosity in AlSi10Mg parts processed by selective laser melting. *Addit. Manuf.* **2014**, *1–4*, 77–86. [CrossRef]
3. Ziegler, J.F.; Ziegler, M.D.; Biersack, J.P. SRIM Code the Stopping and Range of Ions in Matter, version 2013.00. Available online: http://www.srim.org/ (accessed on 1 February 2019).
4. Khandaker, M.U.; Kim, K.; Kim, K.S.; Lee, M.; Kim, G.; Cho, Y.S.; Lee, Y.O. Production cross-section of residual radionuclides from proton-induced reactions on $^{nat}$Ag up to 40 MeV. *Nucl. Instrum. Methods Phys. Res. B* **2008**, *266*, 5101–5106. [CrossRef]
5. International Atomic Energy Agency (IAEA). *Cyclotron Produced Radionuclides: Emerging Positron Emitters for Medical Applications: 64Cu and 124I*; Radioisotopes and Radiopharmaceuticals Report No. 1; International Atomic Energy Agency (IAEA): Viena, Austria, 2016; ISBN 978-92-0-109615-9.

*instruments*

**MDPI**

*Article*

# Boron Nitride Nanotube Cyclotron Targets for Recoil Escape Production of Carbon-11

Johanna Peeples [1],*, Sang-Hyon Chu [2], James P. O'Neil [3], Mustafa Janabi [3], Bruce Wieland [1] and Matthew Stokely [1]

[1]  BTI Targetry LLC, 1939 Evans Road, Cary, NC 27513, USA; wielandb@bellsouth.net (B.W.); stokely@btitargetry.com (M.S.)
[2]  National Institute of Aerospace, 100 Exploration Way, Hampton, VA 23666, USA; schu@nianet.org
[3]  Lawrence Berkeley National Laboratory, 1 Cyclotron Road, Berkeley, CA 94720, USA; jponeil@lbl.gov (J.P.O.); mjanabi@lbl.gov (M.J.)
*   Correspondence: peeples@btitargetry.com; Tel.: +1-919-677-9799

Received: 19 December 2018; Accepted: 24 January 2019; Published: 27 January 2019

**Abstract:** Boron nitride nanotubes (BNNTs) were investigated as a target media for cyclotron production of $^{11}C$ for incident beam energy at or below 11 MeV. Both the $^{11}B(p,n)^{11}C$ and $^{14}N(p,\alpha)^{11}C$ nuclear reactions were utilized. A sweep gas of nitrogen or helium was used to collect recoil escape atoms with a desired form of $^{11}CO_2$. Three prototype targets were tested using an RDS-111 cyclotron. Target geometry and density were shown to impact the saturation yield of $^{11}C$ and percent of yield recovered as carbon dioxide. Physical damage to the BNNT target media was observed at beam currents above 5 µA. Additional studies are needed to identify operating conditions suitable for commercial application of the method.

**Keywords:** cyclotron target; carbon-11; recoil escape; boron nitride nanotubes; BNNTs

## 1. Introduction

Boron nitride nanotube (BNNT) nanomaterials [1–5] can be used in a recoil escape target to produce $^{11}C$ for incident proton energy at or below 11 MeV. This would enable $^{11}C$ production for economical low-energy cyclotrons [6–9], and it could be used to increase production from conventional $^{11}C$ gas targets on MiniTrace and RDS-111 cyclotrons. Preliminary experiments have demonstrated recoil escape production and recovery of small quantities of $^{11}CO_2$, and there has been a continued effort to develop a target design and platform that can be used to produce viable yields for a commercial system.

Conventional gas targets for $^{11}C$ production operate by proton bombardment of nitrogen gas [10–13]. However, due to the low nuclear cross-section of the $^{14}N(p,\alpha)^{11}C$ nuclear reaction below 11 MeV, this production method is not commercially viable for accelerators with proton energy in the range of 7–10 MeV. Adding BNNT nanomaterials to the target allows for an additional production route via the $^{11}B(p,n)^{11}C$ nuclear reaction, which has a higher cross-section at all proton energies, as shown in Figure 1 [14,15]. If the produced $^{11}C$ can be effectively recovered from the target, the total yield of $^{11}C$ will be greater than currently achievable using a conventional gas target.

The nanotube geometry should offer superior recovery of $^{11}C$ recoil atoms to boron powder, due to the smaller mean particle size (nm versus µm) and higher porosity. Nearly complete recoil escape should be possible, because the walls of individual BNNT fibrils are single atomic thickness or at most a few atoms thick. The high porosity of the bulk material should allow sufficient gas flow to facilitate the slowing of recoil ions to thermal energies and their combination with trace oxygen in the sweep gas.

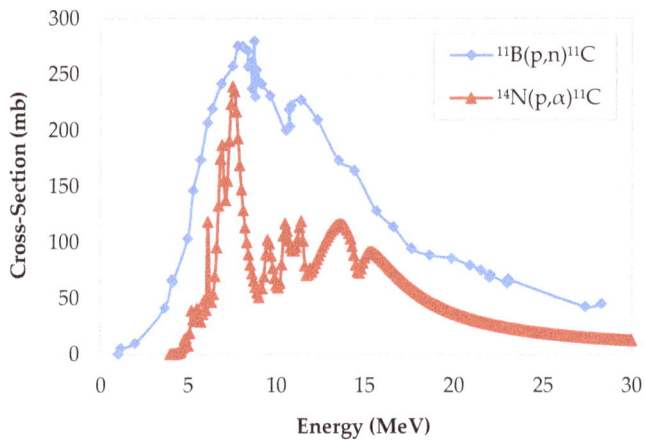

**Figure 1.** Cross-sections for $^{11}$C production from the $^{11}$B(p,n)$^{11}$C and $^{14}$N(p,$\alpha$)$^{11}$C nuclear reactions.

## 2. Materials and Methods

Both natural-abundance boron (19.9% $^{10}$B and 80.1% $^{11}$B) and enriched boron-11 (>98% $^{11}$B) BNNT nanomaterials are commercially available [16]. The bulk material is highly porous, allowing for diffusion of gas, and the nanotube dimensions are compatible with the recoil escape of $^{11}$C. In addition to the nanotubes, impurities of elemental boron and hexagonal boron nitride (*h*-BN) are present. Recovery of $^{11}$C is assumed to be strongly correlated to the impurity content, since larger structures could trap recoiling atoms and prevent recovery in the sweep gas. The BNNT material is supplied at a low density of roughly 1.38 g/cm$^3$ [17]. It undergoes plastic deformation with light pressure and sticks to surfaces both mechanically and due to electrostatic attraction. As a result, loading a target at uniform density is challenging. A photograph of the bulk BNNT material is shown in Figure 2, and individual nanotubes and nontube impurities, such as boron and *h*-BN particles, can be seen in the scanning electron microscope (SEM) images shown in Figure 3.

**Figure 2.** Photograph of 500-mg sample of bulk boron nitride nanotube (BNNT) material from BNNT, LLC, supplier.

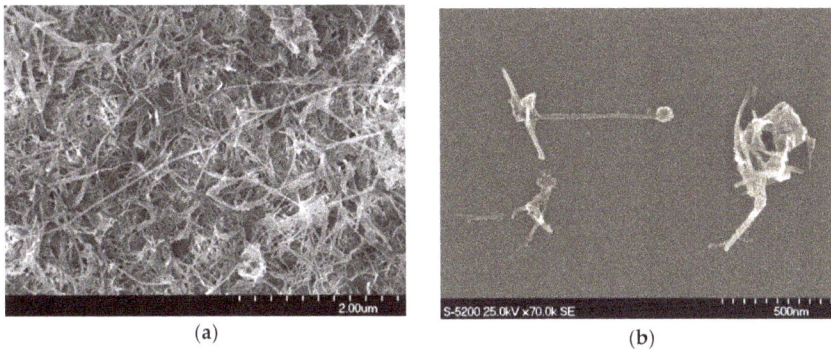

(a)                                              (b)

Figure 3. SEM images of (a) bulk BNNT material and (b) isolated nontube impurities.

Three prototype targets were tested at the RDS-111 cyclotron at Lawrence Berkeley National Laboratory (LBNL) using commercially available BNNT nanomaterials from supplier BNNT, LLC [16]. Preliminary work was performed using BNNT nanomaterials with natural-abundance boron and a target for the Eclipse target changer. The target, shown in Figure 4, features an aluminum target body with a cylindrical target chamber 1.0 cm in diameter by 1.0 cm in length. A window foil of either 25 μm of Havar or 30 μm of aluminum is used to result in a mean incident energy on the BNNT target media of 10.5 MeV or 8.1 MeV, respectively. Annular space between the target and changer creates a flow path for water cooling. Two 1-mm diameter passages, which intercept the target chamber front and rear, allow sweep gas to pass into the target through the BNNT material and out of the target into a delivery line during or after the irradiation process.

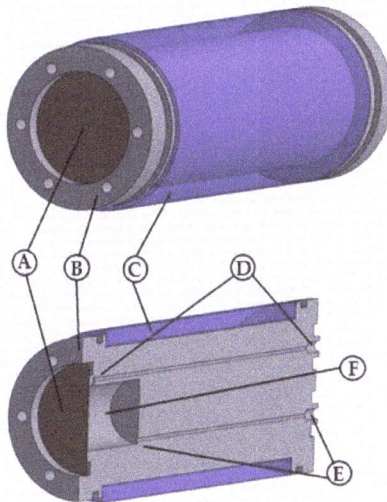

Figure 4. Prototype target for Eclipse target changer including (A) Havar or aluminum window foil, (B) an aluminum target body, (C) cooling water, (D) a front sweep gas flow path, (E) a rear sweep gas flow path, and (F) a target chamber for BNNT.

Using the initial prototype, sufficient quantities of $^{11}C$ were produced and recovered using a range of operating conditions to demonstrate the feasibility of using BNNT as a target material, and more than 80% of the recovered $^{11}C$ was in the desired form of $CO_2$. The principle technical challenge was identified to be minimizing damage to the BNNT target media that occurred during target operation at a high current. Irradiation at 20 µA with 10.5-MeV incident protons resulted in a dramatic reduction in $^{11}$C saturation yield and physical damage to the BNNT material, including the formation of a cavity in the center of the target chamber with translucent glassy crystals along its margins, as shown in Figure 5. The crystalline material was presumed to be boron oxide ($B_2O_3$).

**Figure 5.** Photograph of BNNT material in the prototype target, showing material damage resulting from irradiation at 20-µA beam current.

Two additional prototype targets with larger chamber volumes were developed for the single target station. The aluminum target bodies feature cylindrical target chambers with dimensions of 1.1 cm in diameter by 6.0 cm in length (BN-124 target) and 1.1 cm in diameter by 1.5 cm in length (BN-131 target), with four 6.1-mm in diameter water cooling channels. Helium-cooled Havar vacuum and target windows (25 µm and 38 µm, respectively) resulted in a mean incident energy on the BNNT media of 9.5 MeV. The BN-124 and BN-131 targets are shown in Figures 6 and 7.

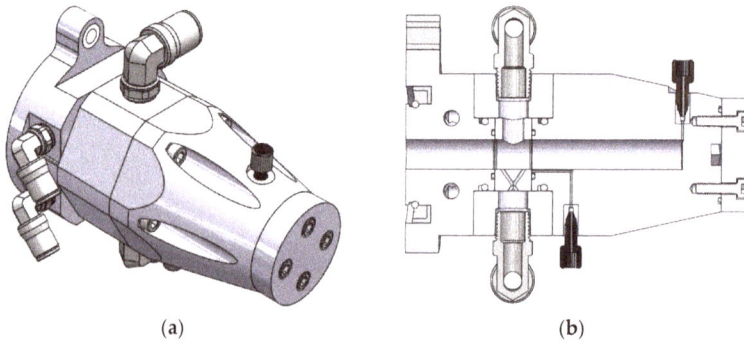

(a)                                          (b)

**Figure 6.** BN-124 target: (**a**) Isometric and (**b**) vertical mid-plane cross-section views.

A custom BTI Targetry target station, shown in Figure 8, was used, consisting of a beam tube, a vacuum isolation valve, a collimator with an 8-mm diameter opening, and target mounting geometry. Because the BN-124 prototype target is considerably longer axially than the prior prototype or commercial $^{18}$F targets, the target–collimator interface is subjected to a greater bending moment while supporting the weight of the target. A pressed stainless steel pin on the vertical mid-plane was added to provide additional mechanical support for this application.

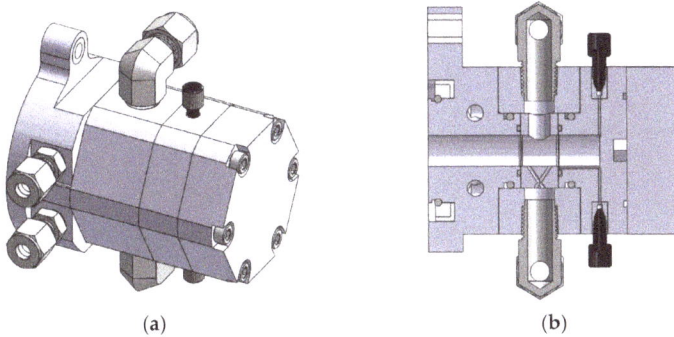

**Figure 7.** BN-131 target: (**a**) Isometric and (**b**) vertical mid-plane cross-section views.

**Figure 8.** Custom target station with mechanical support for additional axial target depth.

A diagram of the experimental setup is shown in Figure 9. The target chamber was filled with BNNT and pressurized by opening the target isolation valve and delivering gas through a flow control valve. For static mode irradiations, the target load valve was closed, and the target was irradiated at a fixed current. After irradiation, the target unload valve was opened, and the target was allowed to depressurize, followed by opening the load valve to sweep gas through the target chamber and remove volatile radioisotopes. For continuous flow mode irradiations, both the load and unload valves were opened with a gas sweep rate established by the flow control valve, and gas was swept through the target chamber throughout the irradiation. In all cases, the same gas was used to both fill the target and for the sweep to remove volatile radioisotopes. Experiments were performed using a fill gas of nitrogen or helium with 1%–10% oxygen.

For both static and continuous flow operation, the target gas was delivered through 10 m of 0.5-mm inside diameter tubing (valve E to trap G in Figure 9) to a hot cell in the laboratory. Once inside the hot cell, the gas was swept through a $^{11}CO_2$ soda lime trapping cartridge, and the untrapped radiolabeled gases and target gas were collected in a Tedlar gas collection bag. The CO and $CO_2$ were identified using an SRI 8610C gas chromatograph (GC) equipped with a Restek ShinCarbon ST 80/100 column (P/N 80486-800, 2 m, 2-mm inside diameter, 1/8-in. outside diameter Silicone). The original GC detector setup was a thermal conductivity detector (TCD) followed by a flame ionization detector (FID). A 1/16-in. outside diameter stainless steel line was added in between the two detectors and formed into a 1-in. loop to redirect flow through an NaI gamma detector with outside dimensions roughly 1 in. outside diameter by 4.75 in. high. A Carroll & Ramsey Associate post-amplifier/integrator (model 105-S) was used.

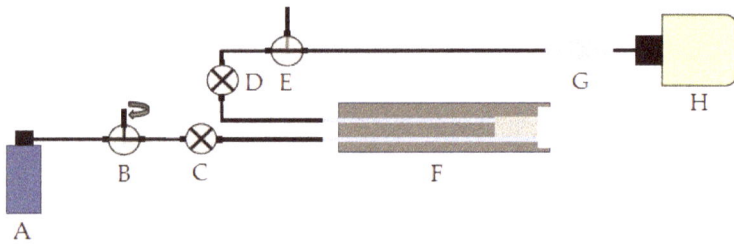

**Figure 9.** Diagram of experimental apparatus for material irradiation and isotope collection, illustrating a (A) target gas tank and regulator, (B) flow control valve, (C) target load valve, (D) target unload valve, (E) delivery/waste valve, (F) target body and target material, (G) soda lime carbon dioxide trap, and (H) radioisotope gas collection bag.

During the collection, the $^{11}CO_2$ trap was held in a dose calibrator (ionization chamber style well detector) to measure the dynamic buildup and final quantity of $^{11}C$ trapped as $^{11}CO_2$ on the cartridge. Subsequently, a 50-mL sample of the volatile gases collected in the sampling bag was analyzed for radioactivity in the dose calibrator, and total activity in the bag was determined by volume ratio between the 50-mL sample and total bag volume. Activities of all samples were measured for a minimum of 5 min and fitted to the known half-lives for $^{11}C$ and $^{13}N$, and the results were decay-corrected to end of bombardment.

The $^{11}C$ was collected in several modes, depending on the experiment (soda lime trap and collection bag, soda lime trap and waste line, or collection bag only). The soda lime trap was made of Fisher Scientific brand (ACS certified, S-196) soda lime placed inside of a Swagelok 1/2–1/4 in. reducing union (PFA-820-6-4) with an approximate volume of 2–4 mL. The collection bag was a 3-L Tedlar bag (SKC Inc., Cat # 232-03). When a GC sample was to be analyzed, the entire target output was collected into the collection bag, and a 1-mL sample was injected onto the GC. Soda lime trap and collection bag mode was used when the $CO_2$ composition was unknown to determine the percentage of recovered activity in the form of $CO_2$ versus other radioactive products. Soda lime trap and waste line mode was used to determine target yield when the $CO_2$ percentage was known to be high.

The recovered target saturation yield of $^{11}C$ ($Y_{sat}$) was calculated by

$$Y_{sat} = \frac{A_{EOB}}{I\left(1 - e^{-\lambda t_{irr}}\right)},\tag{1}$$

using the recovered decay-corrected end-of-bombardment $^{11}C$ activity ($A_{EOB}$), the average beam current ($I$), the $^{11}C$ decay constant ($\lambda = 0.034045$ min$^{-1}$), and a 5-min irradiation time ($t_{irr}$). A portion of the produced $^{11}C$ activity remained trapped in the target media but could not easily be measured.

The BN-124 and BN-131 prototype targets were tested using 98% enriched $^{11}B$ BNNT ($^{11}BNNT$) to increase the yield of $^{11}C$ and reduce the production of $^7Be$, which has a half-life of 53 days. Additional tests were performed for the BN-131 target using $^{11}BNNT$ material from a second commercial supplier, BNNano [18].

## 3. Results

### 3.1. Beam Tests of the BN-124 Target

Proton beam tests were performed on the single target station using the BN-124 prototype, which had a 6.0-cm depth and a 5.7-cm$^3$ volume, for beam currents between 1 and 10 μA, with static operation at load pressures between 200 and 800 psi using a fill gas of nitrogen with 1% oxygen. The target was loaded with 377 mg of $^{11}BNNT$, corresponding to an effective BNNT density of 0.07 g/cm$^3$. A recovered target saturation yield of $^{11}CO_2$ as a function of beam current and load pressure is shown in Figure 10. More than 95% of the recovered activity was in the desired form of $^{11}CO_2$, as indicated by soda lime trapping and confirmed by GC measurements. Target saturation yield increased with pressure between 200 and 600 psi, and then began to drop off. Saturation yield was highest at 1 µA, which corresponded to the lowest target heat input.

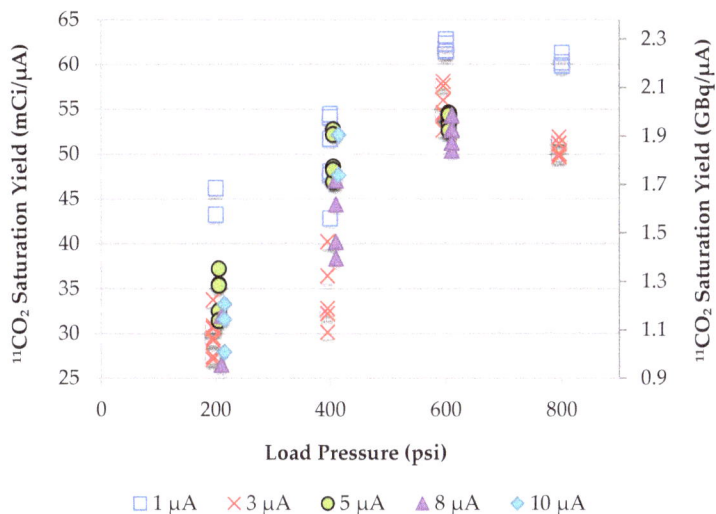

**Figure 10.** $^{11}CO_2$ saturation yield for the BN-124 target using a fill gas of nitrogen with 1% oxygen.

The BN-124 target was opened and visually inspected following a series of irradiations at each beam current. Following irradiation at 1 µA, the material changed color from gray to off-white, which was attributed to the conversion of elemental B impurities into $B_2O_3$. Material shrinkage (volume reduction) of the BNNT target media into the back of the target was observed, with additional shrinkage occurring for higher currents, as summarized in Table 1. After irradiation at 3 µA, the material shrinkage was about 40%, and the color continued to whiten. Following the 8-µA runs, observed changes in the BNNT material included crystallization, gray and black spots, and a white powdery buildup coating the inside walls of the target. Due to the additional material shrinkage, there was a 3.1-cm depth void in the front of the target, which was more than half of the chamber depth.

**Table 1.** Material shrinkage measurements and visual observations for the BN-124 target.

| Beam Current (µA) | Depth of Void (cm) | Depth of $^{11}$BNNT (cm) | Observations |
|---|---|---|---|
| 0 | - | 6.0 | Gray color |
| 1 | - | 6.0 | Off-white color |
| 3 | 2.5 | 3.5 | White color, shrinkage |
| 5 | 2.5 | 3.5 | White color, shrinkage |
| 8 | 3.1 | 2.9 | Crystallization, powdery buildup coating walls |

Photographs of the BN-124 target media post-irradiation at each beam current are shown in Figure 11. It is important to note that shades of gray and white can be very misleading due to variations in the angle of the photograph and lighting conditions.

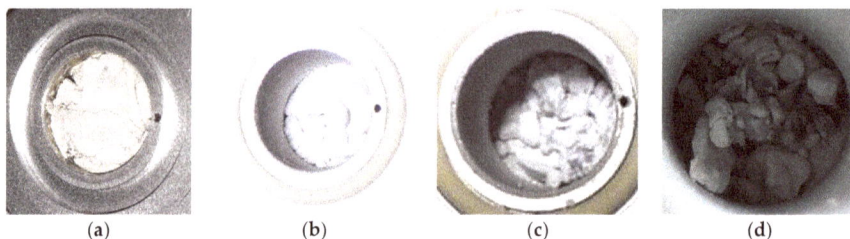

**Figure 11.** Photographs of the BN-124 target post-irradiation at (**a**) 1 µA; (**b**) 3 µA; (**c**) 5 µA; and (**d**) 8 µA.

The void formed in the front of the target during irradiation made interpretation of the data difficult, since most of the recovered yield was likely from the $^{14}N(p,\alpha)^{11}C$ reaction in the nitrogen gas in the voided region. Increasing the effective density in the BN-124 target to mitigate material shrinkage would have required significantly more $^{11}BNNT$ material, which was both cost-prohibitive (>\$1000/g) and inefficient due to the range of the protons. Accounting for the Havar vacuum and target windows (25 µm and 38 µm, respectively), stopping power calculations indicated only 125 mg of BNNT was needed for the target to be axially range thick for 11-MeV protons.

*3.2. Beam Tests of the BN-131 Target*

To address these issues, a second prototype target (BN-131) was designed and fabricated for the single target station with a reduced target depth of 1.5 cm, resulting in a volume of 1.4 cm$^3$. The target was loaded with 400 mg of $^{11}BNNT$, resulting in an effective BNNT density four times higher than that used for the prior prototype. Proton beam irradiations were performed at 5 µA, for static operation at 200–600 psi load pressure, using both nitrogen gas with 1% oxygen and helium gas with 1% oxygen. Negligible material shrinkage was observed for the BN-131 target following irradiation.

Total saturation yield of $^{11}C$ in all forms ($^{11}CO_2$ and $^{11}CO$) for the BN-131 target irradiations is shown in Figure 12. Since helium does not offer a competing production reaction for $^{11}C$, all recovered $^{11}C$ was produced in BNNT for these irradiations. The total yield of $^{11}C$ was roughly equivalent for both gases, suggesting minimal production of $^{11}C$ in the nitrogen fill gas. The fraction of activity recovered in the form of carbon dioxide is shown in Figure 13. Using helium fill gas in the BN-131 target consistently produced less $^{11}C$ as CO$_2$ (35%–55%) compared to using nitrogen gas in the BN-131 target (50%–55%). For both fill gases, the percentage of activity recovered as carbon dioxide was significantly less than that observed when using nitrogen gas in the BN-124 target (>95%).

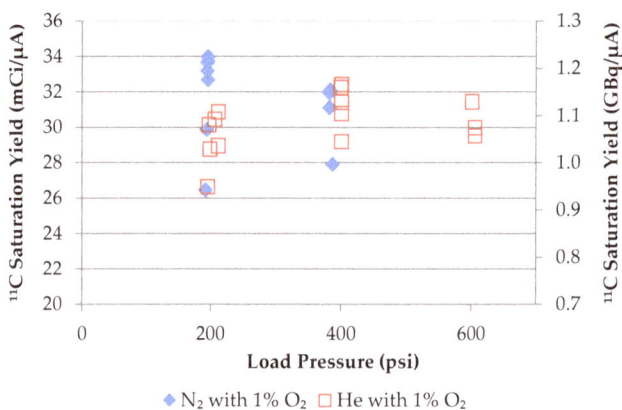

♦ N$_2$ with 1% O$_2$  □ He with 1% O$_2$

**Figure 12.** $^{11}C$ saturation yield for the BN-131 target at 5 µA.

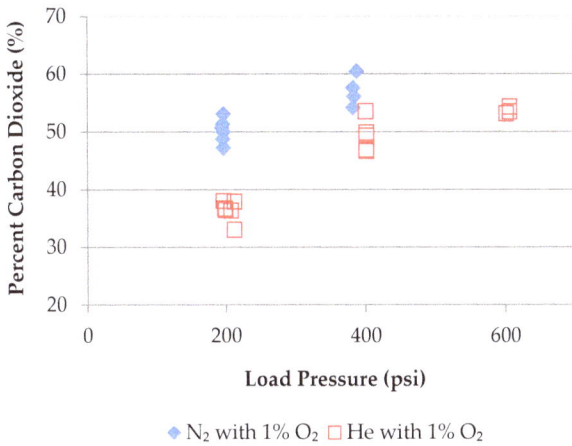

**Figure 13.** Percent of $^{11}$C recovered as $CO_2$ for the BN-131 target at 5 µA.

### 3.3. Comparison of $^{11}$C Saturation Yield for the BN-124 and BN-131 Targets

Total saturation yield of $^{11}$C in all forms ($^{11}CO_2$ and $^{11}CO$) for the BN-124 and BN-131 targets is shown in Figure 14. The results were similar for static operation using a fill gas of nitrogen with 1% oxygen at 200 psi. However, the BN-124 target yield increased with nitrogen gas pressure, while the BN-131 target yield was insensitive to nitrogen gas pressure. This supported the assertion that a significant component of yield from the larger target was due to $^{11}$C production in the voided region and independent of the nanomaterials.

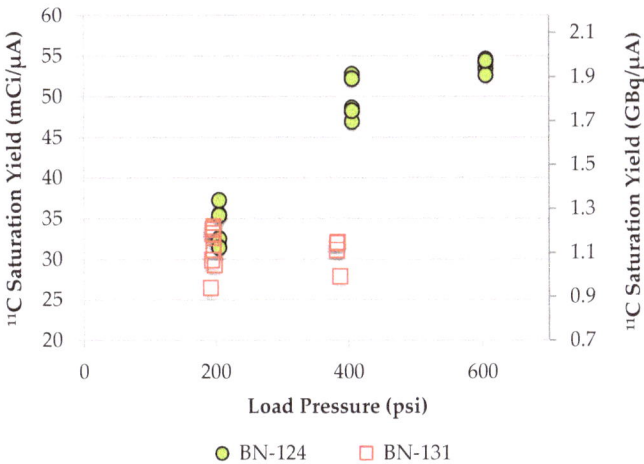

**Figure 14.** $^{11}$C saturation yield for the BN-124 and BN-131 targets at 5 µA using nitrogen with 1% oxygen.

### 3.4. BN-124 Gas-Only Target Operation

To provide a better context for evaluating target performance, the BN-124 target was also operated as a gas target using a fill gas of nitrogen with 1% oxygen with no BNNT nanomaterials present. The recovered target saturation yield of $^{11}CO_2$ for static operation at 200 psi load pressure using 377 mg $^{11}$BNNT and for gas-only static operation at 200 psi load pressure is shown in Figure 15. Monte Carlo radiation transport calculations were performed for the target using Monte Carlo N-Particle eXtended

(MCNPX) [19]. Although MCNPX simulations using the nuclear cross-section data indicated more [11]C was produced when using the [11]BNNT, use of the [11]BNNT resulted in a lower recovered saturation yield for all cases. This suggested a large amount of activity was being trapped in the target.

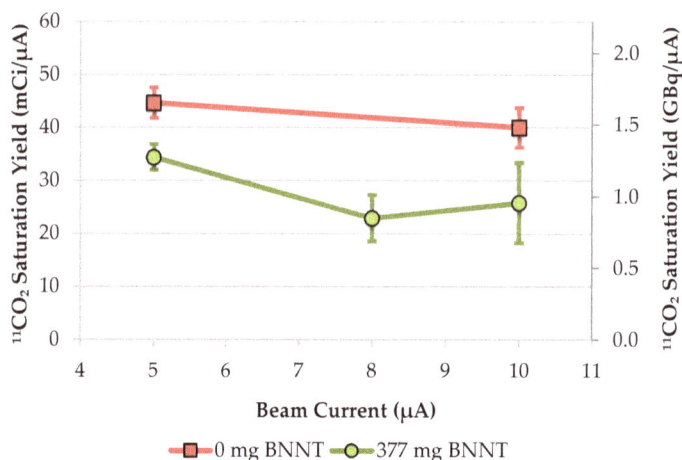

**Figure 15.** [11]$CO_2$ saturation yield for the BN-124 target using a fill gas of nitrogen with 1% oxygen.

*3.5. Evaluation of BNNT Nanomaterials from a Second Supplier*

Additional BNNT nanomaterials were acquired from a second manufacturer, BNNano [18]. Low-purity material (<90% nanotubes) was selected to prevent the material from becoming too brittle to easily load in the cylindrical target chamber. The material was darker in color, with significantly higher density and more mechanical integrity when manipulated, although it was still susceptible to plastic deformation. Photographs of bulk BNNT material from the two suppliers are shown in Figure 16. A limited set of [11]C saturation yield data was collected using the BN-131 target with [11]BNNT nanomaterials from BNNano at the RDS-111 cyclotron at LBNL and at a commercial RDS-111 cyclotron operated by Triad Isotopes. The saturation yield data were 1–2 orders of magnitude lower than what was observed in the prior tests using [11]BNNT from BNNT, LLC. SEM imaging revealed a much higher concentration of nontube impurities in the material, as shown in Figure 17. This further supported the theory that high-purity material was needed to prevent [11]C trapping in targets and achieve a high recovered saturation yield.

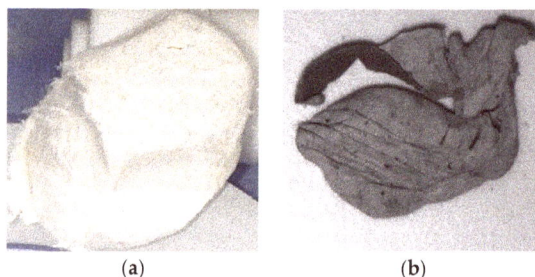

**Figure 16.** Photographs of BNNT nanomaterials from (**a**) BNNT, LLC, and (**b**) BNNano.

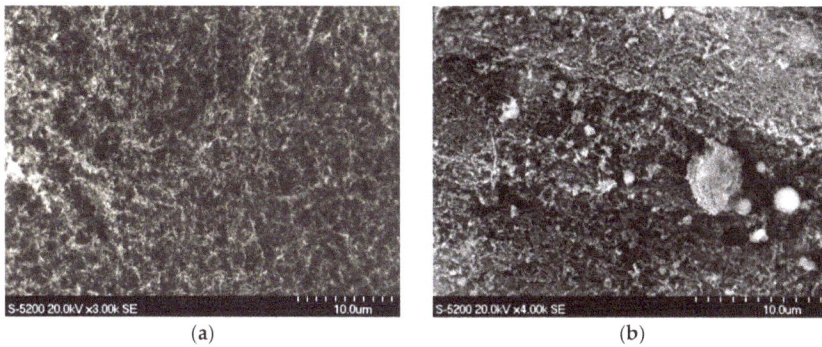

(a)                                            (b)

**Figure 17.** SEM images of BNNT nanomaterials from (**a**) BNNT, LLC, and (**b**) BNNano.

## 4. Discussion

Use of BNNT nanomaterials in a recoil escape target has the potential to increase $^{11}$C yield for proton energies at or below 11 MeV by utilizing both the $^{11}$B(p,n)$^{11}$C and $^{14}$N(p,$\alpha$)$^{11}$C nuclear reactions. However, additional studies are needed to determine operating conditions suitable for a commercial product. Beam irradiation experiments at a low effective BNNT density resulted in material shrinkage, making interpretation of the data difficult. Operating at a higher effective BNNT density prevented material shrinkage, but resulted in a lower saturation yield of $^{11}$C and a lower percentage of activity recovered in the desired form of $CO_2$.

Poor recovery of produced $^{11}$C and material degradation at modest beam currents both pose significant challenges for the development of a commercial system. The trapping of produced $^{11}$C in the target was most likely a function of the nontube impurities present in the commercially sourced BNNT nanomaterials. Impurities could be removed prior to irradiation by thermal oxidation in a furnace, followed by a water rinse. Future work will include beam irradiation tests of a sample of $^{11}$BNNT that has been purified at 600 °C for 1 h at the National Institute of Aerospace (NIA). $^{11}$BNNT material as-received from the BNNT, LLC, supplier is shown next to the purified material (which is layered with wax paper) in Figure 18. The purified material is whiter in color, more brittle, and has increased density of roughly one order of magnitude.

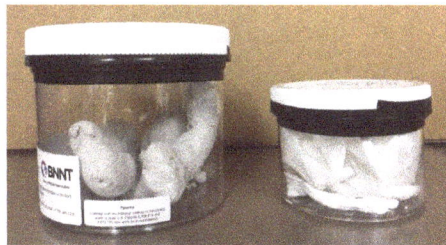

**Figure 18.** Photograph of 500-mg sample of as-received BNNT (left) and 500-mg sample of BNNT following purification at the National Institute of Aerospace (NIA) at 600 °C for 1 h (right).

Material degradation of BNNT materials has been observed as a function of beam current, due to elevated temperature in the target. Bulk BNNT material has low effective thermal conductivity of roughly 1 W/m$^2$-K [20], and significant thermal oxidation has been observed for BNNT targets in the range of 800–900 °C [21]. A new target prototype is currently being designed that aims to reduce peak temperature in the BNNT by utilizing a grazing angle for the beam with only a thin layer of BNNT.

The grazing angle prototype target will use high-purity BNNT target media (99% nanotubes and 1% nontube impurities of B and *h*-BN) from supplier BNNano.

**Author Contributions:** Conceptualization, J.P. and B.W.; methodology, J.P.O. and M.S.; investigation, S.-H.C., J.P.O., M.J., and M.S.; writing—original draft preparation, J.P.; writing—review and editing, M.S.; visualization, J.P. and M.S.; supervision, M.S.; project administration, J.P.; funding acquisition, J.P. and B.W.

**Funding:** This research was funded by the National Science Foundation, grant numbers 1519701 and 1632484.

**Conflicts of Interest:** The authors declare no conflicts of interest.

### References

1. Rubio, A.; Corkill, J.L.; Cohen, M.L. Theory of graphitic boron nitride nanotubes. *Phys. Rev. B* **1994**, *49*, 5081–5084. [CrossRef]
2. Weng-Sieh, Z.; Cherrey, K.; Chopra, N.G.; Blase, X.; Miyamoto, Y.; Rubio, A.; Cohen, M.L.; Louie, S.G.; Zettl, A.; Gronsky, R. Synthesis of $B_xC_yN_x$ nanotubules. *Phys. Rev. B* **1995**, *51*, 11229–11232. [CrossRef]
3. Chopra, N.G.; Luyken, R.J.; Cherrey, K.; Crespi, V.H.; Cohen, M.L.; Louie, S.G.; Zettl, A. Boron nitride nanotubes. *Science* **1995**, *269*, 966–967. [CrossRef] [PubMed]
4. Smith, M.W.; Jordan, K.C.; Park, C.; Kim, J.W.; Lillehei, P.T.; Crooks, R.; Harrison, J.S. Very long single- and few-walled boron nitride nanotubes (BNNTs) via the pressurized vapor/condenser method. *Nanotechnology* **2009**, *20*, 505604. [CrossRef] [PubMed]
5. Kim, K.S.; Kim, M.J.; Park, C.; Fay, C.C.; Chu, S.H.; Kingston, C.T.; Simard, B. Scalable manufacturing of boron nitride nanotubes and their assemblies: A review. *Semicond. Sci. Technol.* **2016**, *32*, 013003. [CrossRef]
6. Jensen, M.; Eriksson, T.; Severin, G.; Parnaste, M.; Norling, J. Experimental yields of PET radioisotopes from a prototype 7.8 MeV cyclotron. In Proceedings of the 15th International Workshop on Targetry and Target Chemistry, Prague, Czech Republic, 18–21 August 2014; pp. 112–113.
7. Khachaturian, M.; Bailey, J. The ABT Molecular Imaging biomarker generator Dose-on-Demand$^{TM}$ high flow tantalum 1.0 water target. In Proceedings of the 16th International Workshop on Targetry and Target Chemistry, Santa Fe, NM, USA, 29 August–1 September 2016; pp. 2–4.
8. Smirnov, V.; Vorozhtsov, S.; Vincent, J. Design study of an ultra-compact superconducting cyclotron for isotope production. *Nucl. Instr. Meth. Phys. Res. A* **2014**, *763*, 6–12. [CrossRef]
9. Podadera, I.; Ahedo, B.; Arce, P.; García-Tabarés, L.; Gavela, D.; Guirao, A.; Lagares, J.I.; Martínez, L.M.; Obradors-Campos, D.; Oliver, C.; et al. Beam diagnostics for commissioning and operation of a novel compact cyclotron for radioisotope production. In Proceedings of the International Beam Instrumentation Conference (IBIC2013), Oxford, UK, 16–19 September 2013; pp. 660–663.
10. Koziorowski, J.; Larsen, P.; Gillings, N. A quartz-lined carbon-11 target: Striving for increased yield and specific activity. *Nucl. Med. Biol.* **2010**, *37*, 943–948. [CrossRef] [PubMed]
11. Savio, E.; García, O.; Trindade, V.; Buccino, P.; Giglio, J.; Balter, H.; Engler, H. Improving production of $^{11}$C to achieve high specific labelled radiopharmaceuticals. *AIP Conf. Proc.* **2012**, *1509*, 185–189. [CrossRef]
12. Fonslet, J.; Itsenko, O.; Koziorowski, J. Indirect measurement of specific activity of [$^{11}$C]$CO_2$ and the effects of target volume fractionation. *AIP Conf. Proc.* **2012**, *1509*, 190–193. [CrossRef]
13. Peeples, J.L.; Magerl, M.; O'Brien, E.M.; Doster, J.M.; Bolotnov, I.A.; Wieland, B.W.; Stokely, M.H. High current C-11 gas target design and optimization using multi-physics coupling. *AIP Conf. Proc.* **2017**, *1845*, 020016. [CrossRef]
14. Firouzbakht, M.L.; Schlyer, D.J.; Wolf, A.P. Yield measurements for the $^{11}$B(p,n)$^{11}$C and $^{10}$B(d,n)$^{11}$C nuclear reactions. *Nucl. Med. Biol.* **1998**, *25*, 161–164. [CrossRef]
15. Takacs, S.; Tarkanyi, F.; Hermanne, A.; Paviotti de Corcuera, R. Validation and upgrade of the recommended cross section data of charged particle reactions used for production PET radioisotopes. *Nucl. Instrum. Methods Phys. Res. B* **2003**, *211*, 169. [CrossRef]
16. BNNT, LLC [US]. Available online: https://www.bnnt.com/products (accessed on 17 December 2018).
17. Zhi, C.; Bando, Y.; Tang, C.; Golberg, D. Specific heat capacity and density of multi-walled boron nitride nanotubes by chemical vapor deposition. *Solid State Commun.* **2012**, *151*, 183–186. [CrossRef]
18. BNNano. Available online: https://www.bnnano.com/products (accessed on 17 December 2018).

19. Pelowitz, D.B. (Ed.) *MCNPX User's Manual Version 2.7.0*; LA-CP-11-00438; Los Alamos National Laboratory: Santa Fe, NM, USA, 2011.
20. Jakubinek, M.B.; Niven, J.F.; Johnson, M.B.; Ashrafi, B.; Kim, K.S.; Simard, B.; White, M.A. Thermal conductivity of bulk boron nitride nanotube sheets and their epoxy-impregnated composites. *Phys. Status Solidi A* **2016**, *213*, 2237–2242. [CrossRef]
21. Chen, Y.; Zou, J.; Campbell, S.J.; Le Caer, G. Boron nitride nanotubes: Pronounced resistance to oxidation. *Appl. Phys. Lett.* **2004**, *84*, 2430. [CrossRef]

![instruments logo] *instruments*

MDPI

*Article*

# Fused Zinc Target for the Production of Gallium Radioisotopes

Stefan Zeisler [1], Alan Limoges [1], Joel Kumlin [1], Jonathan Siikanen [2] and Cornelia Hoehr [1,*]

[1] Life Sciences Division, TRIUMF, Vancouver, BC V6T 2A3, Canada; zeisler@triumf.ca (S.Z.); limogesa@uwindsor.ca (A.L.); kumlin@artms.ca (J.K.)

[2] Medical Radiation Physics and Nuclear Medicine, Karolinska University Hospital, S-171 76 Stockholm, Sweden; jonathan.siikanen@sll.se

* Correspondence: choehr@triumf.ca; Tel.: +1-604-222-1047

Received: 17 December 2018; Accepted: 24 January 2019; Published: 1 February 2019

**Abstract:** Gallium-68 is a popular radioisotope for positron emission tomography. To make gallium-68 more accessible, we developed a new solid target for medical cyclotrons. Fused zinc targets promise a new, efficient, and reliable technique without the downsides of other commonly used time-consuming methods for solid target fabrication, such as electroplating and sputtering. We manufactured targets by fusing small pressed zinc pellets into a recess in aluminum backings. Using a simple hotplate, the fusing could be accomplished in less than two minutes. Subsequently, the targets were cooled, polished, and used successfully for test irradiations at $E_p$ = 12.8 MeV and up to 20 µA proton current.

**Keywords:** PET; medical isotopes; Ga-68; solid target

## 1. Introduction

As positron emission tomography (PET) is becoming more accessible, the use of radiotracers labeled with positron-emitting radiometals has steadily increased. Isotopes such as $^{44}$Sc, $^{68}$Ga, $^{64}$Cu, $^{86}$Y and $^{89}$Zr offer a wide range of half-lives, and chemical and imaging characteristics, thus greatly advancing the development of new radiopharmaceuticals. In recent years, $^{68}$Ga, in particular, has gained much interest in oncological imaging, where it is used in combination with analogous therapeutic radiotracers labelled with $^{90}$Y, $^{177}$Lu or $^{225}$Ac [1].

$^{68}$Ga is commonly eluted from $^{68}$Ge/$^{68}$Ga generators. It can also be produced on small medical cyclotrons via the $^{68}$Zn(p,n)$^{68}$Ga reaction (see Figure 1 for relevant cross sections) in quantities that exceed those available from current generators [2–6]. While other nuclear reactions are also feasible [7], the irradiation of $^{68}$Zn with protons is most convenient and offers high production yield.

Several types of targets, such as foils and coatings, have been described; however, the commercial availability of isotopically enriched foils is rather limited. Zinc can also be sputter coated or electroplated on various substrates. Both processes are time consuming and require specialized equipment and expertise.

Metallic zinc targets necessitate a solid target station on the cyclotron. The target plate may be placed and retrieved manually or via a dedicated transfer system. Manual removal results in radiation exposure to the operator due to radiation fields from the target and activated components in its vicinity. Once the irradiated plate has been moved to the processing hotcell, the target material needs to be dissolved and the gallium radioisotope purified.

A particularly attractive way of extracting gallium radioisotopes from zinc foils is the thermal diffusion method described by Tolmachev and Lundqvist [8]. In their experiments, irradiated foils were heated to approximately 400 °C for an extended period, which caused the gallium isotopes to migrate to the surface of the foil, from which they could be removed by etching with diluted acid.

Reportedly, more than 60% of the produced radioactivity could be separated from the zinc matrix in 15 min, with minimal loss of target material.

Here, we set out to develop an efficient, fast target preparation method to facilitate the production of gallium radioisotopes for research and development of radiopharmaceuticals. Furthermore, we attempted to apply the thermal diffusion method to our fused targets. Finally, we demonstrated the radiochemical purification of gallium radioisotopes from zinc based on the method by Engle et al. [7].

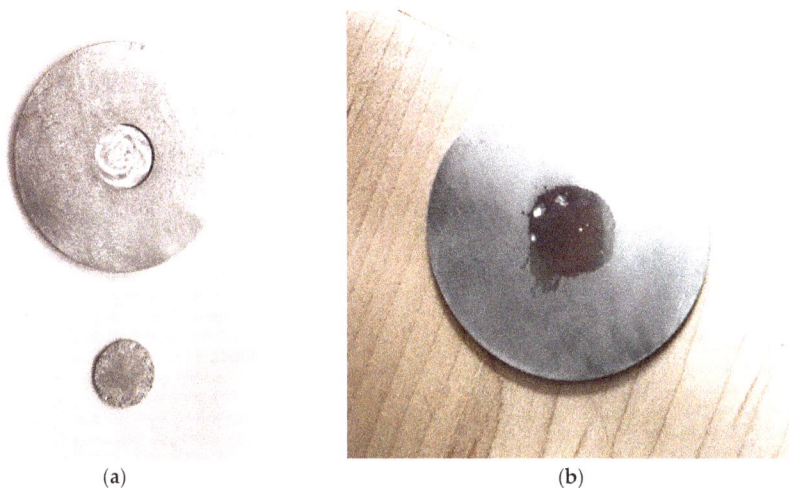

(a)                                    (b)

**Figure 1.** Target plate preparation. (**a**) A Ø35 mm × 1 mm aluminum backing plate with a Ø10 mm × 0.3 mm recess for the target material, and the pellet formed from zinc (150–300 mg) before fusing; (**b**) Zinc pellet fused to the backing plate before polishing.

## 2. Materials and Methods

The target backing was manufactured from 1100 grade aluminum (AL1100, McMaster-Carr) because of its good heat transfer properties, low cost and low activation in the target area of a medical cyclotron. A quantity of 150–300 mg of irregularly shaped natural zinc shavings (99.9%, Sigma-Aldrich) were pressed into flat 10-mm-diameter pellets using a hydraulic press at ~12,500 MPa. A density of 94% of bulk zinc material was achieved. Targets were prepared by fusing the pressed pellet to the aluminum backing (35 mm diameter × 1 mm thickness) by heating the aluminum backing to 450–500 °C on a hotplate (Corning), then placing the pellet into the 0.3 mm deep recess. The fusing process was allowed to proceed for 20–30 s. The target was then removed from the hotplate and quickly cooled by placing it on a cold metal surface. The entire process took about 2 min. After cooling, the targets were manually polished with sandpaper of grades 400, 600, 800 and 1000 (Norton Abrasives), rinsed with methanol and dried.

In this proof-of-principle study we used natural zinc with its isotopic composition shown in Table 1. The prevalent cross sections for the production of gallium radioisotopes (Table 1), scaled by the abundance of the starting zinc isotopes, are shown in Figure 2. The long-lived gallium isotope $^{66}$Ga ($T_{1/2}$ = 9.49 h) was used to study the diffusion of radioactive gallium in the target matrix as well as the radiochemical purification.

**Table 1.** Natural abundance of zinc isotopes [9] and produced gallium isotopes. Half-lives from [10].

| Zn isotopes | $^{64}$Zn | $^{66}$Zn | $^{67}$Zn | $^{68}$Zn | $^{70}$Zn |
|---|---|---|---|---|---|
| Abundance | 48.63% | 27.90% | 4.10% | 18.75% | 0.62% |
| Ga isotope | $^{64}$Ga | $^{66}$Ga | $^{67}$Ga | $^{68}$Ga | $^{70}$Ga |
| Half life | 2.627 min | 9.49 h | 3.2617 d | 67.71 min | 21.14 min |
| Production | $^{64}$Zn(p,n)$^{64}$Ga $^{66}$Zn(p,3n)$^{64}$Ga | $^{66}$Zn(p,n)$^{66}$Ga $^{67}$Zn(p,2n)$^{66}$Ga $^{68}$Zn(p,3n)$^{66}$Ga | $^{67}$Zn(p,n)$^{67}$Ga $^{68}$Zn(p,2n)$^{67}$Ga | $^{68}$Zn(p,n)$^{68}$Ga $^{70}$Zn(p,3n)$^{68}$Ga | $^{70}$Zn(p,n)$^{70}$Ga |

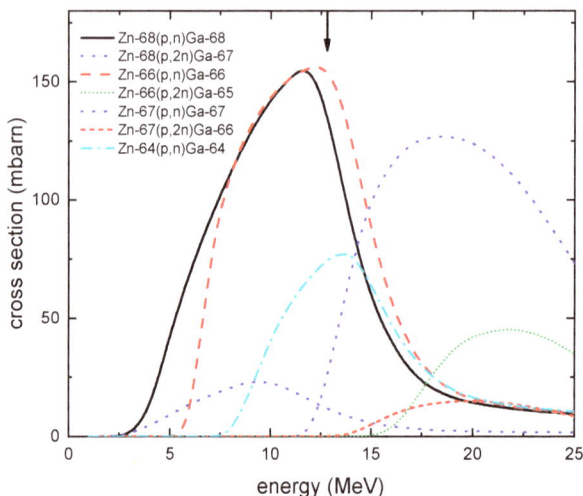

**Figure 2.** Main cross sections for the production of gallium isotopes from zinc, with the zinc isotopes scaled for natural abundance (see Table 1). The entrance energy of the proton beam in our production is marked with an arrow. Data from [11].

All irradiations were performed on TRIUMF's TR13, a 13 MeV self-shielded, negative hydrogen ion cyclotron [12,13]. The target plate was manually mounted in our standard solid target holder, which places the plate perpendicular to the proton beam. The target body and the back of the target plate were water cooled. The front of the target plate was cooled via helium jets impinging onto the plate and onto a 25-μm-thick aluminum foil that separates the target assembly from the cyclotron vacuum. The energy of the proton beam on the target was approximately 12.8 MeV as determined by [14]. As the target body was electrically isolated, the beam current on the target could be directly measured. The maximum beam current was limited to 20 μA due to license restrictions and radiation safety concerns.

In all experiments, a basic target holder was used [15]. The assembly was held together concentrically with four screws. In order to retrieve the irradiated plate, the system needed to be manually removed from the target selector on the cyclotron, which required disconnecting the water and helium cooling lines. The target system was then transported to a fume hood, where it was completely dismantled and the irradiated disc removed. The entire retrieval process took 8–10 min on average, with the operator having to spend approximately 5 min in close proximity to the highly radioactive targets on the selector.

In order to reduce the radiation dose to technical staff, the target was left to decay overnight (16 h). The irradiated plate was then transported to the processing fume hood in a shielded container. Yields were estimated with an ionization chamber (Capintec).

Thermal diffusion experiments were performed using a borosilicate beaker placed on a hotplate. The temperature of the target plate was measured with a thermocouple and displayed on a local readout device. A series of 13 experiments were performed to investigate the effect of various parameters on the diffusion process. Briefly, all targets were heated to a temperature near the melting point of zinc, i.e., up to ~410 °C. Diffusion was allowed to proceed for a period of 60, 120 or 180 min ($n$ = 3, 5, 5, respectively). Targets were heated either in air ($n$ = 10) or under argon flow ($n$ = 3), with the zinc material facing up ($n$ = 11) or down ($n$ = 2).

For the radiochemical purification, the target was placed into a 50-mL beaker and the zinc dissolved with 10–12N HCl (Sigma-Aldrich); lower concentrations proved to dissolve the zinc too slowly. Over 95% of the target material was dissolved in less than 10 min at room temperature, and in 2–5 min at ~50 °C. A small quantity of aluminum (milligrams) was co-dissolved in the process. In order to minimize the dissolution of aluminum, Kapton tape (U-Line) was applied in some experiments to protect the aluminum target face.

Gallium radioisotopes were purified as reported in [7] by ion exchange chromatography on AG50W-X8 cation exchange resin. The gallium was eluted in 4N hydrochloric acid.

## 3. Results and Discussion

### 3.1. Irradiation

Figure 3a shows a fused target after a 15-min irradiation with 12.8 MeV protons and 20 μA beam current. There was no indication of overheating, discoloration or loss of target material. All targets remained intact, with the fused zinc pellet firmly attached to the backing. Further experiments at higher beam currents and with longer irradiation times will be conducted once an improved target system that allows remote handling of the irradiated disc becomes available [15].

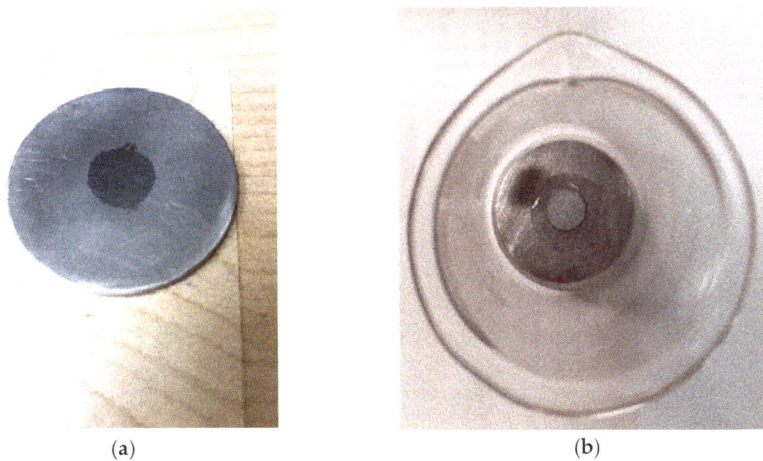

(a)            (b)

**Figure 3.** (**a**) Target plate after irradiation. The streaks of zinc on the aluminum plate are caused by the polishing. (**b**) Target plate being dissolved inside a beaker.

The $^{66}$Ga yield from a 15-min irradiation with 10-μA beam current decay corrected to the end-of-bombardment (EOB) was 400 MBq. Taking the experimental results from [16] and scaling the $^{68}$Ga activity from the $^{66}$Ga activity, we estimate the EOB activity of $^{68}$Ga from our fused target to amount to approximately 3.1 GBq.

## 3.2. Thermal Diffusion Experiments

In summary, none of our diffusion experiments yielded results comparable to those published in [8]. In all cases, the recovery of $^{66}$Ga from the zinc pellet remained below 10% of the total $^{66}$Ga radioactivity produced, which is six to eight times less than the yields reported by Tolmachev et al. [8]. We did not notice any effect of varying process parameters or the orientation of the target plate during the heating cycle or whether the experiment took place in an argon atmosphere or in air. The particular reason for this discrepancy remains unknown. It was speculated that our particular heating apparatus (hot plate/beaker) may have been less than adequate, as the experiments reported in the literature were conducted in a sealed tube furnace under tight temperature control and in high purity argon. Moreover, [8] describes the thermal diffusion of radiogallium in thin, rolled zinc foils. Fused zinc targets may have a different matrix structure or contain a small quantity of zinc oxide that could potentially impede the migration of gallium through the matrix.

## 3.3. Radiochemical Purification

The radiochemical purification of gallium isotopes from zinc is well established, and several procedures to isolate high specific activity $^{66/68}$Ga have been proposed. We chose to demonstrate the general feasibility of extracting $^{66}$Ga from fused zinc targets using the cation exchange method by Engle and co-workers due to its efficiency and convenience.

After irradiation and cooldown, the zinc material was dissolved in an excess of 10N hydrochloric acid (HCl) and passed through a 2-mL column of AG50W-X8 cation exchange resin conditioned with 10N HCl. The column was rinsed with two column volumes each of 10N HCl and 7N HCl. $^{66}$Ga was then eluted with 4N HCl. Approximately 90% of the produced $^{66}$Ga was recovered in the product solution. The entire separation procedure took approximately 30 min.

## 4. Conclusions

We reported on a simple zinc target manufacturing for the production of gallium isotopes. The target plate fabrication took about 2 min using a simple hotplate. Due to the manual operation of our solid target system, the proof-of-principle was conducted on the longer lived $^{66}$Ga to reduce radiation exposure to the cyclotron team and the chemist processing the target. The target was irradiated up to 20 μA at a beam energy on target of 12.8 MeV. No negative effect on the target plate was observed. Heating the irradiated target plate to diffuse the produced $^{66}$Ga to the surface was unsuccessful, maybe due to the heating on a hot plate instead of a more sophisticated tube furnace. After dissolution of the target plate, 400 MBq was measured (decay-corrected). This relates to an estimated activity of 3.1 GBq for $^{68}$Ga at EOB (natural Zn) or 16.7 GBq (99.5% $^{68}$Zn).

Fused targets require little preparation of the target backing, minimal manufacturing time, few materials and tools, thus reducing overall manufacturing cost. Furthermore, the simplistic nature of the method opens up the possibility of large-scale target production. Future experiments will be required to confirm the integrity of fused targets at higher beam currents and extended irradiation periods.

**Author Contributions:** Conceptualization, S.Z. and C.H.; methodology, S.Z., and J.S.; formal analysis, A.L.; investigation, A.L., J.K. and S.Z.; resources, S.Z. and C.H.; writing—original draft preparation, C.H. and S.Z.; writing—S.Z., A.L., J.K., J.S. and C.H.; visualization, X.X.; supervision, S.Z. and C.H.; project administration, C.H.; funding acquisition, C.H.

**Funding:** TRIUMF receives federal funding via a contribution agreement with the National Research Council of Canada. This work was supported by the Natural Sciences and Engineering Research Council (NSERC) via the Discovery Grant program (RGPIN-2016-03972).

**Conflicts of Interest:** The authors declare no conflict of interest.

*Instruments* **2019**, *3*, 10

## References

1. Baum, R.P.; Kulkarni, H.R. THERANOSTICS: From Molecular Imaging Using Ga-68 Labeled Tracers and PET/CT to Personalized Radionuclide Therapy—The Bad Berka Experience. *Theranostics* **2012**, *2*, 437–447. [CrossRef] [PubMed]
2. Alves, F.; Alves, V.H.P.; Do Carmo, S.J.C.; Neves, A.C.B.; Silva, M.; Abrunhosa, A.J. Production of copper-64 and gallium-68 with a medical cyclotron using liquid targets. *Mod. Phys. Lett. A* **2017**, *32*, 1740013. [CrossRef]
3. Nair, M.; Happel, S.; Eriksson, T.; Pandey, M.K.; DeGrado, T.R.; Gagnon, K. Cyclotron production and automated new 2-column processing of [$^{68}$Ga]GaCl$_3$. *Eur. J. Nucl. Med. Mol. Imaging* **2017**, *44* (Suppl. S2), 119.
4. Pandey, M.K.; Byrne, J.F.; Jiang, H.; Packard, A.B.; DeGrado, T.R. Cyclotron production of $^{68}$Ga via the $^{68}$Zn(p,n)$^{68}$Ga reaction in aqueous solution. *Am. J. Nucl. Med. Mol. Imaging.* **2014**, *4*, 303–310. [PubMed]
5. Lin, M.; Waligorski, G.J.; Lepera, C.G. Production of curie quantities of $^{68}$Ga with a medical cyclotron via the $^{68}$Zn(p,n)$^{68}$Ga reaction. *Appl. Radiat. Isot.* **2018**, *133*, 1–3. [CrossRef] [PubMed]
6. Dias, G.; Ramogida, C.; Martini, P.; Hoehr, C.; Kuo-Syanj, L.; Schaffer, P.; Benard, F. Peptide radiolabelling and in vivo imaging using Ga-68 directly produced in liquid targets. *J. Label. Compd. Radiopharm.* **2017**, *6* (Suppl. S1), S400.
7. Engle, J.W.; Lopez-Rodriguez, V.; Gaspar-Carcamo, R.E.; Valdovinos, H.F.; Valle-Gonzalez, M.; Trejo-Ballado, F.; Severin, G.W.; Barnhart, T.E.; Nickles, R.J.; Avila-Rodriguez, M.A. Very high specific activity $^{66/68}$Ga from zinc targets for PET. *Appl. Radiat. Isot.* **2012**, *70*, 1792–1796. [CrossRef] [PubMed]
8. Tolmachev, V.; Lundqvist, H. Rapid separation of gallium from zinc targets by thermal diffusion. *Appl. Radiat. Isot.* **1996**, *47*, 297–299. [CrossRef]
9. The Periodic Table of the Elements. Available online: https://www.webelements.com/ (accessed on 26 January 2019).
10. Interactive Chart of Nuclides. Available online: https://www.nndc.bnl.gov/chart/chartNuc.jsp (accessed on 28 January 2019).
11. TALYS-based Evaluated Nuclear Data Library. Available online: https://tendl.web.psi.ch/tendl_2015 (accessed on 28 January 2019).
12. Laxdal, R.E.; Altman, A.; Kuo, T. Beam measurements on a small commercial cyclotron. In Proceedings of the EPAC94, London, UK, 27 June–1 July 1994; pp. 545–547.
13. Buckley, K.R.; Huser, J.; Jivan, S.; Chun, S.K.; Ruth, T.J. $^{11}$C-methane production in small volume, high pressure gas targets. *Radiochim. Acta* **2000**, *88*, 201. [CrossRef]
14. SRIM—The Stopping and Range of Ions in Matter. Available online: http://www.srim.org/ (accessed on 26 January 2019).
15. Zeisler, S.; Clarke, B.; Kumlin, J.; Hook, B.; Varah, S.; Hoehr, C. A compact quick-release solid target system for the TRIUMF TR-13 cyclotron. *Instruments* **2019**, *3*, 16. [CrossRef]
16. Oehlke, E.; Hoehr, C.; Hou, X.; Hanemaayer, V.; Zeisler, S.; Adam, M.J.; Ruth, T.J.; Celler, A.; Buckely, K.; Benard, F.; et al. Production of Y-86 and other radiometals for research purposes using a solution target system. *Nucl. Med. Biol.* **2015**, *42*, 842–849. [CrossRef] [PubMed]

![instruments logo] *instruments*

MDPI

*Communication*

# Molybdenum Sinter-Cladding of Solid Radioisotope Targets

**William Z. Gelbart [1,*] and Richard R. Johnson [2]**

[1]   Advanced Systems Designs Inc., Garden Bay, BC V0N 1S1, Canada
[2]   University of British Columbia, Vancouver, BC V6T 1Z4, Canada; richardrjohnson81738@gmail.com
*   Correspondence: gelbart@advancesystems.ca or williamgelbart@gmail.com

Received: 28 December 2018; Accepted: 27 January 2019; Published: 2 February 2019

**Abstract:** In solid targets for radioisotope production, the parent materials—mostly metallic—are usually attached to a substrate (metal part, often copper or silver) to support it during handling and irradiation and to facilitate liquid or gas cooling to remove the heat generated by the particle beam. This cladding process is most frequently done by electroplating. One of the biggest challenges of preparing solid, high-current, $^{100}$Mo targets is the difficulty of cladding the substrate with molybdenum—metal that cannot be electroplated. A number of cladding techniques are used with varying degrees of complexity, success, and cost. A simple cladding process, especially suitable for the production of radioisotope targets, was developed. The process uses a metal slurry (metal powder and binder) painted on the substrate and heated in a hydrogen atmosphere where the metal is sintered and diffusion-bound to the substrate in a single step.

**Keywords:** radioisotopes; medical radioisotopes; radioisotope targetry; solid radioisotope targets; radioisotope target processing; medical cyclotrons

## 1. Introduction

In solid targets for radioisotope production, the parent materials—mostly metallic—are usually attached to a substrate (metal part, often copper or silver) to support it during handling and irradiation and to facilitate liquid or gas cooling to remove the heat generated by the particle beam.

This cladding process most frequently employs electroplating. Many metallic elements can be electroplated, but some cannot be plated at all or cannot produce a sufficiently successful cladding of the target substrate. A specific case is molybdenum that so far cannot be successfully electroplated.

In the last few years, we have witnessed an increased interest in accelerator production of Tc-99m using $^{100}$Mo as target material. One of the biggest challenges of preparing a solid, high-current, $^{100}$Mo target is the difficulty of cladding the substrate with molybdenum. For a high-beam current production, a $^{100}$Mo coating in the range of 100 to 500 µm must be bonded to a substrate that provides a rigid support and allows easy manipulation and an efficient liquid cooling. Most metallic element targets are electroplated on copper, silver, or other metal substrates. In the case of molybdenum, a number of cladding techniques were used, including rolling the molybdenum into thin foils and soldering or pressure/diffusion-bonding those to the substrate, laser cladding, electrophoretic deposition, plasma spraying, powder pressing, hot isostatic pressing, among others. Some of these studies are referenced as background material [1–10].

Those processes can often produce successful targets, but there are a number of challenges:

1.   Target preparation time and the cost of each target
2.   Equipment cost
3.   Reliability and reproducibility of the process
4.   Build-up of the required thickness

5. Coating material losses
6. Coating adhesion, especially at a high operating temperature
7. Density of the coating

While this paper focuses on molybdenum cladding, the same technique can likely be used to coat radioisotope production targets with other metals as well. This possibly includes easily electroplated metals that otherwise pose plating difficulties in thicker layers.

Only target cladding is discussed as other stages of processing, dissolution, separation, and recovery are no different than when using other cladding techniques [11].

## 2. Materials and Methods

The process consists of the deposition of the parent material on the substrate in the form of powder mixed with a binder, drying, rolling the deposit to a uniform thickness, and sintering and diffusion-bonding at high temperatures in a hydrogen atmosphere.

Natural molybdenum was used for all the tests. The molybdenum was in powder form; this is in fact the most common for $^{100}$Mo and many other metal isotopes as supplied or as recovered after processing.

Materials employed:

1. Mo powder, <150 µm, 99.99% trace metals basis, Sigma Aldrich 203823
2. Polyvinyl alcohol, Mw 89,000–98,000, 99+% hydrolyzed, Sigma Aldrich 341584
3. C10100 oxygen free copper, OnlineMetals, www.onlinemetals.com
4. 3M 250 Flatback Masking Tape, 3M Company
5. 3M 720 Film Fiber Tape, 3M Company

Equipment used:

1. Hydrogen sintering oven system, Rapidia Inc., Vancouver, BC, Canada

2 g of polyvinyl alcohol was dissolved in 100 cc of water by leaving it overnight at 40 °C while stirring with magnetic stirrer. A metal slurry was prepared by mixing the Mo powder with the polyvinyl alcohol solution in the weight ratio of 2.5 parts Mo to 1.5 parts of polyvinyl alcohol solution.

The process consists of the following steps:

1. The substrates samples, approximately 40 mm × 40 mm, were sheared from a 2 mm thick, C10100, oxygen-free, copper sheet and lightly wet-sanded with #400 silicon carbide paper, followed by a wash with hot water and detergent.
2. The metal slurry was deposited on the substrate. This was done with the substrate on a scale and the weight of slurry deposited and evenly spread over an area of a 25 mm diameter circle that corresponded to final molybdenum thicknesses of 3–5 gm/cm$^2$. Figure 3a.
3. The sample was dried on a hot plate for one hour at 50 °C.
4. The surfaces of the first ten samples were rolled with a silicon carbide ball-bearing (mounted on a handle) using hand pressure. Small rolling mills with steel rollers were used for later samples. The rolling was to consolidate the dry coat and to flatten any small voids that were left as a result of air bubbles. Figure 3b.
5. The sample was placed in a hydrogen oven (Figure 1), the temperature was raised to 980 °C with a ramping rate of 8 °C per minute, and it was kept at 980 °C for a number of hours. Various times were employed from 4 to 8 h. There seemed to be no difference in the results after 4 h.
6. The hydrogen was at atmospheric pressure, and a flow of 0.3 L/min was maintained throughout the process. Since the temperature was close to the melting point of copper and to avoid any deformation, the samples were place on a flat sheet of 99.8% alumina.

7. At the end of the process, samples were allowed to cool while still in the oven (and still under hydrogen atmosphere) to below 100 °C and removed (Figure 2). The cooling time of the oven to this temperature was 12 h, and the cooling was done overnight.

**Figure 1.** Hydrogen oven and $H_2$ generator.

**Figure 2.** Samples after bonding.

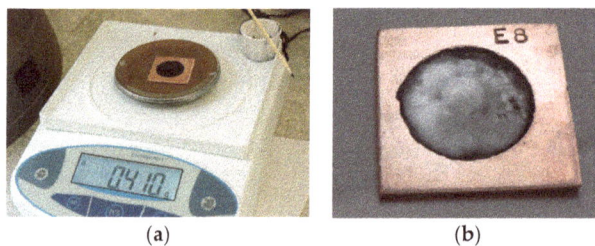

(a)  (b)

**Figure 3.** Application of the slurry: (**a**) Weighting the deposit (**b**) Sample coated, dried, and rolled.

## 3. Results

Sixty-two samples were prepared and tested. The weight of the deposited slurry was chosen to give a final thickness of the metal cladding—after sintering and bonding—to be between 40 μm and 80 μm. This was done by depositing the slurry when the substrate was on a scale and using the known density of the molybdenum in the slurry to determine the molybdenum thickness after the water and the polyvinyl alcohol were baked out from the mixture.

The hydrogen was provided by a hydrogen generator. It was important to keep the hydrogen moisture content low. A desiccator column was used to bring the gas humidity to −60 °C dew-point (10 parts per million).

The finished samples were tested following the *Standard Test Methods for Adhesion of Metallic Coatings*, ASTM B571. The tests selected were the peel test, bend tests, and chisel-knife test. The peel tests were performed using the 3M720 Film Fiber Tape and the 3M 250 Flatback Masking Tape.

Four samples have failed some of the tests (#3. Bend Tests and/or #5. Chisel-Knife Test). Those were from the very first run of six 40 μm samples made before the adequate drying of the hydrogen and before other small corrections in the oven and the technique.

For the rest of the samples, there appears to be no mechanical way to remove the cladding other than machining it off Figure 4.

**Figure 4.** Tests for adhesion of coatings.

Six finished and tested samples of various thicknesses (40–80 μm) were heated in vacuum to 385 °C and retested using the same procedures. All samples passed all of the tests successfully. The cladding remained on the substrate until a chemical dissolution was used to remove it as is done regularly in target processing [11].

## 4. Discussion

In addition to copper substrates, a number of tests were performed on pure silver. The first two samples of the silver were melted at 950 °C. The presence of the molybdenum seemed to lower the melting point of the silver. Samples treated at 900 °C still showed some melting, though the molybdenum cladding was intact and adhering well. More tests will be performed at slightly lower temperatures.

The process of diffusion cladding appeared to work well for the preparation of radioisotope targets. Even in the experimental setup, the preparation and handling time for each target was approximately 5 min—not including the drying or the sintering times. There were no losses of the coating material in the process.

The thickness was measured using Mitutoyo 519-807 Electronic gauge and Mitutoyo 122L indicator. The density of the cladding was calculated from the thickness and weight and was over 90% of full metal density and, if needed, could probably have been consolidated further by pressing. This will be investigated using other thickness-measuring techniques in addition to the currently employed.

*Instruments* **2019**, *3*, 11

Thicker coatings of molybdenum as well as other cladding metals will be tested including yttrium and nickel.

Production targets—with both natural molybdenum and $^{100}$Mo cladding—will be irradiated and evaluated.

## 5. Patents

A United States provisional patent, covering the parts of the process and applications, applied for.

**Author Contributions:** W.Z.G. is the lead scientist of this work and was involved in the sample preparation, testing and analysis. R.R.J. was involved in the analysis.

**Acknowledgments:** The authors wish to thank *Rapidia Inc.* of Vancouver, BC, Canada, for their help and for the use of their equipment.

**Conflicts of Interest:** The authors declare no conflict of interest.

## References

1. Stolarz, A.; Kowalska, J.A.; Jasinski, P.; Janiak, T.; Samorajczyk, J. Molybdenum targets produced by mechanical reshaping. *JRNC* **2015**, *305*, 947–952. [CrossRef] [PubMed]
2. Gelbart, W.; Johnson, R.R.; Abeysekera, B. Solid Target Irradiation and Transfer System. In Proceedings of the 14th International Workshop on Targetry and Target Chemistry, Playa del Carmen, Mexico, 26–29 August 2012.
3. Stolarz, A. Target preparation for research with charged projectiles. *JRNC* **2014**, *299*, 913–931. [CrossRef] [PubMed]
4. Bénard, F.; Buckley, K.R.; Ruth, T.J.; Zeisler, S.K.; Klug, J.; Hanemaayer, V.; Vuckovic, M.; Hou, X.; Celler, A.; Appiah, J.P.; et al. Implementation of Multi-Curie Production of $^{99m}$Tc by Conventional Medical Cyclotrons. *J. Nucl. Med.* **2014**, *55*, 1017–1022. [CrossRef] [PubMed]
5. Gelbart, W.Z.; Johnson, R.R. Irradiation Targets for Accelerator Production of $^{99m}$Tc. In Proceedings of the 14th International Workshop on Targetry and Target Chemistry, Playa del Carmen, Mexico, 26–29 August 2012.
6. Target and Apparatus for Cyclotron Production Of Technetium-99m. EP Patent EP3197246(A1), 26 July 2017.
7. Hanemaayer, V.; Zeisler, S.K.; Buckley, K.R.; Klug, J.; Kovacs, M.; Bérnard, F.; Ruth, T.J.; Schaffer, P. Solid Targets for Tc-99m Production on Medical Cyclotrons. In Proceedings of the 14th International Workshop on Targetry and Target Chemistry, Playa del Carmen, Mexico, 26–29 August 2012.
8. Surette, G.J. Molybdenum 100 Targets for Cyclotron Production of Technetium 99m. Master's Thesis, Department of Mechanical and Aerospace Engineering, Carleton University, Ottawa, ON, Canada, 2015.
9. Thomas, B.A.; Wilson, J.S.; Gagnon, K. Solid $^{100}$Mo target preparation using cold rolling and diffusion bonding. In Proceedings of the 15th International Workshop on Targetry and Target Chemistry, Prague, Czech Republic, 18–21 August 2014.
10. Sklairova, H.; Cisterno, S.; Cicoria, G.; Marengo, M.; Palmieri, V. Innovative Target for Production of Technetium-99m by Biomedical Cyclotron. *Molecules* **2019**, *24*, 25. [CrossRef] [PubMed]
11. Matei, L.; McRae, G.; Galea, R.; Niculae, D.; Craciun, L.; Leonte, R.; Surette, G.; Langille, S.; St. Louis, C.; Gelbart, W.; et al. A new approach for manufacturing and processing targets used to produce $^{99m}$Tc with cyclotrons. *Mod. Phys. Lett. A* **2017**, *32*, 1740011. [CrossRef]

*instruments*

*Article*

# Medical Cyclotron Solid Target Preparation by Ultrathick Film Magnetron Sputtering Deposition

**Hanna Skliarova [1],\*, Sara Cisternino [1],\*, Gianfranco Cicoria [2], Mario Marengo [2], Emiliano Cazzola [3], Giancarlo Gorgoni [3] and Vincenzo Palmieri [1],†**

[1]  Legnaro National Laboratories, Italian National Institute for Nuclear Physics (LNL-INFN), Viale dell'Università, 2, 35020 Legnaro (PD), Italy

[2]  Medical Physics Department University Hospital "S. Orsola–Malpighi", 40138 Bologna, Italy; cicoria.gianfranco@aou.mo.it (G.C.); mario.marengo@unibo.it (M.M.)

[3]  IRCCS Sacro Cuore Don Calabria Hospital, Cyclotron and Radiopharmacy Department, 37024 Negrar (VR), Italy; Emiliano.Cazzola@sacrocuore.it (E.C.); giancarlo.gorgoni@sacrocuore.it (G.G.)

\*   Correspondence: Hanna.Skliarova@lnl.infn.it (H.S.); Sara.Cisternino@lnl.infn.it (S.C.); Tel.: +39-049-806-8416 (H.S. & S.C.)

†   Deceased 16 March 2018.

Received: 22 December 2018; Accepted: 8 March 2019; Published: 13 March 2019

**Abstract:** Magnetron sputtering is proposed here as an innovative method for the deposition of a material layer onto an appropriate backing plate for cyclotron solid targets aimed at medical radioisotopes production. In this study, a method to deposit thick, high-density, high-thickness-uniformity, and stress-free films of high adherence to the backing was developed by optimizing the fundamental deposition parameters: sputtering gas pressure, substrate temperature, and using a multilayer deposition mode, as well. This method was proposed to realize Mo-100 and Y-nat solid targets for biomedical cyclotron production of Tc-99m and Zr-89 radionuclides, respectively. The combination of all three optimized sputtering parameters (i.e., $1.63 \times 10^{-2}$ mbar Ar pressure, 500 °C substrate temperature, and the multilayer mode) allowed us to achieve deposition thickness as high as 100 µm for Mo targets. The 50/70-µm-thick Y targets were instead realized by optimizing the sputtering pressure only ($1.36 \times 10^{-2}$ mbar Ar pressure), without making use of additional substrate heating. These optimized deposition parameters allowed for the production of targets by using different backing materials (e.g., Mo onto copper, sapphire, and synthetic diamond; and Y onto a niobium backing). All target types tested were able to sustain a power density as high as 1 kW/cm$^2$ provided by the proton beam of medical cyclotrons (15.6 MeV for Mo targets and 12.7 MeV for Y targets at up to a 70-µA proton beam current). Both short- and long-time irradiation tests, closer to the real production, have been realized.

**Keywords:** cyclotron solid target; radioisotope production; magnetron sputtering; thick film deposition

## 1. Introduction

A conventional medical cyclotron solid target comprises the target material deposited onto a baking plate that is cooled by water from the back and possibly by helium gas flow from the front. There is a number of techniques for accelerator target preparation based on chemical, mechanical, or physical processes [1]. The list of the most common methods for cyclotron solid target production includes, but is not limited to, rolling or mechanical reshaping, pressing, sintering, electrodeposition, and a set of "physical" methods [2]. Here, we have associated with the group of physical methods different Physical Vapor Deposition (PVD) methods, such as Focused Ion Beam (FIB) or magnetron sputtering, thermal spray deposition, and plasma spray deposition. Table 1 presents a summary of the most commonly used methods for cyclotron solid target preparation and some examples of their application for the preparation of Y and Mo targets, which is the topic of this work. A more detailed overview of Mo cyclotron solid target preparation has been presented recently by the authors [3]. Each method can be used either separately or in a combination with others, for example, pressing or electrophoresis followed by sintering [4], sintering followed by press-bonding [5,6], etc., which can lead to improved thermomechanical performance. The choice of a suitable method for target production is guided by the type of precursor material, target dimensions, desired backing plate, and the dissolution and separation procedures of the irradiated target. The optimal thickness of the target for radionuclides production depends on the preferred particle energy range, chosen in a way to minimize impurities. Usually, it is on the order of hundreds of micrometers or even millimeters. Material losses during the target preparation procedure should be minimized when costly isotopically enriched materials are used for production. Besides that, the target must be mechanically stable and able to withstand the thermodynamic conditions that occur during irradiation: no pealing, sputtering, evaporation, and other thermal damages should occur. Of course, performance under the beam depends on the irradiation parameters (beam energy and beam current).

An "ideal" technique, available for all types of materials and fulfilling all the requirements for the target, does not exist. The choice of technique for target preparation is always a compromise between the approach to fulfilling the particular requirements of each application and the cost of implementation.

**Table 1.** Comparison of the most common target preparation methods.

| Method | Thickness | Deposited Material's Limitations | Backing | Losses | Example Mo, Y |
|---|---|---|---|---|---|
| Rolling (mechanical reshaping) | tens of μm … mm | Metals, sufficiently ductile, not oxidized | Press-bonding to a backing is possible for soft materials. | 10%–20% | [7–9] |
| Pressing | hundreds of μm … mm | Not possible for hard materials without a binder | No backing. Press-bonding or brazing can be used as a second step of target preparation | <5% | [2,5,10] |
| Sintering | hundreds of μm … mm | Oxygen-sensitive materials can be sintered either in a reduced atmosphere or by particular methods | No backing. Press-bonding or brazing can be used as a second step of target preparation | <5% | [11–13] |
| Melting | hundreds of μm … mm | For high-melting-temperature materials, laser melting should be used | Melting temperature of backing is preferred to be higher than precursor material | <5% | [14,15] |
| Sedimentation | tens of μm … hundreds of μm | A binder is needed | Various backing | <5% | [4,16] |
| Electrodeposition | μm … hundreds of μm | Metals or oxides. Metals with high affinity to O cannot be deposited in pure form | Must be electrically conductive | 10%–20% | [17,18] |
| "Physical" deposition * | μm … hundreds of μm | Various materials | Various backing | 70%–80% | [3,19–24] |

* Here, physical deposition methods include different Physical Vapor Deposition (PVD) methods: Focused Ion Beam (FIB) and magnetron sputtering, thermal spray deposition, and plasma spray deposition.

In order to maximize the nuclear reaction yield, production should be performed at maximum proton currents. This means that the target system should provide high efficiency of heat dissipation. In order to achieve this purpose, the materials should have maximum thermal conductivity, including both the target material itself and the target backing plate, and should be connected by a method providing good thermomechanical contact between them. Direct deposition by sputtering may be particularly interesting in cases where: a relatively thin layer of target material is required and obtaining the proper contact between the target material and the backing is critical; the backing is cumbersome (particular, not disklike, shapes, such as microchannel- or metallic-foam-based heat exchanger as a part of the backing, etc.); or the alternative target bonding involves the use of material or processes that may introduce impurities (e.g., brazing, etc.).

The LNL-INFN group, in the framework of the LARAMED (LAboratories for RAdioisotopes of MEDical interest) project [25], has proposed to use magnetron sputtering to deposit the target material onto the appropriate backing plate in order to provide high density, high thickness uniformity, and high adherence to the backing. An innovative study of the Mo cyclotron solid target concept for $^{99m}$Tc production was recently presented in [3]. This included sputter deposition of Mo target material onto a composite high thermal conductivity and chemical resistance backing plate. The previous article [3] was devoted to Mo deposition and the technological aspects of vacuum brazing to realize a composite backing plate. The scope of the present work is instead focused on the deposition method development.

In this work, the method originally developed to produce a Mo cyclotron solid target has been tested for another precursor material, Y for $^{89}$Zr radionuclide production. Some technical details on Mo target preparation are repeated in the current work to compare the main deposition parameters for Y and Mo in order to illustrate the versatility of the developed method, its applicability to different target and backing materials, and its capability to produce targets for different target stations.

Both radionuclides $^{99m}$Tc and $^{89}$Zr have importance for medical applications. The interest in additional/alternative routes for $^{99m}$Tc production has been stimulated by the perceived new $^{99m}$Tc crisis, due to the scheduled shutdown of the Chalk River nuclear power plant in 2018. Cyclotron-based production of $^{99m}$Tc, starting from $^{100}$Mo by $^{100}$Mo(p,2n)$^{99m}$Tc reaction, has been developed and evaluated at the LNL-INFN in the framework of the APOTEMA-TECHN_OSP project [25–30]. Regarding $^{89}$Zr, the main interest in this radionuclide is related to the radiolabeling of slowly accumulating radiopharmaceuticals (*in vivo* imaging of antibodies, nanoparticles, and other large bioactive molecules) for targeting tumor cells [31–34]. The latter goal requires the ready availability of relatively large amounts of $^{89}$Zr with high specific activity: this remains nowadays a challenge.

Magnetron sputtering is a very flexible PVD technique that allows to modify a lot the properties of the deposited film by changing the sputtering parameters. Magnetron sputtering is generally known as a PVD technique for the deposition of thin metallic films. However, it is not used for thick film deposition because of the tensile or compressive stress that is always present in the films [35,36]. Controlling the stress in PVD films is extremely important because of its close relationship to the technological properties of the material; the adhesion strength to the substrate; and the limit of film thickness without cracking, buckling, or delamination.

One of the most challenging issues of this study was to develop a method to deposit dense stress-free films of refractory metal with a thickness of the order of tens or even hundreds of microns. Magnetron sputtering was used to deposit a thick target film directly onto a backing plate. This approach could have a further advantage: to simplify the often-underestimated challenge of establishing good thermal contact between the target and the target backing plate.

Thus, in this work, the validation of the solid target production technology based on the magnetron sputtering technique was realized for both Mo and Y target preparation. Indeed, a set of Mo and Y target prototypes has been realized and successfully tested under the cyclotron's beam. The results have shown that the developed solid target preparation method is attractive for further optimization and implementation in medical radionuclide production.

## 2. Materials and Methods

### 2.1. Materials

Natural molybdenum (99.99% purity, Mateck GmbH, Julich, Germany), natural yttrium (99.9% purity, Gambetti Kenologia Srl, Binasco, MI, Italy), and argon (99.99% purity, SIAD S.p.A., 159 Bergamo, Italy) were used for sputtering deposition as target materials and sputtering gas, respectively.

Different substrate materials were used: Mo was deposited onto copper (Ø32 × 1 mm), sapphire (Ø12.7 × 0.5 mm, Meller Optics Inc., Providence RI, USA), chemical vapor deposited (CVD) synthetic diamond (Ø13.5 × 0.3 mm, II-VI Advanced Materials GmbH, Pine Brook, NJ, USA), and silicon wafers of 50.8 mm diameter and 250–300 μm thickness of semiconductor quality and (100) orientation (Sil'tronix Silicon Technologies, Archamps, France).

Niobium disks of 99.9% purity (Ø24 × 0.5 mm, Goodfellow Cambridge Ltd., Huntingdon, England) were used as the substrates for the yttrium deposition.

Copper substrates were washed in an ultrasonic bath for 20 min with GP 17.40 SUP soap (NGL Cleaning Technology SA, Nyon, Switzerland) and deionized water. This was followed by chemical etching with SUBU5 solution (5 g/L of sulfamic acid, 1 g/L of ammonium citrate, 50 mL/L of butanol, 50 mL/L of H2O2, and 1 L of deionized water) at $72 \pm 4$ °C in order to remove surface oxides, passivation in 20 g/L of sulfamic acid, ultrasonic washing with water for 20 min, rinsing with ethanol, and drying with nitrogen.

Niobium and nonmetallic substrates cleaning procedure included: ultrasonic bath cleaning with Rodaclean® (NGL Cleaning Technology SA, Nyon, Switzerland) soap for 20 min at 40 °C; ultrasonic bath cleaning with deionized water for 20 min at 40 °C; rinsing with ethanol (storage in ethanol in plastic box); mechanical cleaning with ethanol and AlfaWipe® (Texwipe Company, Hoofddorp, The Netherlands); and drying with nitrogen gas immediately before positioning onto the substrate-holder.

## 2.2. Deposition System

The sputtering process was carried out in a cylindrical, stainless-steel vacuum chamber that was 25 cm in diameter and 25 cm in length. A base pressure of $5 \times 10^{-6}$ mbar was reached without backing out (heating the vacuum flanges up to 200 °C to improve degassing during pumping) by means of the Pfeiffer turbo molecular pump of 360 L/min and the Varian Tri Scroll Pump of 210 L/min as a primary.

The films were deposited by DC (direct current) sputtering with a 2-in. planar magnetron cathode source unbalanced of the II Type. The depositions were performed onto planar substrate-holders, with a distance of 6 or 7 cm from the cathode.

The "down-top" deposition configuration, with the magnetron source placed from the downside of the cylindrical chamber and the substrate-holder with substrates from the top of the chamber, was used for the film deposition in order to minimize the film delamination probability caused by the metallic dust particles (Figure 1).

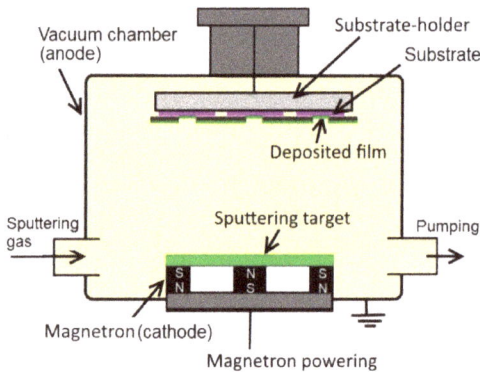

**Figure 1.** Scheme of "down-top" configuration.

The deposition onto 7/8 substrates was realized at the same time. The sputtering materials were deposited on a spot that was 10 mm in diameter in the center of each substrate (backing plate) defined by appropriate masks (Figure 2). For the Mo deposition, a 450-W Infrared (IR) lamp (Helios Italquarz, Cambiago-MI, Italy) was used to heat up the substrate-holder, and a K-type thermocouple, placed inside the furnace, was used to control the temperature with an automatic custom-made infrared lamp backing control system.

(a) (b)

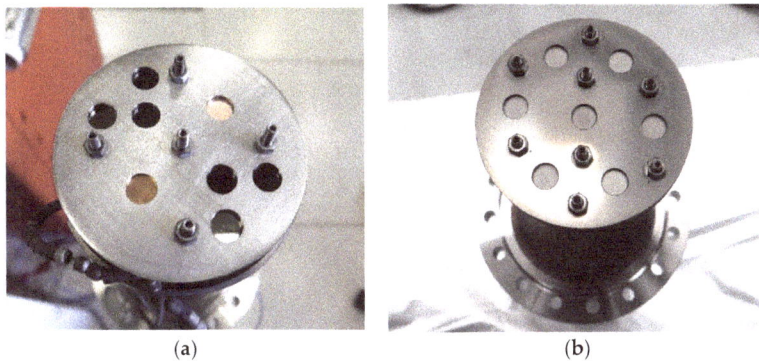

**Figure 2.** Masks assembled on heated substrate-holder for Mo sputtering (**a**) and on nonheated substrate-holder for Y sputtering (**b**).

### 2.3. Deposit Analysis

The evaluation of the film thickness was performed by the contact stylus profiler model Dektak 8 (Veeco, Plainview, NY, USA).

FEI (formerly Philips, OR, USA) Scanning Electron Microscope SEM XL-30 was used for the sputtered film analysis. Samples of 5-μm Mo film onto a silicon wafer substrate and 40-μm Y film onto a copper substrate were prepared in separate runs with the same optimized parameters for SEM cross-section analysis.

### 2.4. Cyclotron Tests

In this study, two different cyclotrons with the corresponding solid target stations were used for testing the Mo and Y sputtered targets. The Mo target irradiation tests were performed at the Medical Physics Department of "S. Orsola-Malpighi" Hospital in Bologna using the PETtrace 800S cyclotron (GE Healthcare, Chicago, IL, USA) equipped with the solid target station prototype of TEMA Sinergie S.P.A. (Faenza, Ra, Italy). The Y targets were tested under the TR19 cyclotron with the corresponding target station (ACSI, Richmond, BC, Canada).

The GE PETtrace 800S cyclotron (GE Healthcare, Chicago, IL, USA) works at a fixed proton energy of 16.5 MeV (deuteron energy 8 MeV) and currents up to 100 μA (the maximum current available practically depends on the source and tuning of the magnets). The solid target station (prototype TEMA Sinergie S.P.A., Faenza, Ra, Italy) is shown in Figure 3a. The target "coin" is cooled directly by the He gas flow in the front and indirectly through contact with the water-cooled aluminum chamber from the back. A detailed description of this irradiation unit was reported previously by Cicoria and co-workers [37,38].

TR19 14–19 MeV is a variable energy proton cyclotron with a high current ion source up to 300 μA. The corresponding ACSI solid target station allows for direct helium gas cooling of the target coin from the front part and water cooling from the back.

The target coin prototypes were realized fitting the design of the corresponding target station, which means disks of 2 mm maximum thickness and diameters of 32 mm (TEMA) and 24 mm (ACSI).

**Figure 3.** Solid target stations used for under-beam target tests: (**a**) TEMA Synergie target station prototype of PETtrace cyclotron (S. Orsola-Malpighi Hospital, Bologna); (**b**) TR19 cyclotron target station (Sacro Cuore Hospital, Negrar, Verona).

The irradiations of the target prototypes for thermomechanical stability control were carried out at 15.6 MeV for the Mo targets and 12.7 MeV for the Y targets at increasing currents. Irradiating for 60 s is sufficient to reach thermal equilibrium in the target. Even such short-time tests are sufficient to reveal the structural characteristics of the target depending on the backing material, the quality of deposition, adherence, etc. Indeed, the target "failing" (i.e., when the deposited layer is detached or cracked) occurs within the first 20 s of irradiation. In practice, an irradiation time of 0.5–2.0 h, or even more, is routinely adopted to allow the production of clinically relevant amounts of radionuclides. Then, the long-term stability of the target withstanding the short-time test is mainly determined by the stability of the beam and the cooling system. For this reason, the irradiation time of 1–2 min was chosen for the initial thermomechanical tests.

After each irradiation, the sample was unloaded to visually inspect the integrity of the target and the adhesion of the Mo or Y film on the backing. Longer irradiations were performed using one of the CVD synthetic-diamond-based Mo target prototypes (30 min at 15.6 MeV, 60 µA, the maximum current reached by the cyclotron at that moment) and two Y targets (5 h, 12.7 MeV, 50 µA).

## 2.5. Estimation of $^{89}Zr$ Expected Yields

In this study, the $^{89}$Zr activity at the end of bombardment (EOB) of Y sputtered targets was not measured experimentally. Instead, it was predicted by means of the Radionuclide Yield Calculator (RYC) 2.0 software [39] containing SRIM (The Stopping and Range of Ions in Matter [40]) modules. The experimental nuclear cross-section data, presented in Experimental Nuclear Reaction Data (EXFOR [41]) and previously reported by Omara et al. [42], Satheesh et al. [43], and Zhao et al. [44] fit by Gaussian generalized distribution (GGD), were utilized for the calculations. In order to validate the calculations obtained using the RYC 2.0 software, the $A_{theor}$ (this work) was compared to the data presented in the literature $A_{theor}$ (Lit.).

In order to compare the produced Y targets with the ones reported in the literature, the EOB thick target yield for 1 h of irradiation was also estimated according to Equation (1), as suggested by Otuka et al. [45]:

$$a(t_{1h}) = \frac{A(t_{irrad})\left(1 - e^{-\lambda t_{1h}}\right)}{I_0 \left(1 - e^{-\lambda t_{irrad}}\right)}, \tag{1}$$

$$\lambda = \frac{ln2}{T_{1/2}} \tag{2}$$

where $A(t_{irrad})$ is the experimentally measured $^{89}$Zr activity at the end of bombardment (mCi) after $t_{irrad}$ (h) irradiation at $I_0$ irradiation current (μA), $t_{1h}$ = 1 h of normalizing irradiation time (h), $\lambda$ is the radioactive decay constant, and $T_{1/2}$ is the $^{89}$Zr radioactive half-life ($T_{1/2}$ = 78.4 h).

While the tests reported in the literature $a(t_{1h})$ were calculated using reported experimental $^{89}$Zr EOB activity $A_{exp}$, for Y-2 and Y-7 targets, instead, $A_{theor}$ (this work) was used to predict the $a(t_{1h})$, since no experimental measurement of produced activity was realized.

## 3. Results and Discussion

### 3.1. Sputtering Parameters Optimization

Besides the classical stress-associated problems (e.g., cracking in the deposit or substrate, cracking at the substrate–deposit interface, and adhesion problems [46]), the stress in deposited films can be a reason for poor adhesion between a film (target material) and a substrate (backing plate). The thermal resistance of this contact can drastically increase, causing a decrease in heat exchange efficiency. Thus, the optimization of the Mo and Y deposition parameters, aiming to reduce stress, was mandatory for the purpose of this work (i.e., cyclotron solid target realization). The sputtering deposition process of Mo and Y, using the same 2-in. magnetron and the same vacuum chamber, is shown in Figure 4.

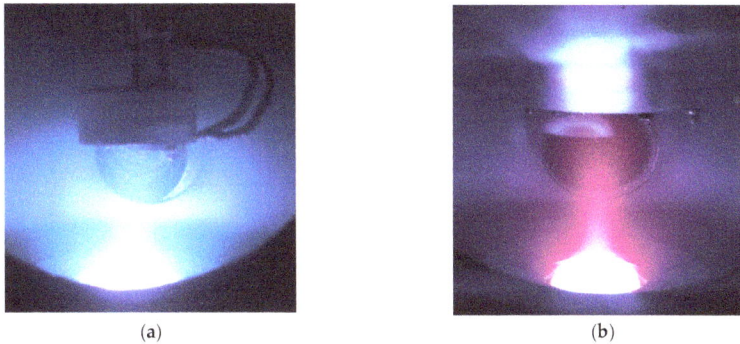

(a)                                                        (b)

**Figure 4.** Plasma of Mo (**a**) and Y (**b**) during the sputtering process.

The intrinsic stress in PVD-deposited films depends on the energy supplied to the growing film surface during the deposition process. The parameters considerably involved in the change of the supplied energy and, thus, in the microstructure growth mechanism are the sputtering gas pressure, the temperature of the holder, bias, etc.

Theoretically, if the other sputtering parameters are kept fixed, there is a particular gas pressure that corresponds to the transition between the tensile and compressive stresses. High pressure corresponds to the decrease of the kinetic energy of sputtered atoms and reflected neutrals bombarding the growing film due to the increased frequency of the collisions with the sputtering gas. In this case, a more porous microstructure is created, which is attributed to tensile intrinsic stress. At low pressure, the arriving particles have higher kinetic energy, and a more dense film, usually with compressive stress, is created [47]. In the current work, the optimal pressure was achieved experimentally by performing short depositions (15 min) of the material of interest (Mo, Y) onto a flexible substrate (Kapton), keeping all the other deposition parameters fixed. The radius of curvature assumed by the Kapton is an indicator of the stress (see Figure 5).

In this way, the "transition" pressure for Mo deposition was found to be $1.63 \times 10^{-2}$ mbar (corresponding to 17 sccm Ar gas flow) and $1.36 \times 10^{-2}$ mbar (corresponding to 19 sccm Ar gas flow) for Y deposition. It should be noted that the distance from the magnetron to the substrate was slightly different in the two cases—6 cm for Mo deposition and 7 cm for Y deposition—due to the difference in the design of the substrate-holders.

Compressive stress    Stress-free    Tensile stress

Ar pressure ⟶

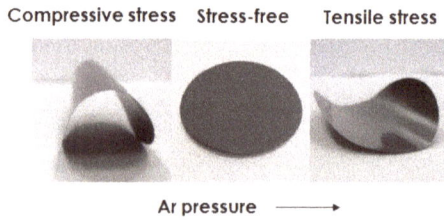

**Figure 5.** Kapton substrate curvature vs. sputtering pressure.

The second parameter, which strongly influences the intrinsic stress in films, is the substrate temperature, since it influences the kinetic energy of the particles that have already arrived at the substrate: the higher the temperature, the higher the density, thanks to renucleation. The transition homologous temperature $T_h = T/T_m = 0.2$–$0.45$ (where $T$ is the temperature during vacuum deposition and $T_m$ is the melting point of a deposited material), presented in the Structure Zone Model as the T-zone [35], corresponds to a transition from a tensile stress, attributed to the porous microstructure, to a near-zero or even low-level compressive stress of the dense bulk-like film. In this work, deposition was realized at the homologous temperature $T_h = T/T_m = 0.2$; this means ~500 °C for Mo and ~250 °C for Y. Indeed, the columnar dense microstructures of Mo and Y films (see Figure 6) obtained at $T_h = 0.2$ corresponded to the standard T-zone in the Structure Zone Model [35].

**Table 2.** Sputtering process parameters.

| Parameters | $^{nat}$Mo | $^{nat}$Y |
|---|---|---|
| Argon flux (sccm) | 17 | 19 |
| Ar pressure (mbar) | $1.63 \times 10^{-2}$ | $1.36 \times 10^{-2}$ |
| Power (W) | 5–550 | 400 |
| Target-substrate distance (cm) | 6 | 7 |
| Substrate temp-re (°C) | 500 | No heating |
| Deposition rate (μm/min) | 11 | 13.3 |
| Multilayer mode | Yes | No |

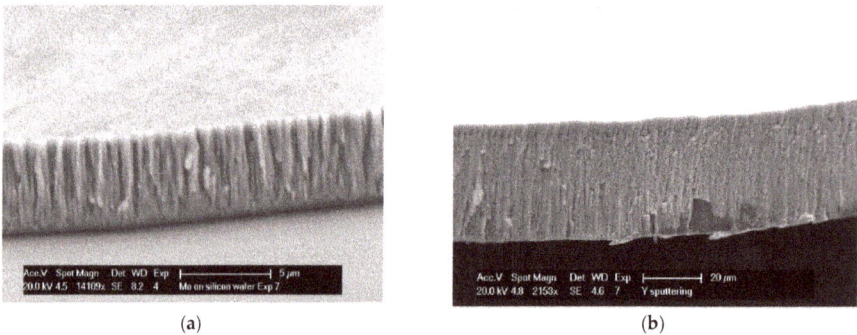

(a)

(b)

**Figure 6.** Cross-section SEM analysis of Mo film deposited onto Si at 500 °C (**a**) and Y film deposited onto copper without forced heating (220–250 °C) (**b**).

Furthermore, a multilayer deposition technique was shown to reduce the stress [48]; thus, the deposition of Mo thick films was fragmented in thousands of subsequent brief depositions of thin films, using an automatic program to control the power. Each deposition was interspersed by a "relaxation time" (80% duty cycle for a 1-min period), in which the film was annealed.

The optimized parameters for magnetron sputtering of both Mo (using copper backing, complex sapphire, or synthetic-diamond-based backing) and Y (on niobium backing) for the described deposition system configuration are shown in Table 2.

The deposition of Mo at $1.63 \times 10^{-2}$ mbar Ar sputtering gas pressure, keeping the substrate-holder heated at 500 °C in a multilayer deposition mode (more details are presented in a previous work by the authors [3]), gave the best over 100-µm-thick Mo films in terms of adhesion, density (more than 95% of the bulk material), and being stress-free. It should be said that, in the past, a much lower Mo thickness of about 0.1 mg/cm$^2$ (~0.1 µm calculated for bulk density Mo), obtained using FIB [24] and ultrahigh vacuum sputtering [23], was reported. Our film thickness is comparable to the 130-µm Mo deposited by thermal spray deposition reported by Jalilian et al. [22]. All eight samples deposited in each sputtering run with the sample-holder (Figure 2a) have been characterized by high film-thickness uniformity.

The fact that ultrathick Mo films were deposited onto ceramic substrates, such as sapphire and CVD synthetic diamond, without stress-induced damage of the substrate demonstrates the versatility of the developed sputtering method. Indeed, the use of chemically inert backing materials (i.e., ceramics) in the dissolution process after target irradiation [30] would avoid radiochemical impurities in the final injectable radiopharmaceutical [3,49].

Instead, the thick stress-free Y films were obtained by merely optimizing the sputtering pressure. Since yttrium is very sensitive to oxidation, the multilayer mode was not applied in order to avoid the introduction of oxide layers between the metallic ones, which might promote the increase of intrinsic stress (and further possible delamination) instead of stress relaxation. Furthermore, the use of the floating temperature of the substrate-holder during the sputtering process simplified the system configuration from the point of view of hardware and safety. On the other hand, forced heating of the substrate-holder was not required in the case of Y sputtering, since the $T_m$ of Y is lower than the one of Mo, and $T_h = 0.2$ was reached thanks to the interaction of the substrate-holder with plasma during the sputtering process.

Seven Y targets were produced in one deposition run. The sputtering deposition of six targets resulted in ~50-µm-thick Y films, with a uniform distribution of film thickness. Only the target placed in the central position during the sputtering process showed a higher but less uniform thickness (70 µm). A representative example of the Y film profile sputtered onto 0.5-mm-thick niobium backing is shown in Figure 7. For Y targets, the niobium backing was chosen since it is inert in concentrated HCl, which was the media used for dissolution after irradiation [34].

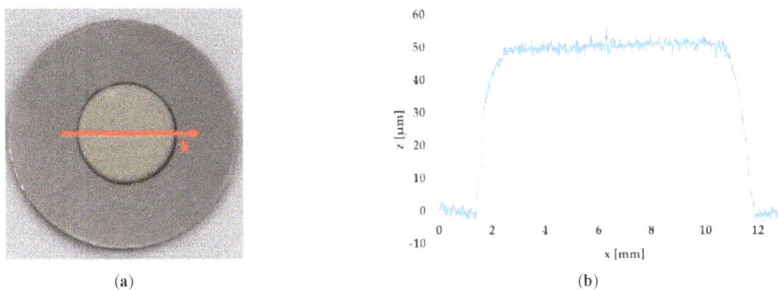

(a)      (b)

**Figure 7.** Deposited target profile measurement. (**a**). Yttrium sputtered target and profile measurement coordinate X (**b**). Typical target profile: X-measurement position, Z-height.

It should be noted that the main defect of the magnetron sputtering deposition technique is the great loss of the deposited material. Therefore, for a very expensive material, such as [100]Mo (starting material for cyclotron-produced [99m]Tc through [100]Mo(p,2n) nuclear reaction), the development of a suitable strategy for the deposition of a small amount of material and an efficient recovery method is necessary. Instead, magnetron sputtering can be a powerful technique for materials with 100% natural abundance, such as [89]Y for the production of [89]Zr when a high yield of production is requested.

### 3.2. Cyclotron Test

The thermomechanical stability of the Mo targets produced by magnetron sputtering was evaluated under the beam of a 16-MeV GE PETtrace cyclotron, in the S. Orsola-Malpighi Hospital, Bologna, at 15.6 MeV, increasing the beam currents with 10-μA steps for 1 min irradiation, starting from 20 up to 70 μA (see Table 3). Visual control of the target after each irradiation was carried out.

All Mo target prototypes, based on about 100-μm-thick Mo film deposited by magnetron sputtering onto copper backing directly and ceramic (sapphire and CVD synthetic diamond) substrates brazed to the copper supports [49], showed good thermomechanical stability under the proton beam. The prototypes could sustain a power density of about 1 kW/cm$^2$, provided by a proton beam of 15.6 MeV, 60 μA, and a spot size of ~11 mm diameter. Excellent adhesion (no delamination) and no film damage were observed after each irradiation. Only one of the sapphire-based samples was cracked, and in our interpretation, this was due to a problem during the brazing process, and not due to the irradiation. In fact, further CVD diamond-based targets were improved by adjusting the parameters of Ti metallization prior to brazing. A more detailed description of the results on different Mo target prototypes is presented in the previous work by the authors [3].

Y targets were irradiated under a ~10 mm in diameter proton beam of the TR19 cyclotron at 12.7 MeV at increasing currents up to 70 μA. The irradiation data are presented in Table 3. It is worth noting that the Y foil targets (thickness from 0.15 to 1 mm) commonly used by different groups to produce [89]Zr [19,38,50–53] can sustain, without any critical damage, only currents under 40 μA. Instead, the targets realized in this work have supported up to 1 kW/cm$^2$ heat power density, corresponding to relatively higher current values, which can increase the [89]Zr radioisotope production yields. Besides that, the thermomechanical performance of the targets realized using the method described in this work is comparable to one of the commercial sputtered Y targets [19]. Indeed, the integrity of the irradiated targets was not compromised, despite a visible dark spot in the center of the target corresponding to the beam profile, as shown in Figure 8. The white spot in the center of some targets can be explained by the creation of some amount of Y oxide/hydroxide due to water leakages in the target station.

| 30 μA for 2 min | 40 μA for 2 min | 60 μA for 2 min | 70 μA for 2 min | 50 μA for 5 h |

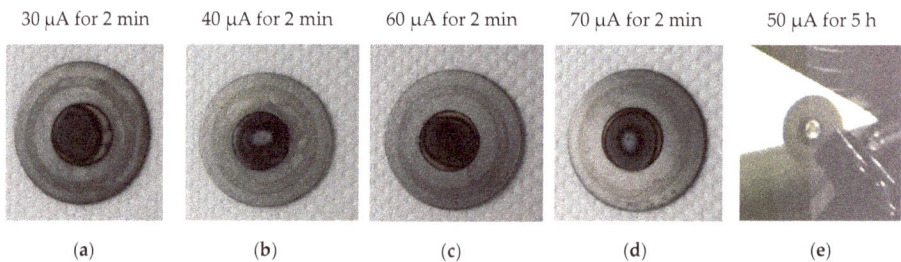

| (a) | (b) | (c) | (d) | (e) |

**Figure 8.** Y sputtered solid targets after irradiation under the 12.7-MeV proton beam of the TR19 ASCI cyclotron (Negrar, Verona, Italy): (**a**) Y-1, (**b**) Y-3, (**c**) Y-4, (**d**) Y-5, and (**e**) Y-7.

**Table 3.** Irradiation tests.

| Target | Deposit Thickness | Backing | Beam Energy | Proton Current | Irradiation Time | Heat Power Density | Result |
|---|---|---|---|---|---|---|---|
| Mo-1 | 110 μm | Cu Ø32 × 1.5 mm | 15.6 MeV | 70 μA | 1 min | 1.2 kW/cm$^2$ | Withstood |
| Mo-2 | 110 μm | Sapphire Ø12.7 × 0.5 mm brazed to Cu Ø32 × 1.5 mm | 15.6 MeV | 60 μA | 1 min | 1 kW/cm$^2$ | Withstood |
| Mo-3 | 125 μm | Diamond Ø13.5 × 0.3 mm | 15.6 MeV | 60 μA | 1 min | 1 kW/cm$^2$ | Withstood |
| Mo-4 | 125 μm | brazed to Cu Ø32 × 1.5 mm | 15.6 MeV | 60 μA | 30 min | 1 kW/cm$^2$ | Withstood |
| Y-1 | 50 μm | | 12.7 MeV | 30 μA | 2 min | 0.5 kW/cm$^2$ | Withstood (Figure 8a) |
| Y-2 | 50 μm | Nb Ø24 × 0.5 mm | 12.7 MeV | 50 μA | 5 h | 0.8 kW/cm$^2$ | Withstood |
| Y-3 | 50 μm | | 12.7 MeV | 40 μA | 2 min | 0.65 kW/cm$^2$ | Withstood (Figure 8b) |
| Y-4 | 50 μm | | 12.7 MeV | 60 μA | 2 min | 1 kW/cm$^2$ | Withstood (Figure 8c) |
| Y-5 | 50 μm | | 12.7 MeV | 70 μA | 2 min | 1.1 kW/cm$^2$ | Withstood (Figure 8d) |
| Y-7 | 70 μm | | 12.7 MeV | 50 μA | 5 h | 0.8 kW/cm$^2$ | Withstood (Figure 8e) |

The expected $^{89}$Zr activity at the EOB estimated using the RYC 2.0 software is reported in Table 4. Since for the experiments reported by Queern et al. [19] some discrepancies were found only for higher-thickness targets of >200 μm (probably due to the chosen nuclear cross-section datasets), the RYC 2.0 was found effective to predict the $^{89}$Zr EOB activity of the Y-2 and Y-7 target irradiation experiments. Thus, the $^{89}$Zr activities of about 41 and 57.2 mCi are expected to be produced irradiating Y-2 of 50 μm and Y-7 of 70 μm sputtered targets for 5 h at 12.7 MeV and 50 μA.

**Table 4.** Comparison of $^{89}$Zr activity produced using sputtered targets.

| Y Thickness, μm | Cyclotron, Target | E, MeV | I, μA | t, h | A$_{exp}$ [1], mCi | A$_{theor}$ (Lit.) [2] mCi | A$_{theor}$ (this work) [3] mCi | a(t$_{1h}$) [4] mCi/μA | Lit. Ref. |
|---|---|---|---|---|---|---|---|---|---|
| 50 | ACSI TR-19 | 12.7 | 50 | 5 | - | - | 41.0 | 0.16 | This work |
| 70 | | 12.7 | 50 | 5 | - | - | 57.2 | 0.22 | |
| 110 | | 12.5 | 10 | 0.5 | 1.63 | 1.82 | 1.8 | 0.33 | |
| 140 | ACSI TR-24, non-inclined target | 12.5 | 21 | 0.5 | 3.94 | 4.86 | 4.8 | 0.37 | [19] |
| 90 | | 12.5 | 30 | 0.5 | 2.6 | 4.46 | 4.42 | 0.17 | |
| 220 | | 12.5 | 40 | 2 | 25.37 | 31.28 | 55.2 | 0.31 | |
| 210 | | 12.8 | 45 | 2 | 43.8 | 45.27 | 60.5 | 0.49 | |
| 90 | | 12.8 | 40 | 2 | 21.9 | 23.09 | 23.8 | 0.28 | |
| 25 (700 eff. [5]) | Philips AVF cyclotron, 1°–2° inclined target | 14 | 100 | 1 | 130 | - | 184.3 | 1.3 | [20] |
| 35 (1000 eff. [5]) | | 14 | 65–80 | 2–3 | 180–360 | - | 243.1–446.9 | 1.39–1.51 | [21] |

[1] A$_{exp}$—$^{89}$Zr measured activity at the end of bombardment (EOB), reported in the literature. [2] A$_{theor}$ (Lit.)—calculated $^{89}$Zr activity at the EOB, reported in corresponding literature reference. [3] A$_{theor}$ (this work)—$^{89}$Zr activity at the EOB, calculated with the RYC 2.0 software. [4] a(t$_{1h}$)—1-h EOB thick target yield. [5] eff.—effective thickness for inclined target calculated as deposit thickness divided into sin (2°).

The $^{89}$Zr 1-h EOB thick target yields a(t$_{1h}$) for Y-2 and Y-7 targets (see Table 4) were a bit lower but of the same order of magnitude with the thick target yields obtained by 90–220 μm ACSI commercial targets [19] and an order of magnitude lower than the ones produced irradiating 25/35-μm Y targets, as reported by Meijs et al. [20] and Verel et al. [21]. This can be explained by higher irradiation energy and the use of a low-angle inclined target configuration, since the effective target thickness is much higher in those cases.

Further depositions are planned to increase the thickness of the Y sputtered film in order to compete better with the commercially available targets. Besides that, new irradiations are required to assess the produced activity and the radionuclide purity in order to confirm the sputtering technique as an alternative route for the realization of Y targets for $^{89}$Zr production [19–21].

## 4. Conclusions

The developed magnetron sputtering technique was successfully applied to the preparation of Mo and Y solid medical cyclotron targets, since this deposition method offers high density of the target material and high adherence to different backing materials. In this way, the good heat transfer allows for increasing the beam current during the irradiation. Indeed, realized solid targets can sustain up to a 1-kW/cm$^2$ proton beam heat power density with no critical damage. In addition, the capability to realize sputtered film onto any substrate gives the possibility of choosing the most suitable backing material for the purpose which each radionuclide production requires (i.e., thermal conductivity, chemical inertness). Thus, sapphire and synthetic diamond materials, inert in $H_2O_2$, which was the dissolution media for irradiated Mo targets, were used as the backing for the Mo targets, and Nb, inert in concentrated HCl used in the case of Y targets, was chosen as the backing for the realization of the Y targets. The performance of the homemade Y sputtered cyclotron solid targets was comparable to the commercial ones.

$^{89}$Zr activity of the order of 40–50 mCi is predicted to be produced by irradiating realized targets for 5 h at 12.7 MeV and 50 µA. The estimated 1-h EOB thick target yield is lower than the one of the ACSI commercial sputtered targets, but can be improved by increasing the Y layer thickness.

The versatility of the developed magnetron sputtering method has been proven in this study, and it can also be a promising alternative for other solid target materials.

## 5. Patents

The method for solid cyclotron target preparation reported in this manuscript was submitted by Istituto Nazionale di Fisica Nucleare on 14.09.17 as an Italian patent application, N. 102017000102990, dep. ref. P1183IT00, inventors V. Palmieri, H. Skliarova, S. Cisternino, M. Marengo, G. Cicoria, entitled "Metodo per l'ottenimento di un target solido per la produzione di radiofarmaci". It was extended to the international patent application PCT/IB2018/056826, dep. ref. P1183PC00, on 07.09.18, entitled "Method for obtaining a solid target for radiopharmaceuticals production".

**Author Contributions:** Conceptualization of the innovative cyclotron solid target prototype described in current work was done by H.S. and V.P.; the methodology for ultrathick Mo film deposition by magnetron sputtering was developed single-handedly by H.S.; further validation of the developed target preparation technique was realized by H.S. and S.C.; investigation of the performance of new target prototypes under cyclotron irradiation was performed by a group of collaborators, including H.S., S.C., M.M., G.C., E.C., and G.G.; resources of material science research lab and cyclotron facilities were provided by V.P., M.M., and G.G., correspondingly; data curation was under the responsibility of H.S., S.C., and E.C.; original draft preparation was performed by H.S. and S.C.; writing—review & editing by H.S., S.C., M.M., and E.C.; work on visualization was realized by H.S. and S.C.; V.P. formally, being responsible for the laboratory, and H.S. informally were in charge of work supervision and administration; funding acquisition was made by V.P.

**Funding:** The part of the research on Mo target preparation was realized in the framework of the TECHN_OSP project funded by CSN5 of the Istituto Nazionale di Fisica Nucleare, Italy for 2015–2017. National responsible: J. Esposito, LNL-INFN. The part on Y target preparation was a part of the Terabio INFN project, funded by the Italian Ministry. National responsible: V. Palmieri.

**Acknowledgments:** Special thanks should be given to Juan Esposito, responsible for the CSN5 INFN project TECHN_OSP, for his precious support, fruitful discussion during paper revision, as well as the funding acquisition. We are grateful to the staff of the LNL-INFN laboratories for the Surface & Material Treatments and for Nuclear Physics and Mechanical workshop for their help. We acknowledge the support and contribution of Stephen Jewkes in reviewing the language of this manuscript. We wish to thank also Anna Taffarello for her contribution to the English language and style editing.

**Conflicts of Interest:** The authors declare no conflict of interest.

## References

1. Stolarz, A. Target preparation for research with charged projectiles. *J. Radioanal. Nucl. Chem.* **2014**, *299*, 913–931. [CrossRef] [PubMed]
2. IAEA. *Cyclotron Based Production of Technetium-99m.*; IAEA: Vienna, Austria, 2017; ISBN 978-92-0-102916-4.

3.  Skliarova, H.; Cisternino, S.; Cicoria, G.; Marengo, M.; Palmieri, V. Innovative Target for Production of Technetium-99m by Biomedical Cyclotron. *Molecules* **2019**, *24*, 25. [CrossRef] [PubMed]

4.  Schaffer, P.; Benard, F.; Buckley, K.R.; Hanemaayer, V.; Manuela, C.H.; Klug, J.A.; Kovacs, M.S.; Morley, T.J.; Ruth, T.J.; Valliant, J.; et al. Processes, systems, and apparatus for cyclotron production of technetium-99m. Patent US 2013/0301769 A1; United States: TRIUMF, 14 November 2013.

5.  Gagnon, K.; Wilson, J.S.; Holt, C.M.B.; Abrams, D.N.; McEwan, A.J.B.; Mitlin, D.; McQuarrie, S.A. Cyclotron production of 99mTc: Recycling of enriched 100Mo metal targets. *Appl. Radiat. Isot.* **2012**, *70*, 1685–1690. [CrossRef] [PubMed]

6.  Wilson, J.; Gagnon, K.; McQuarrie, S. Production of technetium from a molybdenum metal target. Patent US 2014/0029710A1; United States: University of Alberta, 30 January 2014.

7.  Stolarz, A.; Kowalska, J.A.; Jasiński, P.; Janiak, T.; Samorajczyk, J. Molybdenum targets produced by mechanical reshaping. *J. Radioanal. Nucl. Chem.* **2015**, *305*, 947–952. [CrossRef] [PubMed]

8.  Morrall, P.S. The Target Preparation Laboratory at Daresbury. *Nucl. Instrum. Methods Phys. Res. Sect. Accel. Spectrometers Detect. Assoc. Equip.* **2008**, *590*, 118–121. [CrossRef]

9.  Manenti, S.; Holzwarth, U.; Loriggiola, M.; Gini, L.; Esposito, J.; Groppi, F.; Simonelli, F. The excitation functions of 100Mo(p,x)99Mo and 100Mo(p,2n)99mTc. *Appl. Radiat. Isot.* **2014**, *94*, 344–348. [CrossRef] [PubMed]

10. Zweit, J.; Downey, S.; Sharma, H.L. Production of no-carrier-added zirconium-89 for positron emission tomography. *Int. J. Rad. Appl. Instrum. [A]* **1991**, *42*, 199–201. [CrossRef]

11. NISHIKATA, K.; Kimura, A.; Ishida, T.; KITAGISHI, S.; Tsuchiya, K.; Akiyama, H.; Nagakura, M.; Suzuki, K. Method of producing radioactive molybdenum. Patent US 13675769; United States. Japan Atomic Energy Agency, 30 May 2013.

12. Zeisler, S.K.; Hanemaayer, V.; Buckley, K.R. Target system for irradiation of molybdenum with particle beams. Patent US 2017/0048962A1; United States: TRIUMF, 16 February 2017.

13. Schaffer, P.; Bénard, F.; Bernstein, A.; Buckley, K.; Celler, A.; Cockburn, N.; Corsaut, J.; Dodd, M.; Economou, C.; Eriksson, T.; et al. Direct Production of 99mTc via 100Mo(p,2n) on Small Medical Cyclotrons. *Phys. Procedia* **2015**, *66*, 383–395. [CrossRef]

14. Zyuzin, A.; Guérin, B.; van Lier, E.; Tremblay, S.; Rodrigue, S.; Rousseau, J.A.; Dumulon-Perreault, V.; Lecomte, R.; van Lier, J.E. Cyclotron Production of 99mTc. In Proceedings of the WTTC13, Roskilde, Denmark, 26–28 July 2010.

15. Avetisyan, A.; Dallakyan, R.; Sargsyan, R.; Melkonyan, A.; Mkrtchyan, M.; Harutyunyan, G.; Dobrovolsky, N. The powdered molybdenum target preparation technology for 99mTc production on C18 cyclotron. *IJEIT* **2015**, *4*, 8.

16. Taghilo, M. Cyclotron production of 89Zr: A potent radionuclide for positron emission tomography. *Int. J. Phys. Sci.* **2012**, *7*. [CrossRef]

17. Morley, T.J.; Penner, L.; Schaffer, P.; Ruth, T.J.; Bénard, F.; Asselin, E. The deposition of smooth metallic molybdenum from aqueous electrolytes containing molybdate ions. *Electrochem. Commun.* **2012**, *15*, 78–80. [CrossRef]

18. Kazimierczak, H.; Ozga, P.; Socha, R.P. Investigation of electrochemical co-deposition of zinc and molybdenum from citrate solutions. *Electrochimica Acta* **2013**, *104*, 378–390. [CrossRef]

19. Queern, S.L.; Aweda, T.A.; Massicano, A.V.F.; Clanton, N.A.; El Sayed, R.; Sader, J.A.; Zyuzin, A.; Lapi, S.E. Production of Zr-89 using sputtered yttrium coin targets. *Nucl. Med. Biol.* **2017**, *50*, 11–16. [CrossRef] [PubMed]

20. Meijs, W.E.; Herscheid, J.D.M.; Haisma, H.J.; Wijbrandts, R.; van Langevelde, F.; Van Leuffen, P.J.; Mooy, R.; Pinedo, H.M. Production of highly pure no-carrier added 89Zr for the labelling of antibodies with a positron emitter. *Appl. Radiat. Isot.* **1994**, *45*, 1143–1147. [CrossRef]

21. Verel, I.; Visser, G.W.M.; Boellaard, R.; Walsum, M.S.; Snow, G.B.; Dongen, G.A.M.S. van 89Zr Immuno-PET: Comprehensive Procedures for the Production of 89Zr-Labeled Monoclonal Antibodies. *J. Nucl. Med.* **2003**, *44*, 1271–1281.

22. Jalilian, A.; Targholizadeh, H.; Raisali, G.; Zandi, H.; Kamali Dehgan, M. Direct Technetium radiopharmaceuticals production using a 30MeV Cyclotron. *DARU J. Fac. Pharm. Tehran Univ. Med. Sci.* **2011**, *19*, 187–192.

23. Maier, H.J.; Friebel, H.U.; Frischke, D.; Grossmann, R. State of the art of high vacuum sputter deposition of nuclear accelerator targets. *Nucl. Instrum. Methods Phys. Res. Sect. Accel. Spectrometers Detect. Assoc. Equip.* **1993**, *334*, 137–141. [CrossRef]

24. Folger, H.; Klemm, J.; Muller, M. Preparation of Nuclear Accelerator Targets by Focused Ion Beam Sputter Deposition. *IEEE Trans. Nucl. Sci.* **1983**, *30*, 1568–1572. [CrossRef]

25. Esposito, J.; Bettoni, D.; Boschi, A.; Calderolla, M.; Cisternino, S.; Fiorentini, G.; Keppel, G.; Martini, P.; Maggiore, M.; Mou, L.; et al. LARAMED: A Laboratory for Radioisotopes of Medical Interest. *Molecules* **2019**, *24*, 20. [CrossRef]

26. Esposito, J.; Vecchi, G.; Pupillo, G.; Taibi, A.; Uccelli, L.; Boschi, A.; Gambaccini, M. Evaluation of Mo 99 and Tc 99 m Productions Based on a High-Performance Cyclotron. *Sci. Technol. Nucl. Install.* **2013**, *2013*, 1–14. [CrossRef]

27. Boschi, A.; Martini, P.; Pasquali, M.; Uccelli, L. Recent achievements in Tc-99m radiopharmaceutical direct production by medical cyclotrons. *Drug Dev. Ind. Pharm.* **2017**, *43*, 1402–1412. [CrossRef]

28. Martini, P.; Boschi, A.; Cicoria, G.; Zagni, F.; Corazza, A.; Uccelli, L.; Pasquali, M.; Pupillo, G.; Marengo, M.; Loriggiola, M.; et al. In-house cyclotron production of high-purity Tc-99m and Tc-99m radiopharmaceuticals. *Appl. Radiat. Isot.* **2018**, *139*, 325–331. [CrossRef]

29. Uzunov, N.M.; Melendez-Alafort, L.; Bello, M.; Cicoria, G.; Zagni, F.; De Nardo, L.; Selva, A.; Mou, L.; Rossi-Alvarez, C.; Pupillo, G.; et al. Radioisotopic purity and imaging properties of cyclotron-produced $^{99m}$Tc using direct $^{100}$Mo($p$, 2 $n$) reaction. *Phys. Med. Biol.* **2018**, *63*, 185021. [CrossRef]

30. Martini, P.; Boschi, A.; Cicoria, G.; Uccelli, L.; Pasquali, M.; Duatti, A.; Pupillo, G.; Marengo, M.; Loriggiola, M.; Esposito, J. A solvent-extraction module for cyclotron production of high-purity technetium-99m. *Appl. Radiat. Isot.* **2016**, *118*, 302–307. [CrossRef]

31. van de Watering, F.C.J.; Rijpkema, M.; Perk, L.; Brinkmann, U.; Oyen, W.J.G.; Boerman, O.C. Zirconium-89 Labeled Antibodies: A New Tool for Molecular Imaging in Cancer Patients. *BioMed Res. Int.* **2014**, *2014*, 1–13. [CrossRef]

32. Jalilian, A.R.; Osso, J.A. Production, applications and status of zirconium-89 immunoPET agents. *J. Radioanal. Nucl. Chem.* **2017**, *314*, 7–21. [CrossRef]

33. Kasbollah, A.; Eu, P.; Cowell, S.; Deb, P. Review on Production of 89Zr in a Medical Cyclotron for PET Radiopharmaceuticals. *J. Nucl. Med. Technol.* **2013**, *41*, 35–41. [CrossRef]

34. Severin, G.W.; Engle, J.W.; Nickles, R.J.; Barnhart, T.E. 89Zr Radiochemistry for PET. *Med. Chem. Shariqah United Arab Emir.* **2011**, *7*, 389–394.

35. Thornton, J.A. Influence of apparatus geometry and deposition conditions on the structure and topography of thick sputtered coatings. *J. Vac. Sci. Technol.* **1974**, *11*, 666–670. [CrossRef]

36. Hoffman, D.W.; Thornton, J.A. Internal stresses in sputtered chromium. *Thin Solid Films* **1977**, *40*, 355–363. [CrossRef]

37. Cicoria, G.; Pancaldi, D.; Piancastelli, L.; Giovaniello, G.; Bianconi, D.; Bollini, D.; Menapace, E.; Givollani, S.; Pettinato, C.; Spinelli, A.; et al. Marengo Development and Operational Test of a Solid Target for the Pettrace Cyclotron. In Proceedings of the 13th European Symposium on Radiopharmacy and Radiopharamceuticals, Lucca, Italy, 30 March 30–2 April 2006; pp. 3–4.

38. Ciarmatori, A.; Cicoria, G.; Pancaldi, D.; Infantino, A.; Boschi, S.; Fanti, S.; Marengo, M. Some experimental studies on $^{89}$Zr production. *Radiochim. Acta* **2011**, *99*, 631–634. [CrossRef]

39. Radionuclide Yield Calculator - ARRONAX. Available online: http://www.cyclotron-nantes.fr/spip.php?article373 (accessed on 12 March 2019).

40. James Ziegler - SRIM & TRIM. Available online: http://www.srim.org/ (accessed on 12 March 2019).

41. EXFOR: Experimental Nuclear Reaction Data. Available online: https://www.nndc.bnl.gov/exfor/exfor.htm (accessed on 12 March 2019).

42. Omara, H.M.; Hassan, K.F.; Kandil, S.A.; Hegazy, F.E.; Saleh, Z.A. Proton induced reactions on 89Y with particular reference to the production of the medically interesting radionuclide 89Zr. *Radiochim. Acta* **2009**, *97*. [CrossRef]

43. Satheesh, B.; Musthafa, M.M.; Singh, B.P.; Prasad, R. NUCLEAR ISOMERS $^{90m, g}$ Zr, $^{89m, g}$ Zr, $^{89m, g}$ Y AND $^{85m, g}$ Sr FORMED BY BOMBARDMENT OF $^{89}$ Y WITH PROTONS OF ENERGIES FROM 4 TO 40 MeV. *Int. J. Mod. Phys. E* **2011**, *20*, 2119–2131. [CrossRef]

44. Wenrong, Z.; Qingbiao, S.; Hanlin, L.; Weixiang, Y. Investigation of 89Y(p,n)89Zr,89Y(p,2n)88Zr and 89Y(p,pn)88Y reactions up to 22 MeV. *Chin. J. Nucl. Phys.* **1992**, *14*, 7–14.
45. Otuka, N.; Takács, S. Definitions of radioisotope thick target yields. *Radiochim. Acta* **2015**, *103*, 1–6. [CrossRef]
46. Detor, A.J.; Hodge, A.M.; Chason, E.; Wang, Y.; Xu, H.; Conyers, M.; Nikroo, A.; Hamza, A. Stress and microstructure evolution in thick sputtered films. *Acta Mater.* **2009**, *57*, 2055–2065. [CrossRef]
47. Vink, T.J.; Somers, M.A.J.; Daams, J.L.C.; Dirks, A.G. Stress, strain, and microstructure of sputter-deposited Mo thin films. *J. Appl. Phys.* **1991**, *70*, 4301–4308. [CrossRef]
48. Karabacak, T.; Senkevich, J.J.; Wang, G.-C.; Lu, T.-M. Stress Reduction in Sputter Deposited Thin Films Using Physically Self-Assembled Nanostructures as Compliant Layers. *Th Annu. Tech. Conf. Proc.* **2004**, 16.
49. Palmieri, V.; Skliarova, H.; Cisternino, S.; Marengo, M.; Cicoria, G. Method for Obtaining a Solid Target for Radiopharmaceuticals Production. International Patent Application PCT/IB2018/056826, 7 September 2018. National Institute of Nuclear Physics, deposition reference P1183PC00.
50. Dabkowski, A.M.; Paisey, S.J.; Talboys, M.; Marshall, C. Optimization of Cyclotron Production for Radiometal of Zirconium 89. *Acta Phys. Pol. A* **2015**, *127*, 1479–1482. [CrossRef]
51. Wooten, A.; Madrid, E.; Schweitzer, G.; Lawrence, L.; Mebrahtu, E.; Lewis, B.; Lapi, S. Routine Production of 89Zr Using an Automated Module. *Appl. Sci.* **2013**, *3*, 593–613. [CrossRef]
52. Lin, M.; Mukhopadhyay, U.; Waligorski, G.J.; Balatoni, J.A.; González-Lepera, C. Semi-automated production of 89 Zr-oxalate/ 89 Zr-chloride and the potential of 89 Zr-chloride in radiopharmaceutical compounding. *Appl. Radiat. Isot.* **2016**, *107*, 317–322. [CrossRef]
53. Poniger, S.; Tochon-Danguy, H.; Panopoulos, H.; Scott, A. Fully Automated Production of Zr-89 using IBA Nirta and Pinctada Systems. Available online: http://hzdr.qucosa.de/api/qucosa%3A22272/attachment/ATT-0/ (accessed on 12 March 2019).

![instruments logo] *instruments*

MDPI

*Communication*

# Vortex Target: A New Design for a Powder-in-Gas Target for Large-Scale Radionuclide Production

**Gerrie Lange**

GE Healthcare, Cygne Centre, De Rondom 8, 5612 AZ Eindhoven, The Netherlands; gerrie.lange@ge.com;
Tel.: +31-6-53796022

Received: 29 December 2018; Accepted: 31 March 2019; Published: 3 April 2019

**Abstract:** This paper presents a design and working principle for a combined powder-in-gas target. The excellent surface-to-volume ratio of micrometer-sized powder particles injected into a forced carrier-gas-driven environment provides optimal beam power-induced heat relief. Finely dispersed powders can be controlled by a combined pump-driven inward-spiraling gas flow and a fan structure in the center. Known proton-induced nuclear reactions on isotopically enriched materials such as $^{68}$Zn and $^{100}$Mo were taken into account to be conceptually remodeled as a powder-in-gas target assembly, which was compared to thick target designs. The small irradiation chambers that were modeled in our studies for powdery 'thick' targets with a mass thickness (g/cm$^2$) comparable to $^{68}$Zn and $^{100}$Mo resulted in the need to load 2.5 and 12.6 grams of the isotopically enriched target material, respectively, into a convective 7-bar pressured helium cooling circuit for irradiation, with ion currents and entrance energies of 0.8 (13 MeV) and 2 mA (20 MeV), respectively. Current densities of ~2 µA/mm$^2$ (20 MeV), induces power loads of up to 4 kW/cm$^2$. Moreover, the design work showed that this powder-in-gas target concept could potentially be applied to other radionuclide production routes that involve powdery starting materials. Although the modeling work showed good convective heat relief expectations for micrometer-sized powder, more detailed mathematical investigation on the powder-in-gas target restrictions, electrostatic behavior, and erosion effects during irradiation are required for developing a real prototype assembly.

**Keywords:** cyclotron; powder target; thermal study; vortex target; Gallium-68; Technetium-99m

---

## 1. Introduction

Across the world, hundreds of cyclotrons with beam energies of 13 MeV and higher are applied for radionuclide production [1–3]. In the last decade, the cyclotron-based radionuclide production of $^{68}$Ga and $^{99m}$Tc has gained particular interest owing to the growing demand for $^{68}$Ga, and the expected shortages of the most widely used radionuclide, $^{99m}$Tc, obtained from $^{99}$Mo/$^{99m}$Tc generators. The cyclotron-based production of $^{99m}$Tc is an emerging technology that serves as an alternative to reactor-produced $^{99}$Mo. For the Netherlands (population 18 million), the total daily $^{99m}$Tc demand corresponds to 20 MeV proton irradiations of 12,000 µAh in 6-hour-run batches each day. This daily demand can be covered by one or two cyclotrons of 2 mA current. Growing demands for the radioisotope $^{68}$Ga for positron emission tomography can be met by an improved design of the $^{68}$Zn production target.

GE Healthcare uses an IBA Cyclone 30 cyclotron for the single-batch radionuclide production of over 1 TBq of $^{18}$F and $^{123}$I per day by irradiating isotopically enriched target materials [$^{18}$O]-water and $^{124}$Xe using beam currents of 180 µA and 300 µA, respectively [4,5].

A study at activated entrance windows by GE Healthcare and the University of Technology Eindhoven showed acceptable areal beam intensities up to 5 µA/mm$^2$. Correspondence with the cyclotron vendor confirmed that 2.0 milliamp cyclotrons are conceivable [5].

This study establishes a conceptual framework for a powder-in-gas target design with examples. Generally, beam power dissipation causes heat transfer challenges, and thus, the production capacity using a thick solid target is often limited by heat removal restrictions. The (technical) limitations are related to thermal properties, such as the target materials' conductivity and the heat transfer capabilities of the assembly. Prior to irradiation, preparation of target materials and the manufacturing processes of solid targets require pelletizing, sintering, and (multiple) Hydrogen gas reducing steps [1–3].

In this paper, the feasibility of the powder-in-gas target concept (vortex design) with a finely dispersed powder accumulated in an irradiation chamber is discussed.

## 2. Target Design

### 2.1. Vortex Target Design

The assembly design proposed here is based on an inert gas closed-loop circuit for removing the heat induced by the hitting beam to a secondary cooling water circuit. Micrometer-sized powder particles injected into such a gas circuit accumulate inside the circular arranged blade configuration, as indicated by the orange/red zone in Figure 1a. The purpose of the blades and the fan structure is to control the cylindrical-shaped area, where both the centrifugal and inward-directed drag and buoyant forces on the powder are balanced. The blade's front and end (Figure 1a) are conical and inwardly directed to establish small inward axial-directed particle drift. Thus, the conical sections prevent powder accumulation outside the orange/red zone. Figure 1b shows a radial cross-sectional view (A–A) half-way through the powder layer. The corresponding radial profile diagram indicates the angular and tangential gas velocities ($\Omega_{gas}$ and $v_{g.tan}$). The tangential gas velocity $v_{g.tan}$ increased from radial point 1 to a maximum indicated value at radial point 3, which is close to the fin tips of the elongated centered fan structure. The enforced gas spinning (region inside radial points 2 and 3, Figure 1b) and subsequently enhanced centrifugal force lead to the continued accumulation of powder particles in the indicated cylindrical orange/red zone with a length denoted by $L_{layer}$. The product of the dispersed powder density and length $L_{layer}$ is equal to the powder-in-gas mass thickness (i.e., g/cm$^2$) and must correspond to the thick target values.

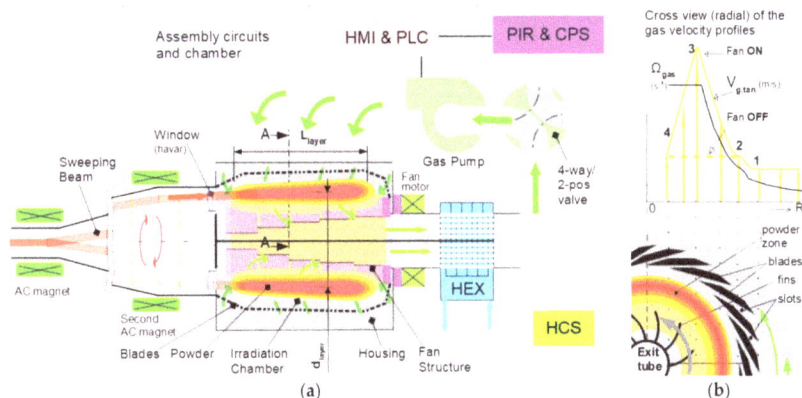

**Figure 1.** (**a**) Sketch of the gas circulation system setup with the beam guiding system, including a dual AC magnet configuration. The volume region in which the powder accumulates is shown in the irradiation chamber. (**b**) Cross-section (A–A) and diagram of the angular and tangential velocities of the gas rotation vs. radius. The numbers in the diagram indicate the velocity profile changes versus radius.

Figure 1a shows the gas circuit components and equipment for process control, including the central parts: cochlea (housing); window section; irradiation chamber with the blades' configuration; and elongated fan structure, which is driven by a magnet-coupled fan motor. The gas pump, shown in green in Figure 1a, generates an inert gas flow that passes the blades and spirals strongly inward to the centered exit tube inside the fan structure.

Other equipment labeled in the figure include:

- Heat exchanger (HEX) for heat removal by the secondary cooling water circuit;
- Helium control system (HCS) for the regulation of the circuit pressure and gas temperatures;
- Four-way/two-position valve for loading and emptying the irradiation chamber;
- Powder injection and recovery (PIR) system and Chemical Processing System (CPS); and
- Process control (PLC) and operator panel (HMI).

Prior to the operation, residual gasses in the target chamber are evacuated. Subsequently, the chamber and circuit are helium pressurized, and the gas pump and the central fan structure are turned on. As a result of the 4w-valve operation, the injected powder is dispersed into the circuit and accumulates in the (indicated orange/red) cylindrical zone.

Subsequently after irradiation, the powder circulating in the irradiation chamber can be scavenged in the PIR system using the 4w-valve operation and reduction of the fan's spinning structure while the gas pump continues inert gas circulation. Reducing the fan's spinning frequency leads to a reduction in centrifugal forces operating on the powder's particles, resulting in a further inward and exiting powder transfer. The decrease in the tangential gas velocity is indicated in Figure 1b by the shift of point 3 to the dashed line level.

## 2.2. Principle of the Vortex

When the operation begins, the powder is dispersed and brought by the carrier gas into the indicated orange/red zone. In this zone, a balance of all forces in the radial direction must be achieved. The carrier gas rotation induces a centrifugal, a buoyant, and a drag force on the particles. The drag force is related to the particle's drift velocity relative to the gas.

For the particles present in the balanced zone, as illustrated in Figure 2, radial forces are expressed by Equation (1):

$$F_{cen} = F_{drag} + F_{buo} \tag{1}$$

where $F_{cen}$ is the centrifugal force, $F_{drag}$ is the drag force, and $F_{buo}$ is the buoyant force.

The main expression for $F_{cen}$ is:

$$F_{cen} = \frac{2 \cdot m_p \cdot v_{g.tan}^2}{d_{layer}} \tag{2}$$

where $m_p$ is the particle's mass (kg), $d_{layer}$ is the average powder layer diameter, and $v_{g.tan}$ is the entering gas velocity equal to the quotient of gas volume flow rate and cross-section of gas inlet.

Calculating drag force $F_{drag}$—on the expected micrometer-sized particles moving in viscous gas—can be described by Stokes' law, which is accurate in a gaseous environment with a Reynolds number of Re ≤ 0.1. For particles having Reynolds numbers of Re ≤ 1.0, Stokes' law remains a proper approximation [6]. Preliminary calculations showed that the range of interest for the particle's size was smaller than 10 μm, while the differential or relative velocities to the carrier gas were expected to be ~1.0 m/s. The Reynolds number verification was carried out for circulating helium gas in the irradiation chamber at a gas density of $\rho_g$ = 1.25 kg/m³ (≈ 7E + 05 Pa, 300 K) and a dynamic viscosity of μ = 2.1E − 05 Pa·s. Particle calculations in the expected ranges of size and velocity ($d_p$ < 10 μm, $v_{p.rel}$ ≈ 1.0 m/s) by Equation (3),

$$Re_p = \frac{\rho_g \cdot d_p \cdot v_{p.rel}}{\mu}, \tag{3}$$

resulted in Reynolds numbers of $Re_p$ < 0.5. Herein, μ is the dynamic viscosity of the carrier gas, $d_p$ is the particle diameter (m), $\rho_g$ is the gas density (kg/m³), and $v_{p.rel}$ is the differential velocity of the particles relative to the gas. The dynamic viscosity's temperature dependency was investigated and determined to be of minor significance for this study.

The drag formula for low differential velocities is expressed by:

$$F_{drag} = 3 \cdot \pi \cdot \mu \cdot f_{eff} \cdot d_p \cdot v_{p.rel} \qquad (4)$$

where $f_{eff}$ is a factor for irregular particle surface condition.

The next equation shows the balance of buoyant force $F_{buo}$ and drag force $F_{drag}$ equal to the centrifugal force $F_{cen}$ by:

$$\frac{\pi \cdot d_p^{3} \cdot \rho_g \cdot v_{g.tan}^{2}}{3 \cdot d_{layer}} + 3 \cdot \pi \cdot \mu \cdot f_{eff} \cdot d_p \cdot v_{p.rel} = \frac{\pi \cdot d_p^{3} \cdot \rho_p \cdot f_p \cdot v_{g.tan}^{2}}{3 \cdot d_{layer}} \qquad (5)$$

where $d_{layer}$ is the average diameter of the intended powder zone inside the blades, and $v_{g.tan}$ is the tangential gas velocity. The expressions $f_p$ and $f_{eff}$ are correction factors for the particle's density and surface roughness, respectively. The variables $\rho_g$ and $\rho_p$ are the densities (kg/m$^3$) of gas and particles, respectively.

The extraction of the particle's velocity relative to the gas results in Equation (6):

$$v_{p.rel} = \frac{d_p^{2} \cdot v_{g.tan}^{2} \cdot (\rho_p \cdot f_p - \rho_g)}{9 \cdot \mu \cdot f_{eff} \cdot d_{dlayer}} \qquad (6)$$

Of course, Equation (6) can be used for areas other than the balanced zone by redefining the quantity $d_{layer}$ by a new expression for the diameter or twice the radius.

Particles not exceeding a certain size or diameter will be transferred inward by the carrier gas, as indicated by the small brown radial resulting velocity vector $v_{p.rad.res}$ (Figure 2, #1).

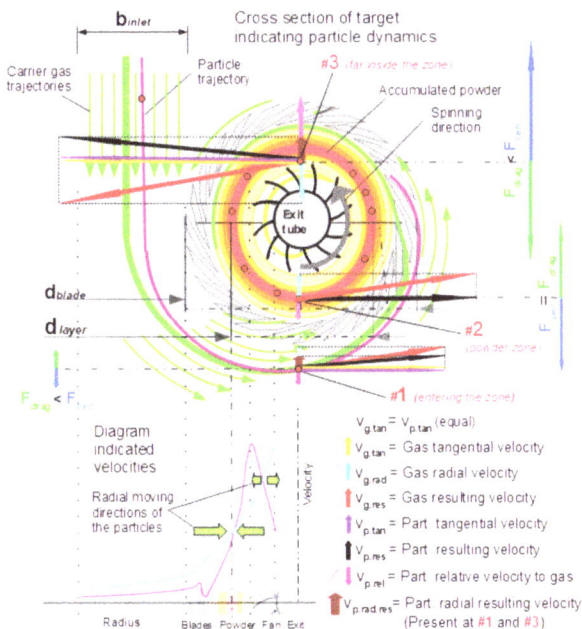

**Figure 2.** Illustrations of the gas trajectories by (curved) green arrows of equal lengths in the housing. The three positions labeled #1, #2, and #3 describe the locations that are outside, inside, and "too far" inside the powder presence zone. The diagram shows that the radius depends on velocity trends for the gas and the particles. Obviously, the particle relative velocity (pink line) rises beyond the radial gas velocity (blue line).

The inward-spiraling carrier gas has a radial velocity component $v_{g.rad}$ (light blue arrow) that is related to gas velocity $v_{g.tan}$, quotient of the indicated housing dimensions $b_{inlet}$, and the circumference of the blade's configuration in the cochlea (Figure 2, #1). Position #2 indicates the location of equal forces without the presence of a brown radial resulting velocity vector $v_{p.rad.res}$. Enhanced gas spinning far inside the zone (position #3) results in a velocity vector $v_{p.rad.res}$ directed outward. Green block arrows in the diagram show the particles' radial moving directions.

Density of the powder material is related to its porosity. Therefore, the factor $f_p$ is defined and estimated to be between 0.5 and 0.9. The shape and (irregular) surface finish correction factor $f_{eff}$ represents the multiplying factor for the diameter $d_p$. The factor $f_{eff}$ varies from 1.0 for a smooth surface to higher values for irregular surfaces. The size, density, and surface finish are surely affected by the preparation of the powder. A preparation procedure must be developed to determine the optimal range of the powder's size that can be applied for injection.

The purpose of carrier gas circuits is to transfer the dissipated heat induced by the beam outside the target system. When the maximum allowed temperature rise, $\Delta T_{gas}$, and circuit pressure of the carrier gas are defined for an expected beam power load $\dot{Q}_{tot}$ (Watt), the mass flow rate $\dot{m}_g$ (kg/s) and volume flow rate $\dot{V}_{fl}$ (m$^3$/s) of the gas cooling can be calculated.

### 2.3. Stopping Power, Ranges, and Beam Features

To model the system, two radionuclide production routes, $^{68}$Zn(p,n)$^{68}$Ga and $^{100}$Mo(p,2n)$^{99m}$Tc, were taken into consideration. For determining the design, the significant factors are the accelerated ion energy at the entrance, the projected ranges related to the electronic and nuclear (minor) stopping powers, and the lower threshold energy for the considered nuclear reaction [7,8].

The ion beam, which enters the assembly nearly parallel to the central assembly's symmetry axis, is intended to pass the full length $L_{layer}$ of the mixed powder-in-gas layer. The powder-in-gas mass thickness (i.e., g/cm$^2$) should correspond closely to the values of known thick targets. The rest of the energy from the ion beam, assumed to be less than the specific nuclear reaction threshold energy, dissipates at the end section of the blades. The ion energy losses due to scattering at the entrance window, the carrier gas, mixed powder-in-gas layer, and blades are defined by Equation (7):

$$E_{ion} = \Delta E_{havar} + \Delta E_{gas.1} + (\Delta E_{mat} + \Delta E_{gas.2})_{mixed} + \Delta E_{blade} \qquad (7)$$

where $\Delta E_{havar}$ and $\Delta E_{gas.1}$ are the energy losses in the window and the first section of the carrier gas, respectively. Generally, beam scattering and energy loss $\Delta E_{gas.1}$ in the carrier gas are expected to be minor. The expression $(\Delta E_{mat} + \Delta E_{gas.2})_{mixed}$ is the ion beam energy loss due to both dispersed powder $\Delta E_{mat}$ and carrier gas $\Delta E_{gas.2}$ in the same volumetric region. The relative contribution of the latter is much lower than the former.

For a technical assessment of the concept, the maximum temperature rise of the powder particles was estimated. Temperature rise depends on the energy level of the local proton beam hitting the particles. For a particle at the start of the layer near the target's entrance, energy loss will be significantly lower than that for a particle at the end of the passed powder-in-gas layer ($L_{layer}$). Otherwise, beam intensity ($\mu$A/mm$^2$) at the entrance is significantly higher compared to when it is further "away" inside the irradiation chamber. To account for worst-case scenario, we calculated the temperature rise of a cylindrical-shaped particle (Figure 3b), with an energy loss at the maximum stopping power at the top of the Bragg peak. Therefore, the beam intensity was calculated at the end of the powder-in-gas layer and supported by ion range and scattering (SRIM) calculations [7–9].

A particle's passage (in a static beam) driven by the tangential velocity $v_{g.tan}$ occurs in a few milliseconds, while particle heating occurs instantly in tens of microseconds. The heating and convective cooling of a particle reach equilibrium at a differential temperature $\Delta T_{tr}$ relative to the gas. The maximum dissipated ion energy $\Delta E_{max}$ in MeV per powder particle (1 eV = 1.602E − 19 J), with the diameter $d_p$ and density $\rho_p$, is by approximation:

$$\Delta E_{max} = SP_{max} \cdot \rho_p \cdot d_p \tag{8}$$

where $SP_{max}$ is the maximum mass stopping power at the Bragg peak in MeV·cm$^2$/gr. The dissipated beam power $\dot{Q}_p$ in a powder particle is:

$$\dot{Q}_p = \pi/4 \cdot d_p^2 \cdot I_{int.static} \cdot \Delta E_{max} \tag{9}$$

where $I_{int.static}$ is the beam intensity in μA/mm$^2$ (or μC/(s·mm$^2$)), which corresponds to the accelerated ion particle's 'flow rate' per square millimeter.

The maximum powder particle differential temperature $\Delta T_{tr.staic}$ relative to the gas is then:

$$\Delta T_{tr.static} = \frac{\dot{Q}_p}{h_{He} \cdot A_p/2} \tag{10}$$

where $A_p$ is the cylindrical particle surface divided by two, given the assumption that only the front-half of the particle's surface is cooled.

Otherwise, when the maximum for the differential temperature $\Delta T_{tr.max}$ is set for the worst-case scenario of energy loss due to the mass stopping power $SP_{max}$ at the "Bragg peak" for a known particle diameter $d_p$ and density $\rho_p$, the maximum allowed beam intensity can be calculated by the following formula:

$$I_{int.static.max} = \frac{3 \cdot h_{He} \cdot \Delta T_{tr.max}}{d_p \cdot \rho_p \cdot SP_{max}} \tag{11}$$

where $I_{int.static.max}$ is the maximum for the static flat-top beam profile, and $h_{He}$ is the heat transfer coefficient (W/m$^2$·K), which is determined by gas flow data and explained in the discussion.

The flat-top beam profile reduces the damaging effects of hotspots on the (2 × 15 μm Havar) windows and allows a higher beam power while keeping the maximum allowed peak current density noted in the introduction (5 μA/mm$^2$) unchanged. Further decrease in beam intensity or a higher allowed total beam power can be established by sweeping the flat-top beam around the assembly's symmetrical center. Sweeping around the center further reduces the window's heat stress as well as the particle heating. Preliminary calculations showed that due to the instantaneous heating of powder particles, a beam sweeping frequency of 1 kHz results in a significant reduction in the particle's differential temperature $\Delta T_{tr}$. Figure 3a shows an impression:

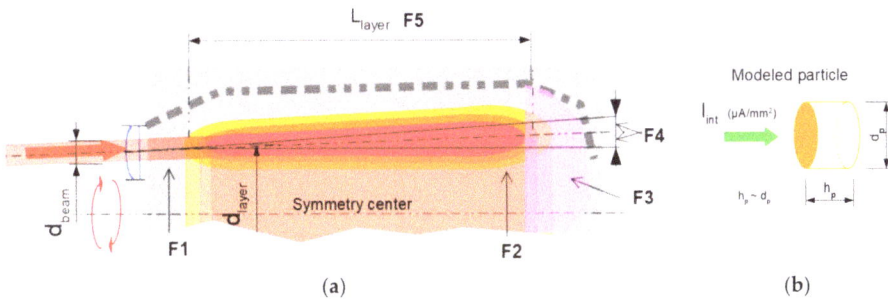

**Figure 3.** (**a**) Cross-sectional view of the powder layer in the irradiation chamber, where F1 is the entering beam at an energy of 14–30 MeV; F2 is the expected beam area at a threshold energy of 4–10 MeV; F3 is the area with the highest stopping power (Bragg Peak); F4 is the scattering-induced lateral range and straggling; and F5 is the length of the powder layer corresponding to thick targets setups. (**b**) Modeled particle.

## 3. Results

To optimize the vortex target design, an iterative modeling process using ion range and scattering (SRIM) data as well as heat transfer calculations was carried out for two nuclear reaction routes [8–10].

Initial data processing involved determining (in the order presented): beam features, ion range and scattering, and particle features and specifications.

Next, particle velocity calculations were performed for the numbered positions indicated in Figure 2 of the following:

- Position #1: the particle in the entering zone and confirmation of the inward-directed transfer;
- Position #2: the particle in the powder zone and confirmation of the balanced particle presence; and
- Position #3: calculation of 'too far' inside the powder zone and confirmation of the outward drifting of the particles.

Confirmation by approximated calculations of particle heating and convective heat relief.

Table 1 shows the input and calculated results of the modeling work of a high-capacity $^{68}$Zn and $^{100}$Mo powder-in-gas target. The beam enters the target chamber almost horizontally in a sweeping mode by the beam guiding system. Beam energy loss due to interaction with the powder layer over the full length is equal to the difference between $E_{ion}$ and $E_{threshold}$. Particle size in the table indicates a certain accuracy range for the operation. Particles outside this range will be transferred to the PIR unit. Total powder mass for 'thick targets' depends on the flat-top ion beam size $d_{beam}$, the nominal diameter of the powder layer $d_{layer}$, powder's density, and beam scatter (Figure 3a).

**Table 1.** Input data and results of $^{68}$Zn and $^{100}$Mo modeling and calculations.

|   | Input data and results | Quantities | Unit | $^{68}$Zn(p,n)$^{68}$Ga | $^{100}$Mo(p,2n)$^{99m}$Tc | Remarks |
|---|---|---|---|---|---|---|
|   | Proton Beam energy | $E_{ion}$ | MeV | 13 | 20 |   |
|   | Beam current | $I_{tar}$ | mA | 0.8 | 2.0 |   |
|   | Threshold energy | $E_{threshold}$ | MeV | 3.8 | 7.8 |   |
|   | Particle size | $d_p$ | μm | 4 ± 1.5 | 3 ± 1 |   |
|   | Target thickness | $L_{matter}$ | mm | 0.38 | 0.60 |   |
|   | Mass thickness (average) | Mass thickness | gr/cm$^2$ | 0.27 | 0.62 |   |
|   | Powder layer length | $L_{layer}$ | mm | 40 | 60 |   |
|   | Diameter powder zone | $d_{layer}$ | mm | 42 | 56 |   |
|   | Beam areal intensity | $I_{sweep}$ | μA/mm$^2$ | 1.35 | 1.81 |   |
|   | Powder's total mass | $m_{matter}$ | gr | 2.5 | 12.6 |   |
| #1 | Particle relative velocity | $v_{p.rel}$ (#1) | m/s | 0.44 | 0.44 |   |
|   | Gas radial velocity | $v_{g.rad}$ (#1) | m/s | 1.69 | 1.82 |   |
|   | Particle incidence angle | $\psi_{fl}$ (#1) | deg | 8.06 | 6.87 | (>0 = Ok) |
| #2 | Particle relative velocity | $v_{p.rel}$ (#2) | m/s | 2.70 | 2.93 |   |
|   | Gas radial velocity | $v_{g.rad}$ (#2) | m/s | 2.71 | 2.91 |   |
|   | Particle incidence angle | $\psi_{fl}$ (#2) | deg | −0.00 | −0.03 | (≈0 = Ok) |
| #3 | Particle relative velocity | $v_{p.rel}$ (#3) | m/s | 8.45 | 10.30 |   |
|   | Gas radial velocity | $v_{g.rad}$ (#3) | m/s | 4.06 | 4.76 |   |
|   | Particle incidence angle | $\psi_{fl}$ (#3) | deg | −9.02 | −7.61 | (<0 = Ok) |
|   | Diff. temp. static beam | $\Delta T_{tr.static}$ | K | 624 | 670 | Both not allowed |
|   | Diff. temp. sweeping beam | $\Delta T_{tr.sweep}$ | K | 270 | 300 | Freq. 1 kHz |
|   | Max. sweeping beam intensity | $I_{int.sweep.max}$ | μA/mm$^2$ | 2.13 | 2.05 |   |

The differential temperatures $\Delta T_{tr.static}$ and $\Delta T_{tr.sweep}$ (1 kHz sweeping beam) were calculated at the maximum mass stopping power (Bragg Peak). The maximum sweeping beam intensity $I_{int.sweep.max}$ at the end of the powder layer was calculated for a scattered beam on a modeled cylindrical particle (Figure 3b). The values for the differential temperature $\Delta T_{tr.static}$ are interpreted as not acceptable.

## 4. Discussion

The design work showed an interesting route toward achieving a powder-in-gas vortex target. However, several processes, such as preventing adverse powder accumulation outside the intended layer, must be investigated in detail. The presence of the fan structure is essential for supporting the layer's stability by the carrier gas's enhanced tangential velocity in the irradiation chamber.

Generally, prior chemical processing results in different shapes, sizes, and surface conditions of particles. An apparatus, possibly identical to a vortex assembly, must be developed for powder particle size selection, and reprocessing powder particles that exceed the desired 'range'.

The heat transfer coefficient $h_{He}$ and gas flow velocity $v_{He}$ at the contact surface were examined in this study using the pressure and gas flow data of existing and designed assemblies. The heat transfer coefficient has close to a linear dependency on the circuit pressure $P_{cir}$ and gas velocity $v_{He}$. An empirical expression $\xi_{He}$ was determined and is shown in the following formula:

$$h_{He} = \xi_{He} \cdot P_{cir} \cdot v_{He} \tag{12}$$

where $\xi_{He}$ is defined as the 'heat transfer constant' $\xi_{He}$ between $1 \times 10^{-4}$ and $4 \times 10^{-4}$ W·s/(Pa·K·m$^3$). Further investigation of the basic heat transfer coefficient is recommended to determine helium flow angle dependency relative to the (particle's) surface.

Each powder has a certain level of hardness. This potentially induces erosion effects on the blades and other components. Candidate materials for internal structures must have excellent properties for thermal and electrical conductivity, negligible long-term radio-activation profiles, and significant chemical differences from the produced radionuclide.

When a beam passes the target containing the powder and carrier gas, they become partly ionized. The gas will contain a dilute plasma of free electrons and positive ions determined by a balance between beam-induced ionization and recombination processes. The powder particles become positively charged primarily by secondary electron emission induced by the impinging accelerated ions. The target's design is intended to cause particles, when attracted to the blades, to discharge and re-enter the powder layer by the intensive tangential gas flow. Further research into the particle's behavior during irradiation conditions is addressed in Section 5. Further, quantities such as the net ion current, as well as differential gas circuit pressures and various temperature positions in the assembly's structure, must be monitored.

Our calculations showed that the instantaneous heating of particles occurs in tens of microseconds during beam passage. The average differential temperatures, as shown in Table 1, were calculated for a beam sweeping frequency of 1 kHz. When the beam sweeps in the opposite direction relative to the powder rotation, the particles' beam passage, heating, and cooling occur in tens of microseconds. Further increasing this frequency leads to more averaging of particle heating, resulting in reduced temperature of individual particles.

## 5. Conclusions

This study suggests that the excellent surface-to-volume ratio of micrometer-sized particles in a carrier-gas-driven environment leads to optimal heat relief. Gas flow induced by the gas pump transfers the powder inside the blade's structure. Inside, the fan's structure maintains the accumulated powder near the blade's structure via enhanced tangential gas velocity.

The advantages of the powder-in-gas target design are:

- Shortened production cycles when higher beam intensities are applied;
- Expected shortened target material preparation procedures; and
- Faster recovery, dilution procedures, and reprocessing to powdery material.

Assemblies for nuclear reactions on $^{68}$Zn and $^{100}$Mo materials were modeled as a powder-in-gas target. The conceptual design of small-diameter chambers for $^{68}$Zn and $^{100}$Mo revealed the need for 2.5 and 12.6 grams of material, respectively, for a beam current operation of 0.8 and 2 mA, respectively. Calculations were carried out for the worst-case scenario of particle heating. Beam current densities were calculated close to 2 µA/mm$^2$ and corresponded to energy-dependent power loads of ~4 kW/cm$^2$.

Further detailed examinations are required to:

- Determine the optimal cochlea, blade shape, and fan structures;
- Reduce and control erosion effects;
- Optimize the preparation of powder and its injection into the target system;
- Determine the tendency of powder to form dendrites or small deposits at the blades;
- Avoid agglomeration of sponged particle formation due to discharging effects;
- Control particle drift to the blade's surfaces by, for instance, (proton) current measurement; and
- Determine the effect of applying a (positive) electrical voltage to the fan structure and control powder's behavior possibilities during irradiation.

Summarizing the conceptual design, the nuclear reactions: $^{68}Zn(p,n)^{68}Ga$ and $^{99}Mo(p,2n)^{99m}Tc$, could be good candidates for large-scale production because of the broad interest and their use across the world. This study shows that the concept might be also applicable to other production routes.

**Funding:** This research received no external funding.

**Conflicts of Interest:** The author declares no conflict of interest.

## References

1. Schaffer, P.; Benard, F.; Bernstein, A.; Buckley, K.; Celler, A.; Cockburn, N.; Corsaut, J.; Dodd, M.; Economou, C.; Eriksson, T.; et al. Direct Production of 99mTc via 100Mo(p,2n) on Small Medical Cyclotrons. *Phys. Procedia* **2015**, *66*, 383–395. Available online: https://www.triumf.ca/sites/default/files/20140815_Direct_Production_of_Tc99monSmallMedicalCyclotrons_0.pdf (accessed on 20 November 2018). [CrossRef]
2. IAEA. Cyclotron Based Production of Tecgnetium-99m. Available online: https://www.iaea.org/publications/10990/cyclotron-based-production-of-technetium-99m (accessed on 15 December 2018).
3. Sklairova, H.; Cisternino, S.; Cicoria, G.; Marengo, M.; Palmieri, V. Innovative Target for Production of Technetium-99m by Biomedical Cyclotron. *Molecules* **2019**, *24*, 25. Available online: https://www.mdpi.com/1420-3049/24/1/25 (accessed on 26 February 2019). [CrossRef] [PubMed]
4. University of Technology Eindhoven. Available online: http://www.acctec.nl/acctec.nl/products/wobbling-system.html (accessed on 1 November 2018).
5. IBA. Available online: http://www.iba-radiopharmasolutions.com/products/cyclotrons (accessed on 15 December 2018).
6. Veldman, A.E.P.; Velicka, A. University Groningen; Report; Fluid Dynamics ('Stromingsleer'). Available online: http://www.math.rug.nl/~{}veldman/Colleges/stromingsleer/Stromingsleer1011.pdf (accessed on 18 November 2018).
7. Zeisler, S.; Limoges, A.; Kumlin, J.; Siikanen, J.; Hoehr, C. Fused Zinc Target for the Production of Gallium Radioisotopes. *Instruments* **2019**, *3*, 10. [CrossRef]
8. National Institute of Standards and Technology. Available online: https://physics.nist.gov/PhysRefData/Star/Text/intro.html (accessed on 10 November 2018).
9. Ziegler, J.F.; Ziegler, M.D.; Biersack, J.P. SRIM—The Stopping and Range of Ions in Matter. Available online: https://www.srim.orgZinc (accessed on 15 February 2018).
10. Engineering Tool Box. Available online: https://www.engineeringtoolbox.com/gases-absolute-dynamic-viscosity-d1888.html (accessed on 16 November 2018).

![instruments logo] *instruments*

MDPI

*Article*

# Automated Purification of Radiometals Produced by Liquid Targets

Vítor H. Alves [1,*], Sérgio J. C. do Carmo [1], Francisco Alves [2] and Antero J. Abrunhosa [2]

[1]  ICNAS—Produção, University of Coimbra, Pólo das Ciências da Saúde, Azinhaga de Santa Comba, 3000-548 Coimbra, Portugal; sergiocarmo@uc.pt
[2]  ICNAS, University of Coimbra, Pólo das Ciências da Saúde, Azinhaga de Santa Comba, 3000-548 Coimbra, Portugal; franciscoalves@uc.pt (F.A.); antero@pet.uc.pt (A.J.A.)
*   Correspondence: vitoralves@uc.pt; Tel.: +351-239-488-510

Received: 13 August 2018; Accepted: 12 September 2018; Published: 14 September 2018

**Abstract:** An automated process for the production and purification of radiometals produced by irradiating liquid targets in a medical cyclotron, using a commercially available module, has been developed. The method is suitable for the production and purification of radiometals such as $^{68}$Ga, $^{64}$Cu and $^{61}$Cu through irradiation of liquid targets and is important for producing high specific activity radioisotopes with a substantial reduction in processing time and cost when compared with the solid target approach. The "liquid target" process also eliminates the need for pre- and post-irradiation target preparation and simplifies the transfer of irradiated material from target to hotcell. A $^{68}$GaCl$_3$ solution can be obtained in about 35 min with an average yield of 73.9 ± 6.7% in less than 10 mL of volume. $^{64}$CuCl$_2$ solutions can be obtained with an average yield of 81.2 ± 7.8% in about 1 h of processing time. A dedicated single-use disposable kit is used on a commercial IBA Synthera® extension module.

**Keywords:** liquid targets; medical cyclotron; radiometals; gallium-68; copper-64; copper-61; purification; disposable kit; radiopharmaceuticals

## 1. Introduction

The interest of radiometals in Nuclear Medicine has increased dramatically over the last decade fostered by the successful clinical use of metal-based radiopharmaceuticals in combined targeted diagnosis and therapy (the so-called theragnostic concept) [1–5]. To produce these radiometals, most hospitals would require the purchase of isotope generators, when available, or to make a substantial investment in a medical cyclotron with a solid target system. This is not a trivial option as most cyclotrons typically handle liquid and gas targets only and are used to produce non-metallic isotopes such as $^{18}$F, $^{11}$C and $^{13}$N. Therefore, the possibility to produce metal isotopes using a medical cyclotron without the investment in a solid-target system provides an easy and accessible way to produce these isotopes within a wide range of accelerator facilities [6–11]. Recent developments concerning the production of radiometals using liquid targets have been published by our group [12,13], paving the way for a new, safer and simplified procedure for automated loading and transfer of target solution to an automated chemistry module inside of a shielded hot-cell, and helping compliance with current Good Manufacturing Practices (GMP) regulations [10].

The methods described allow the production of radioisotopes—such as $^{68}$Ga, $^{64}$Cu, $^{61}$Cu and others—through the irradiation of liquid targets, with a substantial reduction in processing time and cost when compared with the solid target approach. The process also eliminates the need for pre- and post-irradiation target preparation and simplifies the transfer of irradiated material from target to hotcell (Figure 1).

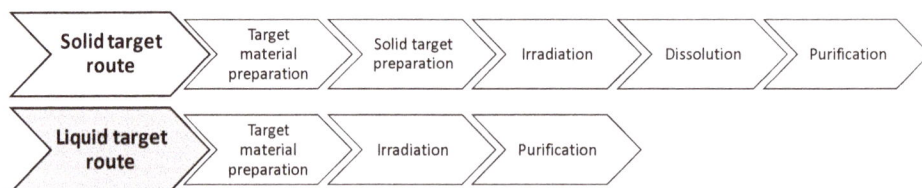

**Figure 1.** Solid and liquid routes to produce radiometals in a medical low energy cyclotron.

Based on the potential of fast and cost-effective production of radiometals in medical cyclotrons, we present a fully automated process, using a commercially available module, for the purification of metal radioisotopes produced by cyclotron irradiation of liquid targets. This work describes the fully automated separation of $^{68}$Ga and $^{64}$Cu or $^{61}$Cu from target material and formulation in a solution for radiolabelling in compliance with European Pharmacopoeia (Ph. Eur.) requirements [14]. The purified chloride solution can be used for labelling molecules using a conventional automated procedure in a reactor vial followed by post purification by a C18 cartridge [15–17] or by means of a cold-kit based method [18–20].

The process described can easily be extended to other metal radioisotopes. Irradiation of liquid targets in medical cyclotrons involves the previous preparation of a target solution containing the enriched (when needed) material in a process that benefits from the high yields provided by the same nuclear reactions used in the solid target while avoiding the inherent limitations of using such targets. In addition to the post-irradiation handling and transport of the solid target to a processing unit (shielded hotcell), such solid targets require a large amount of expensive enriched material (hundreds of mg are necessary) and such a long and complicated process is also associated with inevitable contamination with other metal ions due to the use of higher volumes of strong acids for the dissolution.

## 2. Materials and Methods

All steps required for the production and separation of a metal radioisotope from a liquid target are implemented in a fully integrated system. For each isotope, a dedicated IBA Nirta Conical® target system (IBA, Louvain-la-Neuve, Belgium) is used. To separate the metal isotopes from the target solution and reformulate them in a ready-to-use chloride solution, a commercially available IBA Synthera® Extension module (IBA, Louvain-la-Neuve, Belgium) is used with single-use kits. For the radiolabelling step, an IBA Synthera® Extension is used to label compounds with $^{64}$Cu/$^{61}$Cu (e.g., bis(4-methyl-3-thiosemicarbazone), PTSM; diacetyl-2,3-bis(N4-methyl-3-thiosemicarbazone), ATSM) and an IBA Synthera@ for $^{68}$Ga-based compounds (e.g., 1,4,7,10-tetraazacyclododecane-1,4,7,10-tetraacetic acid (DOTA) peptides, *N,N'*-bis(2-hydroxybenzyl)ethylenediamine-*N,N'*-diacetic acid (HBED) peptides. For metal trace analysis of samples, is used an inductively coupled plasma mass spectrometry (ICP-MS) equipment: Thermo Scientific iCAP Qc (Thermo Fisher Scientific, Waltham, MA, USA). To measure the activities of samples, an ISOMED 2010 (Nuklear-medizintechnik, Dresden, Germany) is used.

All chemicals and solvents used are trace-metal grade.

### 2.1. Targetry/Irradiation

As target material, the enriched isotopes are diluted in a 0.01 M nitric acid solution (Table 1). Concentrations are adjusted to produce a maximum of required activity while avoiding precipitation and providing stability of the solution over time for storage and better behaviour under the cyclotron beam without corrosion of target support materials [6,7].

**Table 1.** Composition of various solutions used.

| Isotope | Target Material | Reaction | Chemical form Solution |
|---|---|---|---|
| Gallium-68 ($^{68}$Ga) | Zinc-68 ($^{68}$Zn)–$^A$E = 99.5% Enrich. | $^{68}$Zn(p,n)$^{68}$Ga | $^{68}$Zn(NO$_3$)$_2$·6H$_2$O |
| Copper-64 ($^{64}$Cu) | Nickel-66 ($^{64}$Ni)–95% Enrich. | $^{64}$Ni(p,n)$^{64}$Cu | $^{64}$Ni(NO$_3$)$_2$·6H$_2$O |
| Copper-61 ($^{61}$Cu) | Natural Zinc ($^{nat}$Zn) | $^{nat}$Zn(p,$\alpha$)$^{61}$Cu | $^{nat}$Zn(NO$_3$)$_2$·6H$_2$O |

$^{64}$Ni and $^{68}$Zn targets are typically irradiated with a beam current of about 70 µA and 45 µA, respectively, using an IBA 18/9 Cyclone cyclotron. The amount of enriched material on target varies from 10–100 mg of $^{64}$Ni and 100–400 mg of $^{68}$Zn, depending on the required activity. After irradiation, solutions are transferred to a processing hot-cell under nitrogen pressure.

### 2.2. Post-Processing

For the Gallium-68 production, the irradiated $^{68}$Zn target solution is dissolved multiple times in water and the solution is passed through a cation exchange resin (SCX; DOWEX 50W, 200–400 mesh, H+ form, treated with 10 mL of 3 M HCl followed by 10 mL of water) loaded on a 1 mL catridge. The cartridge is then washed with 30 mL of Acetone/HBr mixture to remove zinc ions as described by Strelow [21,22]. The adsorbed $^{68}$Ga cations are eluted from the SCX cartridge with 6 mL of HCl 3 M mixed with 10 mL of HCl 30% (to increase the molarity of HCl) to an intermediate reservoir (Figure 2) and passed through an anion exchange resin (SAX; Biorad AG1 100 mesh, treated with 10 mL of water followed by 10 mL of HCl 8 M) loaded on 0.5 mL size-cartridge where the anionic complex $[^{68}$GaCl$_4]^-$ remained strongly adsorbed [23,24]. A flow of inert gas is then applied to dry the column and remove any traces of HCl. Finally, $^{68}$Ga is eluted from the column with water into a final collection vial in the form of $^{68}$GaCl$_3$ solution in 0.1–0.25 M HCl. The $^{68}$Zn ions are collected on a separate vial and can be recycled to be reused as target material.

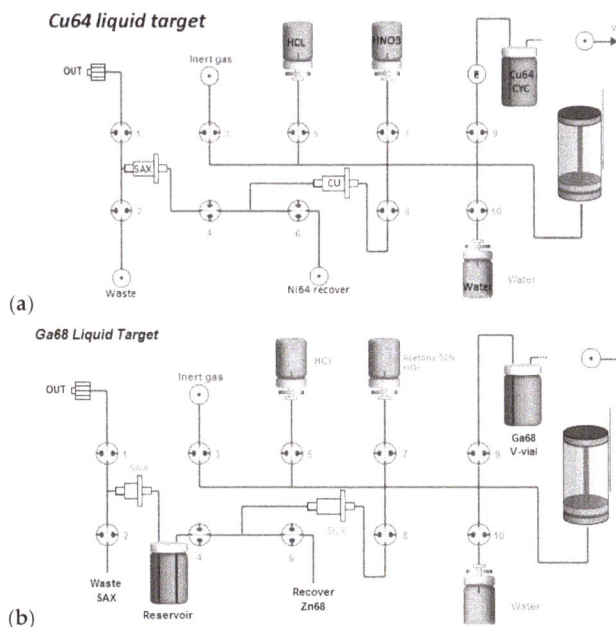

**Figure 2.** Schematic diagram of IBA Synthera® Extension synthesizer software to purify and prepare $^{64}$Cu/$^{61}$Cu-chloride solution (**a**) and to purify and prepare $^{68}$Ga-chloride solution (**b**).

The entire purification process takes about 35 min from end of bombardment (EOB).

Conversely, for the production of copper radioisotopes ([61]Cu and [64]Cu) the irradiated [nat]Zn or [64]Ni liquid target solution is dissolved multiple times in water to bring the pH to a suitable range for the adsorption of the copper ions onto a highly selective Cu resin (TrisKem International, Bruz, France) loaded on 2 mL cartridge, as described by Dirks [25]. The pH adjusted solution is then passed through the resin (pre-conditioned with 10 mL of water) that is then washed with 10 mL of $HNO_3$ 1 mM to remove any traces of non-copper ions. The adsorbed [64]Cu/[61]Cu cations are eluted from the cartridge with 5 mL of HCl 3 M, directly to an anion exchange resin (SAX; TrisKem International, treated with 10 mL of water followed by 10 mL of HCl 8 M) (Figure 2) loaded on 0.5 mL cartridge size where the anionic complex $[^{64}CuCl_4]^-/[^{61}CuCl_4]^-$ remains strongly adsorbed. A flow of inert gas is then applied to dry the column and remove any traces of HCl. Finally, copper is eluted from the column with water into a final collection vial in the form of a copper chloride solution. In the case of [64]Cu production, [64]Ni ions are recovered on a separated vial and can be recovered to be recycled. As for [nat]Zn, there is no need to recover, as natural zinc is quite inexpensive.

The entire purification process takes about 1 h from EOB.

## 2.3. Specific Activity and Trace Metal Analysis

Specific activities (TBq/μg) of [68]Ga and [64]Cu were calculated by measuring the total Ga and Cu present in the final chloride solution after purification using inductively coupled plasma mass spectrometry (ICP-MS). Other metal contaminants including Al, Co, Cu, Ga, Fe, Ni and Zn were also analysed by ICP-MS.

## 3. Results

Figure 3 shows the successful separation of [68]Ga, [61]Cu and [64]Cu from their target nuclides using the methods described. The presented procedure for processing radiometals is able to recover $81.2 \pm 7.8\%$ (n = 10, average of 10 runs) of Copper-64 chloride solution in a small volume (4 mL) using the cartridge-based purification with a disposable kit on a commercial IBA Synthera® extension module. Using an almost identical process, we recovered $73.9 \pm 6.7\%$ (n = 33, average of 33 runs) of Gallium-68 chloride solution in 5–10 mL of volume using the ionic exchange principle applied on the same synthesizer module with a dedicated disposable tubing kit. The efficiency of our separation (Figure 3) is consistent with the previously reported purification yields [7,9,26].

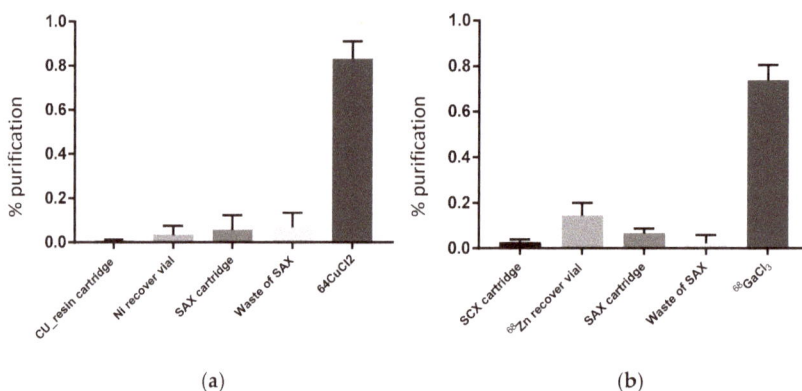

(a)   (b)

**Figure 3.** Average yields and activity loss in each step. Purification of [64]CuCl$_2$ (**a**), with $81.2 \pm 7.8\%$ (decay corrected) yield in 1 h of process. Purification of [68]GaCl$_3$ (**b**), with $73.9 \pm 6.7\%$ (decay corrected) yield in 35 min of process.

Average production yields and respective specific molarities are summarized on Table 2. In Figure 4, the results of ICP-MS analysis for determination of metal impurities in final solutions are presented.

**Table 2.** Purified activities obtained and respective specific activity.

| Isotope | Target Material Amount | Irradiation and Purification Time | Activity at End of Purification (EOP) (GBq) | Specific Activity (TBq/µg) |
|---------|------------------------|-----------------------------------|---------------------------------------------|----------------------------|
| Gallium-68 | 100 mg | 1 h 35 min | 1.5–2.7 | 0.3–24 ($^{68}$Ga/Ga) |
|  | 200 mg | 1 h 35 min | 4.4–5.1 | 0.3–24 ($^{68}$Ga/Ga) |
| Copper-64 | 10–100 mg | 1 h 30 min–9 h 30 min | 0.54–4.6 | 5.0–122.8 ($^{64}$Cu/Cu) |

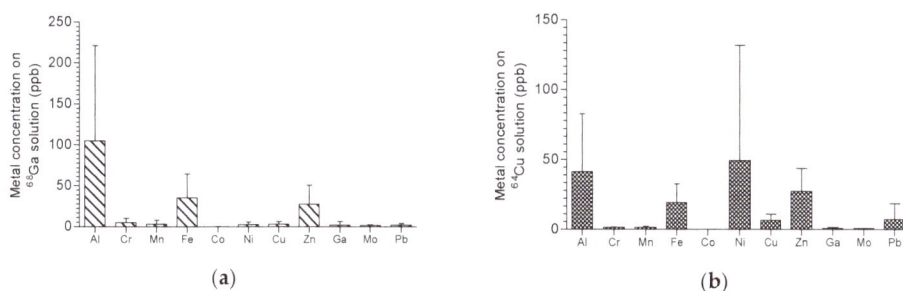

(a)　(b)

**Figure 4.** Concentration of Ga, $^{64}$Cu and other contaminants in the final product solution. Final volume: 10 mL for Ga and 4 mL for Cu. (number of values $n = 25$ for Ga (**a**) and $n = 3$ for Cu (**b**)).

$^{68}$Ga solutions produced were tested for the presence of iron and zinc using ICP-MS. Results are shown in Figure 5 and are in accordance with the Ph. Eur. requirement of a maximum of 10 µg/GBq [14] up to 4 h after the end of purification.

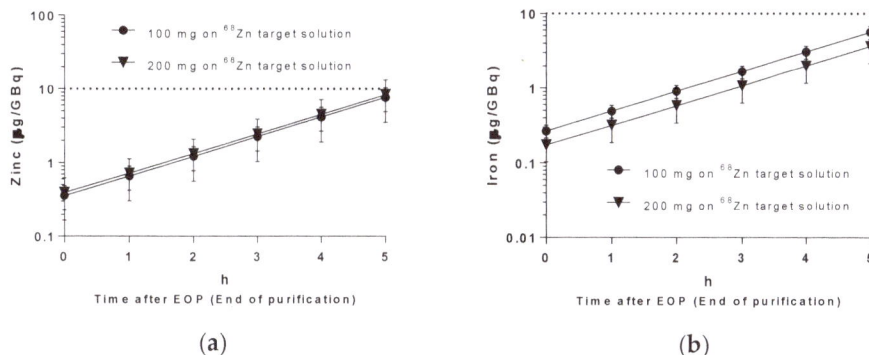

(a)　(b)

**Figure 5.** Chemical purity of final $^{68}$Ga-peptide formulation. The product complies with the Ph. Eur. regarding the maximum amount in µg of Zinc (**a**) and Iron (**b**) per GBq of activity. Values are decay corrected.

An iTLC analysis was made for all chloride solutions to confirm the presence of the ionic forms of the radiometal isotopes and the absence of colloidal complexes (Figure 6).

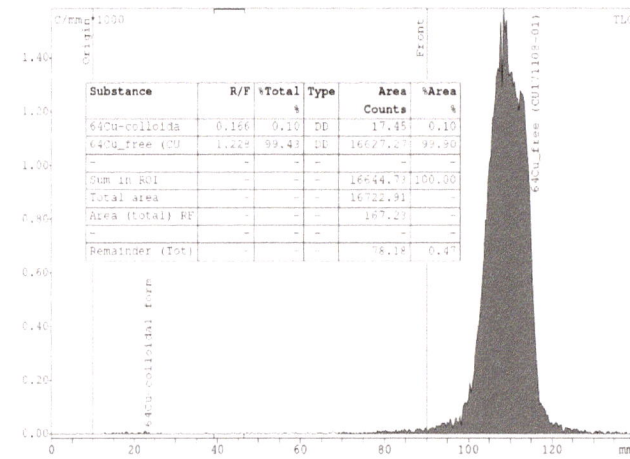

(a)

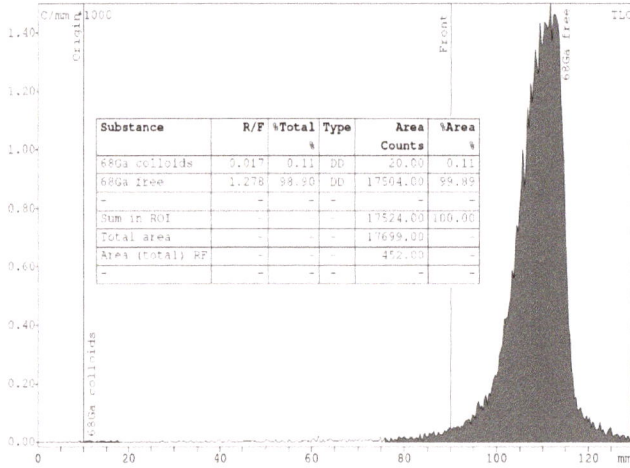

(b)

**Figure 6.** iTLC analysis of purified $^{64}CuCl_2$ (**a**) and $^{68}GaCl_3$ (**b**) using a Raytest miniGita detector. Stationary phase: iTLC-SG strips; mobile phase: 0.1 M sodium citrate (pH adjusted to 4–4.5). Rf = 0.1–0.2 for colloidal form and Rf = 1.0 for free radioisotope.

## 4. Discussion

A complete setup for radiometal production and purification based on the irradiation of a liquid target was implemented using an IBA target and an IBA Synthera® Extension module. Using commercially available disposable kits (Fluidomica, Coimbra, Portugal) the system is able to recover 81.2 ± 7.8% of copper-64 chloride solution for radiolabelling in less than one hour of processing time, and 73.9 ± 6.7% of gallium-68 chloride solution for radiolabelling in less than 35 min processing time. The activity of $^{68}Ga$ lost to zinc-68 recover vial is explained with amount of SCX resin used on first purification step. It was decided to keep this to ensure the proper clean of zinc ions from resin without increasing the volume of washing solution and subsequent increasing of processing time. This is a significant improvement, with less processing time and considerably lower costs, when compared with a conventional solid target system [27] or the published results from other liquid target systems [11,26]. The approach described here enables the production of purified radiometal solutions ready to be used for labelling radiopharmaceuticals for human use with final activities large enough for multiple doses and/or for distribution to other positron emission tomography (PET) facilities.

Specific activities were in the range of 0.3–24 TBq/μg for $^{68}$Ga and 5.0–122.8 TBq/μg for $^{64}$Cu. The presence of metal contaminants, especially iron and zinc, were very low and in compliance with the Ph. Eur. regarding the maximum amount permitted per GBq of activity in the final product vial. As Figure 4 shows, for gallium and copper purification processes, it was found in the final chloride solutions, some traces (part per billion level) of aluminium that was explained by the glass storage container, and iron coming from all the reagents used by the "concentration" effect of process once iron has the same behavior of gallium and will be 'carried' together with to final purified solutions. Zinc presence on gallium-68 and copper-61 chloride solutions can be observed in small concentration at end once are the start target material. The same is said for nickel on copper-64 purification. The zinc present on copper-64 chloride solution was not explained and was assumed as possible contamination from used reagents.

Since the amount of expensive enriched material can be chosen and optimized to suit the requirements for each production, a very substantial cost reduction is achieved when compared to the solid target technique. The purified radiometal solution is ready to be used for radiolabelling in about 30 min for $^{68}$Ga and 1 h for $^{64}$Cu after EOB which is a significant improvement considering the inevitable time-consuming post-irradiation processing associated with the solid target technique.

This improvement is even more important for the case of $^{68}$Ga where the purity of the product is maintained for up to 5 h after EOB. When compared with generator obtained $^{68}$Ga, two major advantages emerged: (1) it is possible to make more consecutive runs as only 1 h 35 min is necessary to produce $^{68}$GaCl$_3$, from the beginning of irradiation till the end of purification, compared with the generators' $^{68}$Ga-grown waiting time and (2) no risk of contamination with long-lived impurities, as occurs with $^{68}$Ge/$^{68}$Ga generators, where there is a significant risk of $^{68}$Ge breakthrough.

## 5. Conclusions

The described process makes feasible the production of metal radioisotopes, such as $^{68}$Ga, $^{64}$Cu and $^{61}$Cu, through the irradiation of a liquid target, using a medical cyclotron, with a considerable reduction in processing time and cost when compared with the traditional solid target approach. The process also eliminates the complex and time-consuming tasks associated with pre- and post-irradiation target preparation and simplifies the transfer of irradiated material from target to hot-cells.

Additionally, the automated process with disposable cassettes reduces radiation exposure to the operator, improves robustness of the production and provides documentation of the manufacturing process that can be used to fulfil GMP requirements.

Considering that virtually all medical cyclotrons installed worldwide are using liquid targets for routine production of PET radiopharmaceuticals, this approach provides an easier and accessible way to produce medical radioisotopes for human use in a wide range of accelerator facilities.

## 6. Patents

EP20150170854. Process for producing gallium-68 through the irradiation of a solution target. (Grant 2017-08-09; Publication 2017-08-09)

US15172905. Process for producing gallium-68 through the irradiation of a solution target. (Pending).

**Author Contributions:** Conceptualization, V.H.A.; Data curation, V.H.A.; Formal analysis, V.H.A.; Investigation, V.H.A.; Methodology, V.H.A.; Project administration, F.A. and A.J.A.; Resources, S.J.C.d.C., F.A. and A.J.A.; Supervision, F.A. and A.J.A.; Validation, F.A. and A.J.A.; Visualization, V.H.A.; Writing—original draft, V.H.A.; Writing—review & editing, F.A. and A.J.A.

**Funding:** This research received no external funding.

**Conflicts of Interest:** The authors declare no conflict of interest.

## References

1. Yordanova, A.; Feldmann, G.; Ahmadzadehfar, H.; Essler, M.; Eppard, E.; Kürpig, S.; Schönberger, S.; Gonzalez-Carmona, M.; Feldmann, G.; Ahmadzadehfar, H.; et al. Theranostics in Nuclear Medicine Practice. *Preprints* **2017**, 2017010094. [CrossRef]

2. Velikyan, I. Prospective of $^{68}$Ga-radiopharmaceutical development. *Theranostics* **2013**, *4*, 47–80. [CrossRef] [PubMed]

3. Velikyan, I. Continued rapid growth in $^{68}$Ga applications: Update 2013 to June 2014. *J. Label. Compd. Radiopharm.* **2015**, *58*, 99–121. [CrossRef] [PubMed]

4. Brasse, D.; Nonat, A. Radiometals: Towards a new success story in nuclear imaging? *Dalt. Trans.* **2015**, *44*, 4845–4858. [CrossRef] [PubMed]

5. Cutler, C.S.; Hennkens, H.M.; Sisay, N.; Huclier-Markai, S.; Jurisson, S.S. Radiometals for combined imaging and therapy. *Chem. Rev.* **2013**, *113*, 858–883. [CrossRef] [PubMed]

6. Do Carmo, S.J.C.; Alves, V.; Alves, F.; Abrunhosa, A.J. Fast and cost-effective cyclotron production of $^{61}$Cu using a $^{nat}$Zn liquid target: An opportunity for radiopharmaceutical production and R&D. *Dalt. Trans.* **2017**, *46*, 14556–14560. [CrossRef]

7. Alves, F.; Alves, V.H.P.; Do Carmo, S.J.C.; Neves, A.C.B.; Silva, M.; Abrunhosa, A.J. Production of copper-64 and gallium-68 with a medical cyclotron using liquid targets. *Mod. Phys. Lett. A* **2017**, *32*, 1740013. [CrossRef]

8. Hoehr, C.; Badesso, B.; Morley, T.; Trinczek, M.; Buckley, K.; Klug, J.; Zeisler, S.; Hanemaayer, V.; Ruth, T.R.; Benard, F.; et al. Producing radiometals in liquid targets: Proof of feasibility with 94mTc. *AIP Conf. Proc.* **2012**, *1509*, 56–60. [CrossRef]

9. Hoehr, C.; Oehlke, E.; Hou, X.; Zeisler, S.; Adam, M.; Ruth, T.; Buckley, K.; Celler, A.; Benard, F.; Schaffer, P. Production of Radiometals in a Liquid Target. Available online: http://hzdr.qucosa.de/api/qucosa%3A22237/attachment/ATT-0/ (accessed on 13 September 2018).

10. Alves, F.; Alves, V.H.; Neves, A.C.B.; do Carmo, S.J.C.; Nactergal, B.; Hellas, V.; Kral, E.; Gonçalves-Gameiro, C.; Abrunhosa, A.J.; Gonçalves-gameiro, C.; et al. Cyclotron production of Ga-68 for human use from liquid targets: From theory to practice. *AIP Conf. Proc.* **2017**, *1845*, 20001–20005. [CrossRef]

11. Oehlke, E.; Hoehr, C.; Hou, X.; Hanemaayer, V.; Zeisler, S.; Adam, M.J.; Ruth, T.J.; Celler, A.; Buckley, K.; Benard, F.; et al. Production of Y-86 and other radiometals for research purposes using a solution target system. *Nucl. Med. Biol.* **2015**, *42*, 842–849. [CrossRef] [PubMed]

12. Abrunhosa, A.; Alves, V.; Alves, F. Process for Producing Gallium-68 Through the Irradiation of a Solution Target 2016. U.S. Patent US15172905, 7 December 2016.

13. Alves, V.H.; Abrunhosa, A.J.; Alves, F. Process for Producing Gallium-68 Through the Irradiation of a Solution Target 2015. U.S. Patent EP20150170854, 5 June 2015.

14. European Pharmacopeia. Gallium ($^{68}$Ga) Chloride Solution for Radiolabelling. Available online: https://www.edqm.eu/sites/default/files/content-list-90.pdf (accessed on 13 September 2018).

15. Alves, V.H.; Prata, M.I.M.; Abrunhosa, A.J.; Castelo-Branco, M. GMP production of $^{68}$Ga-labelled DOTA-NOC on IBA Synthera. *J. Radioanal. Nucl. Chem.* **2015**. [CrossRef]

16. Ocak, M.; Antretter, M.; Knopp, R.; Kunkel, F.; Petrik, M.; Bergisadi, N.; Decristoforo, C. Full automation of $^{68}$Ga labelling of DOTA-peptides including cation exchange prepurification. *Appl. Radiat. Isot.* **2010**, *68*, 297–302. [CrossRef] [PubMed]

17. Matarrese, M.; Bedeschi, P.; Scardaoni, R.; Sudati, F.; Savi, A.; Pepe, A.; Masiello, V.; Todde, S.; Gianolli, L.; Messa, C.; et al. Automated production of copper radioisotopes and preparation of high specific activity [$^{64}$Cu]Cu-ATSM for PET studies. *Appl. Radiat. Isot.* **2010**, *68*, 5–13. [CrossRef] [PubMed]

18. Mukherjee, A.; Pandey, U.; Chakravarty, R.; Sarma, H.D.; Dash, A. Development of single vial kits for preparation of $^{68}$Ga-labelled peptides for PET imaging of neuroendocrine tumours. *Mol. Imaging Biol.* **2014**, *16*, 550–557. [CrossRef] [PubMed]

19. Ma, M.T.; Cullinane, C.; Waldeck, K.; Roselt, P.; Hicks, R.J.; Blower, P.J. Rapid kit-based (68)Ga-labelling and PET imaging with THP-Tyr(3)-octreotate: A preliminary comparison with DOTA-Tyr(3)-octreotate. *EJNMMI Res.* **2015**, *5*, 52. [CrossRef] [PubMed]

20. Asti, M.; Iori, M.; Capponi, P.C.; Rubagotti, S.; Fraternali, A.; Versari, A. Development of a simple kit-based method for preparation of pharmaceutical-grade [68]Ga-DOTATOC. *Nucl. Med. Commun.* **2015**, *36*, 502–510. [CrossRef] [PubMed]

21. Strelow, F.W. Quantitative separation of gallium from zinc, copper, indium, iron(III) and other elements by cation-exchange chromatography in hydrobromic acid-acetone medium. *Talanta* **1980**, *27*, 231–236. [CrossRef]

22. Van der Walt, T.N.; Strelow, F.W.E. Quantitative Separation of Gallium from Other Elements by Cation-Exchange Chromatography. *Anal. Chem.* **1983**, *55*, 212–216. [CrossRef]

23. Velikyan, I. *Synthesis, Characterization and Application of Ga-Labelled Peptides and Oligonucleotides*; Licentiate Dissertation, Institute of Chemistry, Department of Organic Chemistry: Uppsala, Sweden, 2004.

24. Meyer, G.J.; Macke, H.; Schuhmacher, J.; Knapp, W.H.; Hofmann, M. [68]Ga-labelled DOTA-derivatised peptide ligands. *Eur. J. Nucl. Med. Mol. Imaging* **2004**, *31*, 1097–1104. [CrossRef] [PubMed]

25. Dirks-Fandrei, C. *Entwicklung von Methoden zur Selektiven Trennung von Scandium, Zirkonium und Zinn für Radiopharmazeutische Anwendungen*; Philipps-Universität Marburg: Marburg, Germany, 2014; p. 20.

26. Pandey, M.K.; Byrne, J.F.; Jiang, H.; Packard, A.B.; DeGrado, T.R. Cyclotron production of (68)Ga via the (68)Zn(p,n)(68)Ga reaction in aqueous solution. *Am. J. Nucl. Med. Mol. Imaging* **2014**, *4*, 303–310. [PubMed]

27. Malinconico, M.; Asp, J.; Lang, C.; Boschi, F.; Guidi, G.; Takhar, P. Radiometals Production by Only One Solid Target System. Available online: https://www.comecer.com/radiometals-production-by-only-one-solid-target-system/ (accessed on 10 July 2018).

*instruments*

MDPI

*Article*

# Optimized Treatment and Recovery of Irradiated [$^{18}$O]-Water in the Production of [$^{18}$F]-Fluoride

**Antje Uhlending \*, Harald Henneken, Verena Hugenberg and Wolfgang Burchert**

Institute for Radiology, Nuclear Medicine and Molecular Imaging, Heart and Diabetes Center North Rhine Westphalia, University Hospital, Ruhr University Bochum, 32545 Bad Oeynhausen, Germany; hhenneken@hdz-nrw.de (H.H.); vhugenberg@hdz-nrw.de (V.H.); wburchert@hdz-nrw.de (W.B.)
\* Correspondence: auhlending@hdz-nrw.de; Tel.: +49-5731-97-3532

Received: 28 May 2018; Accepted: 22 June 2018; Published: 4 July 2018

**Abstract:** Enriched [$^{18}$O]-water is the target material for [$^{18}$F]-fluoride production. Due to its high price and scarce availability, an increased interest and necessity has arisen to recycle the used water, in order to use it multiple times as a target material for [$^{18}$F]-fluoride production. This paper presents an efficient treatment and reprocessing procedure giving rise to high chemical quality [$^{18}$O]-water, thereby maintaining its enrichment grade. The reprocessing is subdivided into two main steps. In the first step, the [$^{18}$F]-FDG (fluorodeoxyglucose) synthesis preparation was modified to preserve the enrichment grade. Anhydrous acetonitrile is used to dry tubing systems and cartridges in the synthesis module. Applying this procedure, the loss in the enrichment throughout the reprocessing is <1%. The second step involves a fractional distillation in which the major part of the [$^{18}$O]-water was recycled. Impurities such as solvents, ions, and radioactive nuclides were almost completely separated. Due to the modified synthesis preparation using acetonitrile, the first distillation fraction contains a larger amount of an azeotropic [$^{18}$O]-water/acetonitrile mixture. This fraction is not further distillable. Contents of the remaining [$^{18}$O]-water were separated from the azeotropic mixture by using a molecular sieve desiccant. This process represents a fast, easy, and inexpensive method for reprocessing used [$^{18}$O]-water into new [$^{18}$O]-water quality for further application.

**Keywords:** recycling of [$^{18}$O]-water; enrichment grade; fractional distillation; azeotropic mixtures; molecular sieve desiccant

---

## 1. Introduction

The [$^{18}$O]-water recovered from the synthesis of [$^{18}$F]-based radiopharmaceuticals could in principle be reused for further irradiation. During synthesis and irradiation, the [$^{18}$O]-water gets contaminated with solvents and ions from the synthesis module and the target [1,2]. The synthesis of [$^{18}$F]-based radiopharmaceuticals involves an [$^{18}$F]-fluoride separation step, in which the irradiated [$^{18}$O]-water is eluted through an anion exchange cartridge, which is commonly preconditioned with [$^{16}$O]-water [3]. Most of the end users also rinse the target and the [$^{18}$F]-fluoride transfer line with [$^{16}$O]-water [4]. Hence, the degree of enrichment drops through intermixing with [$^{16}$O]-water. For a further irradiation, the recycled [$^{18}$O]-water has to be of high chemical quality, and of the highest possible enrichment grade. A method for recycling the [$^{18}$O]-water must also be able to separate all impurities [5]. The recycling processes using condensation and/or UV-irradiation did not lead to good [$^{18}$O]-water quality, sufficient quantity, or a stable enrichment grade [6–8]. Most of the end users collect the irradiated [$^{18}$O]-water samples and return them to the manufacturers for recycling. The aim of this work was to develop a cost-effective, coherent reprocessing concept for [$^{18}$O]-water suitable for routine production. The desired goal was to recycle >90% of the [$^{18}$O]-water, while maintaining the enrichment grade with the prescribed method. For further radiation, the [$^{18}$O]-water quality needs to possess a high chemical quality, thereby attaining the release specifications [9].

## 2. Materials and Methods

### 2.1. Overview of the Optimized Recovery Method

For use in the target, the [$^{18}$O]-water has to be present with high chemical and microbiological purity, as well as sufficient enrichment grade. After the bombardment, the irradiated [$^{18}$O]-water is usually transferred via [$^{18}$F]-transfer lines into the synthesis modules. For a meaningful recovery, an additional rinse of the transfer lines and target was conducted exclusively with [$^{18}$O]-water. During the synthesis of [$^{18}$F]-radiopharmaceutical, the irradiated [$^{18}$O]-water was passed through an anion exchange cartridge to separate the [$^{18}$F]-fluoride from the [$^{18}$O]-water, which was collected separately. Residues of [$^{16}$O]-water on the cartridge and in the tubing system of the module would lead to a significant contamination of the [$^{18}$O]-water with [$^{16}$O]-water. Therefore, an improved synthesis module preparation ensures the maintenance of the enrichment grade of the [$^{18}$O]-water. A fractional distillation was performed for the purification of the [$^{18}$O]-water. The first collected fraction contained an azeotropic [$^{18}$O]-water/acetonitrile mixture (10–15% of the preparation). The [$^{18}$O]-water was separated from this azeotropic mixture via an optimized molecular sieve procedure. Microbiological contaminants were removed via UV-irradiation before the final distillation, followed by a quality control of the recycled [$^{18}$O]-water before its release for the next irradiation. A schematic overview of the whole recovery method is shown below in Figure 1.

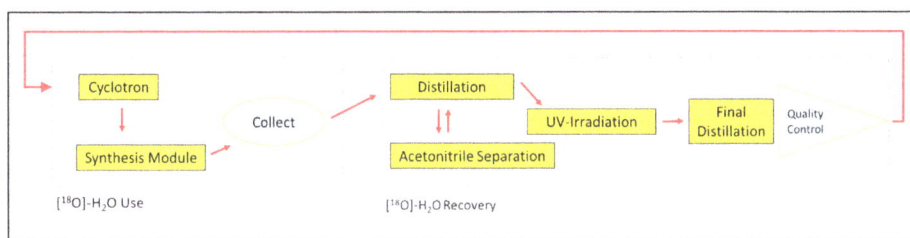

**Figure 1.** Overview of the optimized recovery method.

### 2.2. Improved Synthesis Preparation for Maintaining the Enrichment Grade

The use of [$^{16}$O]-water to rinse the transfer lines during the cleaning process and the preparation of the synthesis modules resulted in a significant loss of the enrichment grade in the collected [$^{18}$O]-water. During the synthesis, [$^{18}$O]-water was recovered by trapping [$^{18}$F]-fluoride on an anion exchange cartridge (Chromafix 30-PS-HCO$_3$ by Macherey-Nagel) and subsequent elution with an aqueous potassium carbonate solution [10]. The cartridge itself was also preconditioned with 5 mL of [$^{16}$O]-water. Residues of the [$^{16}$O]-water reduce the enrichment grade of the recollected [$^{18}$O]-water. First attempts to remove the water residues from the cartridge with a stream of helium gas still revealed a significant loss of the enrichment grade (Table 1). A clear improvement in the retention of the enrichment grade was achieved by an additional flushing of the red tubing system in Figure 2 with anhydrous acetonitrile [11]. Additionally, the anion exchange cartridge was rinsed with 1 mL of anhydrous acetonitrile after conditioning. After performing the above-mentioned steps, the synthesis module was prepared according to common literature procedures [12]. A flow chart representing the tubing system is shown in Figure 2.

**Table 1.** Influence of the synthesis preparation on the enrichment grade.

| Enrichment Grade | Common Synthesis Preparation | Drying Module + Cartridge with Helium Flow | Drying Module + Cartridge with Acetonitrile |
|---|---|---|---|
| [%] | 74.0–80.0 | 84.0–88.0 | 95.0–96.5 |

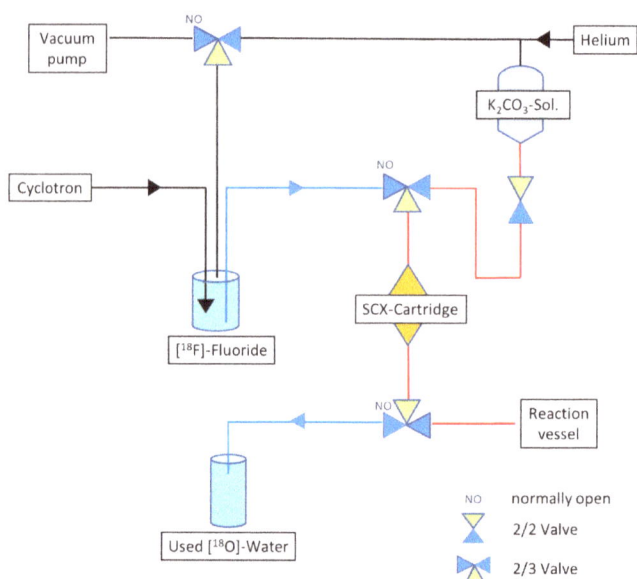

**Figure 2.** Schematic partial representation of the tubing system.

## 2.3. Purification by Fractional Distillation

The distillation is a simple separation process that can be used for the cleaning and separation of liquids with boiling points below 150 °C. A fractionated distillation in commercial glassware is used for the purification of the collected [$^{18}$O]-water [13]. The distillation unit (Figure 3a) consists of a heating jacket, a 500 mL flask (a 1–2 mL flask can also be used), a 30 cm vacuum-coated Vigreux column, a thermometer (20–150 °C), a 100 mL distillate receiver with vent stopcock, and a special distillation cooling attachment (Figure 3b). This special semi-micro scale apparatus allows an improved separation of the fractions, thereby obtaining a lower volume loss.

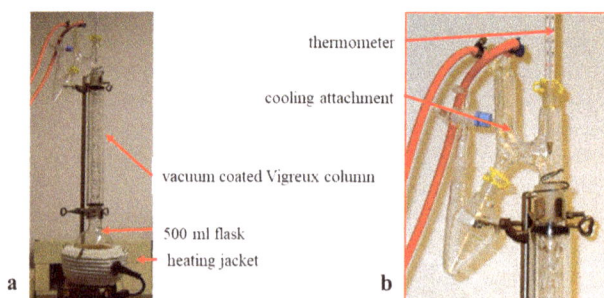

**Figure 3.** (**a**) Distillation unit; (**b**) special distillation cooling attachment.

## 2.4. Separation of the [$^{18}$O]-Water from the Azeotropic Mixtures by Means of a Molecular Sieve

The idea of the separation of [$^{18}$O]-water from organic solvents using a desiccant (including molecular sieves) has already been described in the literature [14]. We adapted this method for the purification of the first distilled fractions, which contain an [$^{18}$O]-water/acetonitrile mixture with an acetonitrile content of 50–80% (10–15% of the preparation). Separation of the [$^{18}$O]-water by further distillation is not possible [15]. A simple and suitable routine method for an almost complete recovery of the [$^{18}$O]-water from azeotropic mixtures using molecular sieve 3Å is briefly described below [16]. Special attention had been paid to the preservation of the [$^{18}$O]-enrichment grade.

A condensation unit (Figure 4) consisting of a three-necked flask with a thermometer, a stopcock, an inlet pipe with a diameter of 10 mm, and a heating jacket has been constructed. A cooling trap with ice cooling was connected to the diaphragm vacuum pump via PTFE tubing connections with a piston and a three-way stopcock.

The use of a molecular sieve containing [$^{16}$O]-substitute groups led to a significant loss of enrichment grade. Therefore, the molecular sieve has to be pretreated before use. Dust and breakage from the molecular sieve 3Å was removed by washing with double distilled water. For activation, the molecular sieve was dried under argon flow in the condensation unit at a maximum temperature of 300 °C. Subsequently, the molecular sieve was mixed with [$^{18}$O]-water of about 50% enrichment grade, and dried again at about 200 °C. Further handling and storage was carried out under argon.

For the separation of the [$^{18}$O]-water from the azeotropic [$^{18}$O]-water/acetonitrile mixture, 150 g of the preconditioned 3Å molecular sieve was added to 150 g of the mixture, and allowed to stand under argon for 24 h. After 24 h, most of the [$^{18}$O]-water molecules had been absorbed by the molecular sieve. After decantation of the acetonitrile, the residual amount of acetonitrile was removed by drying for 1 h at 40 °C under a membrane pump vacuum, and subsequently under a weak argon stream for another 5 min. Following this, the cold trap was cleaned and placed in ice water. For the recovery of the [$^{18}$O]-water, the molecular sieve was heated for 3 h under atmospheric pressure in a weak stream of argon at a maximum temperature of 200 °C. The pure [$^{18}$O]-water was collected in the cold trap. The solvent content of the recollected [$^{18}$O]-water is usually >1%, and can be returned to the distillation.

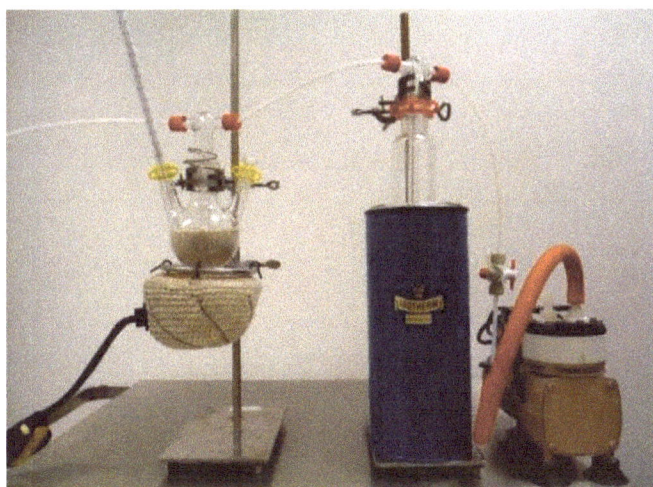

**Figure 4.** Condensation unit for the recovery of [$^{18}$O]-water from azeotropic mixtures.

*2.5. Determination of the Degree of Enrichment via Pycnometry*

The effects on the enrichment grade of the individual process steps were examined via pycnometry [17,18]. With an increased amount of [$^{18}$O] in [$^{16}$O]-water, an increase of the density of the [$^{18}$O]-water mixture is observed. The increase of the density is dependent on the increase of the amount of [$^{18}$O] present in the [$^{18}$O]-water.

$$\rho[^{18}O]\text{-water}(T) = \rho[^{16}O]\text{-water}(T) \times (20.0153/18.0153) \text{ [g/mL] for pure } [^{18}O]\text{-water} \qquad ()$$

$(20.0153/18.0153)$ is the molecular weight ratio of $[^{18}O]$-water to $[^{16}O]$-water.

Since the number of atoms is proportional to the volume, the enrichment grade is dependent on the density ($\rho = m/V$) at constant temperature.

$$\rho_{\text{sample}} = (w[^{18}O]\text{water} \times \rho[^{18}O]\text{water}) + (w[^{16}O]\text{water} \times \rho[^{16}O]\text{water}) \qquad ()$$

At a constant volume measurement (using the same pipette), the enrichment grade can be determined by the mass ratio of $[^{18}O]$-water to $[^{16}O]$-water. The density of the $[^{18}O]$-water can be determined experimentally, and the enrichment grade can be calculated using the following formula:

$$w_{\text{sample}}\left[^{18}O\right] = 9 \times \frac{m\left[^{18}O\right]\text{water}}{m\left[^{16}O\right]\text{water}} - 9[\%]$$

$m\,[^{16}O]$-water (T) = mass of $[^{16}O]$-water at certain temperature and pipette

$m\,[^{18}O]$-water (T) = mass of $[^{18}O]$-water sample at certain temperature and pipette

w: enrichment grade in %

$9 = 1/K$, $K = (20.0153/18.0153) - 1$;

K: constant from the difference of the molecular weights.

Hence, the enrichment grade is determined by simple weighing in combination with volume measurement, whereas the density is temperature dependent.

## 3. Results

For the validation of the purification procedures and the quality control of the purified $[^{18}O]$-water, the following parameters (Table 2) have been investigated.

**Table 2.** Examination parameters for the quality control of $[^{18}O]$-water.

| Parameter | Method | Device |
|---|---|---|
| $[^{18}O]$-Enrichment grade | Relative pycnometry | Libra, pipette |
| Solvent content | Gas chromatography | Trace 1310 Thermo Fischer, FID |
| Aromatic compounds | UV-Spectroscopy | Genesys 10 S Thermo Fisher |
| Ions | Conductivity measurement | Vario Cond, WTW, 0.001-200 µS/cm |
| Radioactive nuclide | Gamma measurement, multi-channel-analyzer | Germanium Detector GC 1200-7500 SL Canberra |

Using fractionated distillation, a separation of the $[^{18}O]$-water from the solvent mixture in the first step was possible, in both good quantity and sufficient quality (Table 3). About 80% of the used $[^{18}O]$-water can be recycled in the first step. The process was established in common laboratory scales of 1 to 2 L. No differences in the yields of activity during $[^{18}F]$-fluoride production and the radiochemical yields of $[^{18}F]$-FDG in the following radiosynthesis were observed, comparing the use of recycled $[^{18}O]$-water to fresh $[^{18}O]$-water. After more than 10 times of use and purification, no accumulation of any impurity, in particular of radionuclides, has been detected. Furthermore, during irradiation in the cyclotron, the recycled $[^{18}O]$-water showed no pressure increase in the target, normally observed when additional solvent residues are present. It has also been found that purification by fractional distillation does not diminish the enrichment grade. In summary, following the above-mentioned recycling process, used $[^{18}O]$-water can be recovered without any loss in the enrichment grade.

**Table 3.** Quality of the recovered $[^{18}O]$-water.

| Parameter | Unit | Collected-$[^{18}O]$-H$_2$O | Main Fraction | First Fraction | Residue | Specification |
|---|---|---|---|---|---|---|
| Volume | % | 100 | 80 | 15 | 5 | - |
| Conductivity | µS/cm | 1200 | 0.9 | 133 | 3000 | <10 |
| Acetonitrile | % | 8–10 | <0.00001 | 40–80 | <0.1 | <0.0001 |
| Ethanol | µg/mL | 126 | <0.1 | 1600 | <0.1 | <100 |
| Acetone | µg/mL | 19.2 | <0.1 | 260 | <0.1 | <100 |
| UV-Spectrum | 220 nm | 0.120 | 0.0005 | 0.052 | 2.095 | <0.2 |
| | 280 nm | 0.040 | 0.0014 | 0.010 | 0.400 | <0.1 |
| MCA | keV | 810; 846; 1238; 1460; 1770 | not detectable | 136; 1460 | 122; 320; 744; 810; 846; 1037; 1238; 1770 | not detectable |
| $[^{18}O]$-Enrichment | % | 100 | <0.1 | 0.3 | 99.6 | - |
| | % | 95.0 | 94.5 | - | - | 95.5 |

Gamma spectroscopy, in particular a multi channel analyzer (MCA) was used to determine the radionuclidic purity of the recovered $[^{18}O]$-water fractions. Specifications for $[^{18}O]$-water request no detectable energy lines. Aliquots of the different $[^{18}O]$-water fractions, comprising the same volume and geometry, were examined in regards to their relative nuclidic purity (Table 3).

During the development of an improved synthesis procedure, the effects on the enrichment grade were investigated. Conventional cleaning of the synthesis module reduces the enrichment grade of the $[^{18}O]$-water by about 25% per use. As part of the further optimization, with $[^{16}O]$-water-contaminated collection vessels, tubing and cartridges were identified as the origin of the contaminations. By excluding these contaminants, the loss in the enrichment grade was reduced to 15%. Further helium drying reduced the enrichment loss to 10%. However, a constant enrichment grade (98.3% to 98.8%) could be obtained by further rising of the whole system with acetonitrile. The results of the optimization steps in the improved preparation and handling protocol that preserve the enrichment grade of the $[^{18}O]$-water are summarized in Table 1.

The synthesis preparation with acetonitrile drying increased the acetonitrile content in the used $[^{18}O]$-water from approximately 1% to 8–10%. As a result, a volume increase of the first distillation fraction from around 10% to 25% was obtained. The $[^{18}O]$-water in this azeotropic mixture can be separated by the molecular sieve process, without any loss in the enrichment grade (Table 4).

**Table 4.** Influence of the recovery on the enrichment grade.

| Enrichment Grade | $[^{18}O]$-Water Original | $[^{18}O]$-Water after Distillation | $[^{18}O]$-Water after Molecular Sieve |
|---|---|---|---|
| [%] | 97.0 | 96.5 | 96.5 |

The use of a new molecular sieve could possibly lead to a reduction in the enrichment grade of the $[^{18}O]$-water, due to $[^{16}O]$ substituted groups in the resin. Therefore, the molecular sieve has to be preconditioned before use. The quantitative determination of $[^{16}O]$-substituted groups was conducted by isotope dilution analysis (Tables 5 and 6).

**Table 5.** Parameters of the molecular sieve separation method.

| Molecular Sieve | Amount | Equates to X mL H$_2$O |
|---|---|---|
| | 150 g | 200 mL |
| Absorbed water | 0.2 mL/g | 30 mL |
| Substituted O-Functions | 3.3 mmol/g | 7.6 mL |

**Table 6.** Temperature range of the separation method using a molecular sieve.

| Separation Temp. for Acetonitrile | 40 °C | Desorption Temp. [$^{18}$O]-H$_2$O | 200 °C |
|---|---|---|---|
| Recovery rate | >95% | time | 30 min |

The untreated molecular sieve contains 3.3 mmol/g substitutable O-functions, which reduces the enrichment grade, when occupied by [$^{16}$O]. In our case, that correlates to 7.6 mL of water per 150 g, which would reduce the enrichment grade from 90% to 72%. After conditioning and multiple uses of the molecular sieve, the separation of the [$^{18}$O]-water was achieved without detectable reduction of the enrichment grade.

The molecular sieve was left in the apparatus under argon, and can be reused several times. Therefore, the disadvantages of using a molecular sieve are negligible, or at least more than offset by the considerable robustness of the process.

## 4. Discussion

The developed recycling concept allows the repetitive recovery of >90% of the [$^{18}$O]-water, while retaining the enrichment grade and the required chemical quality, which corresponds to the release specifications, and allows for reuse without any loss in performance of the overall process. The purification procedure can be performed in conventional glassware in usual laboratory scale, and is both time-saving and cost-effective. The recycling process also allows the desired rinsing of the [$^{18}$F]-fluoride transfer line with [$^{18}$O]-water, since the optimized treatment of the [$^{18}$O]-water-circuit has been adjusted to maintain the enrichment grade. Additionally, the optimization of the cleaning and preparation of the synthesis module enabled a recovery of the [$^{18}$O]-water without any loss of the enrichment grade.

Further use of the recovered [$^{18}$O]-water did not reveal any interfering effects during [$^{18}$F]-fluoride production and subsequent syntheses. Compared to the use of fresh [$^{18}$F]-water, no difference in the yields of [$^{18}$F]-fluoride production and the radiochemical yields of subsequent radiosyntheses were observed.

The developed optimization steps can be easily established in the routine process, and thus allow for considerable cost savings.

**Author Contributions:** A.U. participated in the design of the study, did the majority of the chemical experiments, designed and composed most of the draft manuscript and participated in its finalization. H.H. was involved in the initial investigations of the study and experiments. W.B. supervised the study. V.H. participated in the design of the manuscript and finalized it. All authors read and approved the final manuscript.

**Funding:** This research received no external funding.

**Conflicts of Interest:** The authors declare no conflicts of interest.

## References

1. Uhlending, H.; Burchert, W. Einfluss von FKM oder Nitril-Target-Dichtungen auf radiochemische Ausbeuten. *Nuklearmedizin* **2017**, *56*, A84.
2. Bowden, L.; Vintro, L.L.; Mitchell, P.I.; O'Donnell, R.G.; Seymour, A.M.; Duffy, G.J. Radionuclide impurities in proton-irradiated [$^{18}$O]H$_2$O for the production of $^{18}$F-: Activities and distribution in the [$^{18}$F]FDG synthesis process. *Appl. Radiat. Isot.* **2009**, *67*, 248–255. [CrossRef] [PubMed]
3. Schleyer, D.J.; Bastos, M.A.V.; Alexoff, D.; Wolf, A.P. Separation of [$^{18}$F]fluoride from [$^{18}$O]water using anion exchange resin. *Int. J. Radiat. Appl. Instrum. A* **1990**, *41*, 531–533. [CrossRef]
4. GE Healthcare. PETtrace OPERATOR GUIDE; DIRECTION 2131768-100 REVISION 15. Unpublished work. 2013; 102.
5. Schlyer, D.J.; Firouzbakht, M.L.; Wolf, A.P. Impurities in the [$^{18}$O]water target and their effect on the yield of an aromatic displacement reaction with [$^{18}$F]fluoride. *Appl. Radiat. Isot.* **1993**, *44*, 1459–1465. [CrossRef]

6. Mangner, T.J.; Mulholland, G.K.; Toorongian, S.A.; Jewett, D.M.; Kilbourn, M.R. Purification of used O-18 target water by photochemical combustion. *J. Nucl. Med.* **1992**, *33*, 982–983.

7. Kitano, H.; Magata, Y.; Tanaka, A.; Mukai, T.; Kuge, Y.; Nagatsu, K.; Konishi, J.; Saji, H. Performance assessment of O-18 water purifier. *Ann. Nucl. Med.* **2001**, *15*, 75–78. [CrossRef] [PubMed]

8. Asti, M.; Grassi, E.; Sghedoni, R.; De Pietri, G.; Fioroni, F.; Versari, A.; Borasi, G.; Salvo, D. Purification by ozonolysis of (18)O enriched water after cyclotron irradiation and the utilization of the purified water for the production of [18F]-FDG (2-deoxy-2-[18F]-fluoro-D-glucose). *Appl. Radiat. Isot.* **2007**, *65*, 831–835. [CrossRef] [PubMed]

9. Guillaume, M.; Luxen, A.; Nebeling, B.; Argentini, M.; Clark, J.C.; Pike, V.W. Recommendations for fluorine-18 production. *Int. J. Radiat. Appl. Instrum. A* **1990**, *42*, 749–762. [CrossRef]

10. Moon, B.S.; Lee, K.C.; An, G.I.; Chi, D.Y.; Yang, S.D.; Choi, C.W.; Lim, S.M.; Chun, K.S. Preparation of 3'-deoxy-3'-[18F]fluorothymidine ([18F]FLT) in ionic liquid, [bmim][OTf]. *J. Label. Compd. Radiopharm.* **2006**, *49*, 287–293. [CrossRef]

11. Uhlending, A.; Henneken, H.; Burchert, W. Abreicherungsarme [18O]-Wasserrückgewinnung durch optimierte Synthesevorbereitung. *Nuklearmedizin* **2013**, *52*, A98.

12. Krasikova, R. Synthesis Modules and Automation in F-18 Labeling. In *PET Chemistry*; Schubiger, P.A., Lehrmann, L., Friebe, M., Eds.; Ernst Schering Research Foundation Workshop; Springer: Berlin/Heidelberg, Germany, 2007; Volume 64.

13. Uhlending, A.; Henneken, H.; Burchert, W. Destillation als Reinigungsverfahren zur Mehrfachverwendung von [18O]-angereichertem Wasser bei der 18F-Fluorid-Produktion. *Nuklearmedizin* **2005**, *44*, A175.

14. Gail, R.; Hamacher, K. Verfahren zur Abtrennung von H218O Aus Einem Organischen Lösungsmittel. DE19948427A1, 3 May 2001.

15. Müller, E.; Bayer, O.; Meerwein, H.; Ziegler, K. *Methoden der Organischen Chemie (Houben-Weyl)—Band I/2: Allgemeine Laboratoriumspraxis*; Thieme: Stuttgart, Germany, 1959; p. 828.

16. Uhlending, A.; Henneken, H.; Burchert, W. Routinetaugliche Abtrennung von [18O]-Wasser aus azeotropen [18O]-Wasser/Acetonitril-Mischungen unter Einsatz von Molsieb. *Nuklearmedizin* **2010**, *49*, A112.

17. Fawdry, R.M. A simple effective method for estimating the [18O] enrichment of water mixtures. *Appl. Radiat. Isot.* **2004**, *60*, 23–26. [CrossRef] [PubMed]

18. Uhlending, A.; Henneken, H.; Burchert, W. Pyknometrische Bestimmung des 18O-Isotopengehaltes von 18O-angereicherten Wasser: Validierung und Anwendung in der Routine. *Nuklearmedizin* **2004**, *43*, A155.

*instruments*

MDPI

Article

# Recovery of Molybdenum Precursor Material in the Cyclotron-Based Technetium-99m Production Cycle

**Hanna Skliarova [1,\*], Paolo Buso [1], Sara Carturan [1,2], Carlos Rossi Alvarez [1], Sara Cisternino [1], Petra Martini [1,3], Alessandra Boschi [3] and Juan Esposito [1]**

[1]   Legnaro National Laboratories, Italian National Institute for Nuclear Physics (LNL-INFN), Viale dell'Università, 35020 Legnaro, Italy; paolo.buso@lnl.infn.it (P.B.) Sara.Carturan@lnl.infn.it (S.C.); Carlos.Rossi.Alvarez@lnl.infn.it (C.R.A.); Sara.Cisternino@lnl.infn.it (S.C.); petra.martini@lnl.infn.it (P.M.); Juan.Esposito@lnl.infn.it (J.E.)

[2]   Department of Physics and Astronomy, University of Padua, 35121 Padua, Italy

[3]   Department of Morphology, Surgery and Experimental Medicine, University of Ferrara, 44121 Ferrara, Italy; bsclsn@unife.it

\*   Correspondence: Hanna.Skliarova@lnl.infn.it; Tel.: +39-049-806-8416

Received: 22 December 2018; Accepted: 5 February 2019; Published: 13 February 2019

**Abstract:** A closed-loop technology aiming at recycling the highly $^{100}$Mo-enriched molybdenum target material has been developed in the framework of the international research efforts on the alternative, cyclotron-based $^{99m}$Tc radionuclide production. The main procedure steps include (i) $^{100}$Mo-based target manufacturing; (ii) irradiation under proton beam; (iii) dissolution of $^{100}$Mo layer containing $^{9\times}$Tc radionuclides (produced by opened nuclear reaction routes) in concentrated $H_2O_2$ solution; and (iv) Mo/Tc separation by the developed radiochemical module, from which the original $^{100}$Mo comes as the "waste" alkaline aqueous fraction. Conversion of the residual $^{100}$Mo molybdates in this fraction into molybdic acids and $MoO_3$ has been pursued by refluxing in excess of $HNO_3$. After evaporation of the solvent to dryness, the molybdic acids and $MoO_3$ may be isolated from $NaNO_3$ by exploiting their different solubility in water. When dried in vacuum at 40 °C, the combined aqueous fractions provided $MoO_3$ as a white powder. In the last recovery step $MoO_3$ has been reduced using a temperature-controlled reactor under hydrogen overpressure. An overall recovery yield of ~90% has been established.

**Keywords:** radioisotope production; Molybdenum-100; cyclotron; molybdenum material recovery

---

## 1. Introduction

$^{99m}$Tc is a radionuclide widely used in Conventional Nuclear Medicine diagnostic examinations. It is routinely eluted from portable generators containing the parent radionuclide $^{99}$Mo, coming from highly-enriched (>80 wt.% $^{235}$U) uranium targets irradiated in nuclear fission reactors. Once separated from the $^{235}$U fission products and further purified, $^{99}$Mo is loaded onto an alumina column inside the portable generator. $^{99m}$Tc-radiopharmaceuticals are then directly prepared in hospital radiopharmacies by adding the $^{99m}$Tc-pertechnetate eluted from generators directly in lyophilized "kits", following the manufacturer instructions and quality control specifications to get an injectable product. Approximately 95% of the world's production of $^{99}$Mo is provided by few ageing nuclear reactors, whose unplanned outages have already caused global shortages in the last decade, even recently (November 2018) [1]. In order to mitigate this problem in the mid-long term, alternative $^{99m}$Tc production routes have been intensively investigated all over the world.

A dedicated CRP (Coordinated Research Project) program launched by the International Atomic Energy Agency (IAEA) in 2011–2015 [2,3] was dedicated to the $^{99m}$Tc/$^{99}$Mo accelerator-based production routes. Main CRP outcomes have revealed that the direct $^{99m}$Tc production, through the $^{100}$Mo(p,2n) nuclear reaction, starting from highly $^{100}$Mo-enriched (i.e., enrichment >99.5%) molybdenum targets is the most promising approach [3–7]. In this context, the optimal proton energy range 10–25 MeV, able to provide sufficient amounts of the radioisotope with impurity level within the limits defined by corresponding European Pharmacopeia [8], has been identified [6,9]. Furthermore, recommended target thickness and irradiation times for optimized production at different irradiation energies (15, 20 and 25 MeV) have been calculated by Esposito et al. (2013) [10].

The purpose of the TECHnetium direct-production in hOSPital (TECHN-OSP) research project at INFN [4,10–12] was the development of a technology able to produce GBq amounts of $^{99m}$Tc through the (p,2n) nuclear reaction route on $^{100}$Mo-enriched metal targets. This approach could provide the daily routine supply by exploiting the existing medical cyclotron network in Italy.

A cyclotron-based $^{99m}$Tc closed-loop production cycle includes the following steps, as shown in Figure 1. (i) $^{100}$Mo target manufacturing, (ii) target irradiation with a proton cyclotron, (iii) dissolution of $^{100}$Mo layer containing $^{9x}$Tc radionuclides (plus Nb, Zr impurities) in the concentrated $H_2O_2$ solution, (iv) separation/purification of $^{99m}$Tc from $^{100}$Mo and by-products by a developed dedicated separation module, and (v) recovery of the costly $^{100}$Mo material remaining as the "waste" after the separation procedure [4,12].

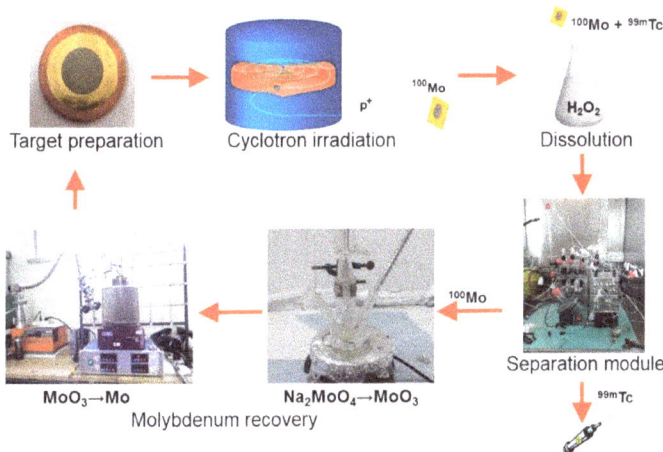

**Figure 1.** TECHnetium direct-production in hOSPital (TECHN-OSP) $^{99m}$Tc production cycle.

As regards the $^{100}$Mo metal target, several configurations have been developed and tested in the framework of the TECHN-OSP project:

- Mo-sputtered layer onto a complex backing plate (patent PCT/IB2018/056826) [13,14].
- HIVIPP (HIgh energy VIbrational Powders Plating) electrostatic deposition [15] of Mo onto metallic backing with a high yield deposition efficiency (>95%) and uniformity, but limited thickness (2–3 μm).
- Stacked-foils clamped into a dedicated target holder.
- Spark Plasma Sintering (SPS) of $^{100}$Mo powders onto an inert backing plate.

In order to meet the requirements for a fast and efficient dissolution, extraction and purification of $^{99m}$Tc yielded from $^{100}$Mo-enriched molybdenum metallic target irradiation, a remotely controlled module based on the Solvent Extraction (SE) technique has been developed. After the irradiation, the targets were processed according to a protocol presented in Figure 2 and described in details by Martini et al. [12]. All quality control procedures on the final products, radiolabeling, in vivo and on phantom imaging studies with a clinical gamma camera, have been conducted as well [16].

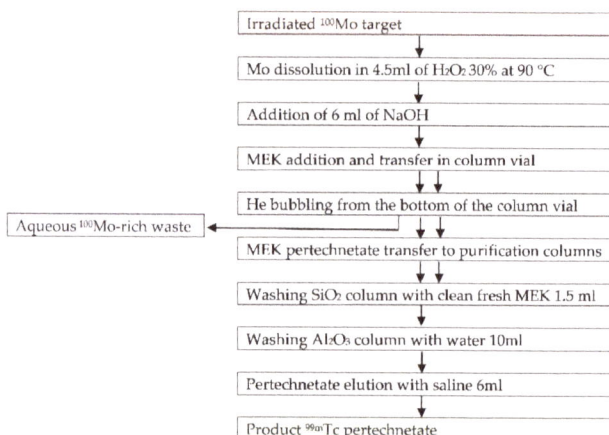

**Figure 2.** Target processing and $^{100}$Mo/$^{99m}$Tc separation–purification protocol.

In order to make the cyclotron-based $^{99m}$Tc production costs affordable, the expensive $^{100}$Mo-enriched material should be recovered from the separation module waste with the aim to be reused for the preparation of new targets, thus closing the production cycle.

Therefore, the goal of this work was to develop a molybdenum material recovery procedure under metallic form, starting from the isotope-rich "waste" fraction from the dedicated separation module [4,12], which has been cited as the most efficient among the ones reviewed in the work by Gumiela et al. [17]. We applied a two-step procedure, which consists of the conversion of sodium molybdates present in the waste fraction into $MoO_3$, then followed by further reduction of molybdenum oxide into metallic molybdenum.

## 2. Materials and Methods

In order to simulate the Mo-rich waste coming from the radiochemical separation module [12], $304.2 \pm 3.3$ mg of natural Mo (100 μm thickness foils, 99.95% purity purchased from Advent Research Materials Ltd., Eynsham, Oxford, England) went through the automatic radiochemical processing as described in Figure 2. Briefly, molybdenum was dissolved in 4.5 mL 30% $H_2O_2$ at 90 °C in ~30 min; then 6 mL of 6M NaOH were added to the solution, and the same procedure applied in the case of SE of technetium from the solution was repeated twice using methyl ethyl ketone (MEK) [12]. The residual aqueous Mo-rich solution, containing molybdate and polymolybdate species, was stored in glass vials, and water was removed by evaporation just before using it for the recovery studies. The residue was then washed with ethanol, filtered and dried at room temperature obtaining $0.79 \pm 0.18$ g of precipitate. In the following sections, it will be mentioned as molybdates mixture.

### 2.1. MoO₃ Recovery from Mo-Rich Module "Waste"

The first step of molybdenum recovery is the conversion of sodium molybdates mixture to molybdic acid and $MoO_3$. The experiment has been performed using Mo-rich "waste" coming from the separation module [12], standard chemical glassware, and commercially available chemical compounds: HCl, 37%, p.a. grade, Aldrich (Darmstadt, Germany); $HNO_3$, 65%, p.a. grade, Sigma Aldrich S.r.l. (Milan, Italy); and $NH_4OH$, 30%, p.a. grade, Carlo Erba (Cornaredo, Italy).

In order to achieve the highest recovery yield and purity, different techniques were tested (recipes 1–4) as described in the following section. The main approach used for $MoO_3$ recovery is known as the Mo-based "spent catalyst regeneration" procedures. Two methods were proposed by Park [18] (Figure 3a) and Kar [19] (Figure 3b). In both processes, the key steps interesting for the present recovery study, start from the conversion of ammonium molybdate in $MoCl_6$ by HCl treatment (step

4 on Figure 3a and steps 7 and 12 on Figure 3b). Then, ammonium hydroxide at controlled pH is added in order to obtain ammonium molybdate (step 5 on Figure 3a and steps 8 and 14 on Figure 3b). Afterwards, ammonium molybdate is precipitated by varying the solution pH (step 6 on Figure 3a and step 15 on Figure 3b) and calcinated at 450 °C to finally produce the pure MoO$_3$ product (step 7 on Figure 3a and steps 9 and 16 on Figure 3b).

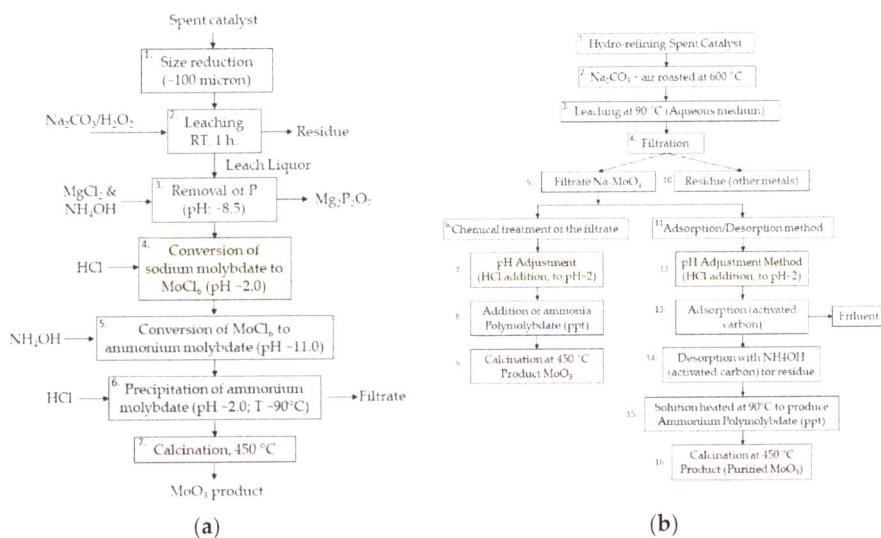

**Figure 3.** MoO$_3$ recovery as regeneration of a spent catalyst described by Park [18] (**a**) and Kar [19] (**b**).

### 2.1.1. Spent Catalyst Approach

Recipe MoO$_3$-1. Into ~0.5 g of the dry Mo-rich "waste" coming from the separation module, 15 mL of 0.1M hydrochloric acid were added. The solution became yellow as a result of MoCl$_6$ production at pH = 2. Then, 50 mL of 1M NH$_4$OH were added to the solution, and the color turned blue/transparent at pH = 10.6. After heating the solution to 80 °C, the pH was observed to decrease to 9.5, owing to ammonia release. The precipitate was centrifuged for 8 min, recovered, and finally dried in vacuum overnight at 40 °C. The weight of the obtained pale yellow powder was 0.4 g.

### 2.1.2. Spent Catalyst Approach, Nitric Acid

Recipe MoO$_3$-2. Into ~0.5 g of the product coming from the separation module 2–3 mL of 5M HNO$_3$ and 10 mL of deionized water were added. The solution became yellow at pH = 0 (lower limit of pH-meter). Then, concentrated NH$_4$OH was added dropwise to the solution until pH = 2, and the color turned yellow-green. The last drop of ammonium hydroxide changed the pH to 5.3 (desired pH = 2.5–3). The solution became pale blue after heating at 80 °C for 2 h and precipitation of white flakes occurred. The precipitate was centrifuged for 8 min, washed twice with 10 mL of ethanol, and dried in vacuum overnight at 40 °C. The weight of the obtained white powder was 0.11 g.

### 2.1.3. Direct Precipitation with Nitric Acid

Recipe MoO$_3$-3. Into ~0.13 g of the product coming from the separation module 3–5 mL of 5M HNO$_3$ (excess) were added. The mixture was heated up in reflux mode at 95 °C for 7 h. Then, the liquid was evaporated and the solid was washed twice with 15 mL of a mixture of EtOH:H$_2$O (3:1 volumetric), prior to centrifugation at 4000 rpm for 8 min. The precipitate was dried in vacuum overnight at 40 °C, giving 0.082 g of white powder.

Recipe MoO₃-4. 0.7 g natMo foil of 250 μm thick (99.99% purity, Goodfellow, Cambridge Ltd., Huntingdon, England) was dissolved in 15 mL of 30% $H_2O_2$ at 90 °C. The dissolution went on for ~20 min. When the solution was cooled down to 50 °C, 5 mL of 6 M NaOH was added to the solution and it was maintained for 30 min in order to simulate the process in a module. Hence, 25 mL of 5 M $HNO_3$ (excess) was added to the solution. The mixture was heated up in reflux mode at 95 °C overnight. After evaporation, a white precipitate was obtained and washed with several portions of distilled water (15 mL, 3 × 7.5 mL). Then, the precipitate was dried in vacuum overnight at 40 °C giving 0.986 g of white powder.

## 2.2. MoO₃ Reduction System

The second step of the closed-loop technology (i.e., the enriched molybdenum recovery process) is the molybdenum oxide reduction to metal, as described above. It was carried out using a dedicated hydrogenation reactor system. The apparatus is schematized in Figure 4a and it is composed of a vacuum system, a furnace, a tungsten reduction cell filled with hydrogen, a liquid nitrogen Dewar, and a process data acquisition system.

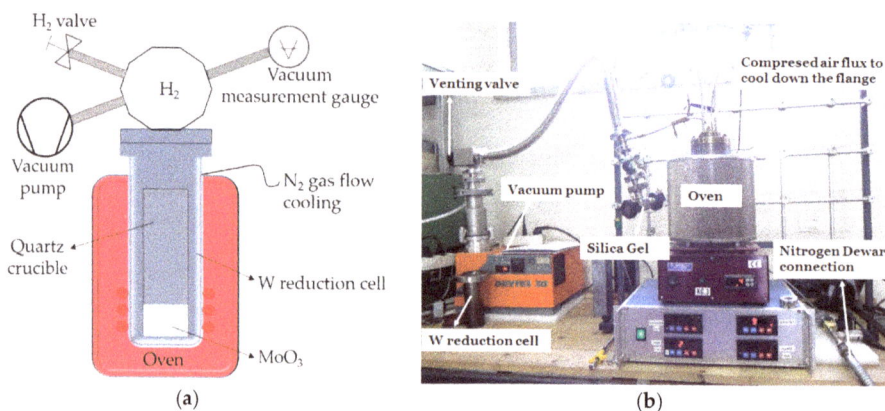

**Figure 4.** Reduction system scheme (**a**) and the laboratory prototype (**b**).

While other groups have used laboratory-scale hydrogenation systems [20] where the reduction process requires a constant hydrogen flow (i.e., an open system), in the present work we applied a different recovery method, using a batch overpressure reactor (i.e., a closed system). This approach has been chosen to fulfil safety regulations operating at LNL. Moreover, it is expected to be effective since the amount of oxide to be treated is lower with respect to the experiments presented by other groups. In order to shift the equilibrium towards the products in the chemical reaction, two techniques were simultaneously used: (i) carrying out the experiment in a hydrogen overpressure condition and (ii) condensing and capturing the water released during reduction by an appropriate amount of silica gel. Therefore, the reduction capability of the system depends on the maximum amount of hydrogen that can be inserted in safety conditions and on the water adsorption capability of the dried silica gel trap.

The full reduction system prototype developed in present work is schematically shown in Figure 4b. During reduction experiments, MoO₃ was placed in a quartz crucible inside the tungsten reduction reactor with the total volume (including the volume inside the tubes before the valves) of ~1.2 L. The system was pumped down to remove air, and then hydrogen in overpressure was inserted by the gas valve.

The resistive furnace was used to heat up the reduction cell. Nitrogen gas released from the liquid nitrogen (LN₂) Dewar was used for cooling down both the tube to condensate the water produced during the reaction and the tungsten reduction cell external walls. The nitrogen flux valve opened automatically when the oven temperature exceeded 250 °C. Silica gel (~10 g) was placed in a box inside the vacuum system for trapping the produced water. Compressed air was used to cool down the flange that closes the reduction cell on the top.

The Alcatel Drytel 30 (Anaheim, CA, USA) oil-free multistage high vacuum pumping system composed of a 7.5 L/s turbo drag pump and a 1 m$^3$/h diaphragm pump was used for preliminary evacuation of the reduction cell. The Drytel 30 can operate from atmosphere to ~1 × 10$^{-6}$ mbar, with a maximum pumping speed of 16 cfm (~27 m$^3$/h) in the high vacuum range.

The 1400 W resistive melting furnace (Giuseppe Mealli S.r.l., Firenze, Italy) was used as a heating source for the reduction process. It is equipped with Kanthal (Hallstahammar, Sweden) resistors to ensure high temperatures and long duration. The thermal insulation is made of ceramic fiber in order to achieve rapid heating in combination with low power consumption. The oven can reach a maximum temperature of 1150 °C.

The reduction reactor was realized in a tungsten–copper-sintered composite material allowing to combine the high melting temperature of tungsten with the good machinability of copper. Nitrogen gas flow was passing around the external walls of the reduction cell in order to minimize its oxidation at elevated temperatures.

The main parameters controlled during the process included

- oven temperature (0–1100 °C);
- flange temperature (0–500 °C);
- hydrogen pressure (0–8000 mbar absolute);
- nitrogen pressure (2–2000 mbar absolute).

A dedicated program written in Java language (version 1.7) was used to control the reduction system. It allowed to set up the multistep heating process. The heating parameters are shown in Table 1.

**Table 1.** Heating parameters for reduction control.

| Step | Temperature Range (°C) | Programmed Temperature Rate (°/min) |
|------|------------------------|-------------------------------------|
| 1 | Room temperature—350 | 5 |
| 2 | 350–750 | 2 |
| 3 | 750–950 | 5 |
| 4 | 950 | Constant for X h |
| 5 | 950—room temperature | Natural convection cooling |

In step four, the temperature was kept constant for X hours, depending on the starting amount of $MoO_3$. The heating parameters were modified, directly by the user during the process, through the GUI (graphical user interface). In case of emergency, a button allowed to immediately stop the process. If the flange temperature exceeds 350 °C, the program interrupts the process and the cooling starts.

The maximum amount of $MoO_3$ powder that can be reduced using 2500 mbar $H_2$ (270 mg) is 2.7 g (18 mmol). In this case, the amount of $H_2$ used for reduction is 108 mg. When the process is finished, $H_2$ final pressure of 1500 mbar (~1685 mbar theoretical) is obtained, thus maintaining the requirement of overpressure. The increase of the $MoO_3$ load is possible: in this case, the volume of the apparatus should be increased proportionally in order to maintain the corresponding hydrogen overpressure after the reduction. It should be also taken into account that the maximum pressure achieved inside the new apparatus during the process has to be controlled for the safety issue.

*2.3. Analysis of the Products*

The applied analytical techniques to characterize the recovery and reduction products comprised X-ray diffraction (XRD) and scanning electron microscopy equipped with energy-dispersive spectroscopy (EDS-SEM).

XRD analysis has been performed with X'Pert Philips PW3040/60 diffractometer equipped with 1.54 Å Cu-Kα X-ray in θ/θ scanning mode. The scans were taken at 40 kV, 40 mA of Cu-Kα X-ray gun. The PW3071/xx Bracket sample stage was used in reflection mode.

For powders analysis, previously grinded in agate mortar Si zero-background sample holder was used. A semiquantitative analysis with X'Pert HighScore software was applied without any additional calibration. The outcome of XRD analysis is supported and completed by SEM-EDS investigation in order to fully investigate both crystalline and amorphous phases. Moreover, the results in terms of structure, composition and quantification have been compared with those expected from the involved chemical reactions.

Fei (former Philips) Scanning Electron Microscope SEM XL-30 equipped with the QUATNTAX energy-dispersive X-ray spectrometer of BRUKER (Billerica, Massachusetts, United States) were used for analysis of powders. It should be mentioned that the inaccuracy of the quantitative elemental analysis by EDX method is not less than 5% [21].

Furthermore, the quantification due to mass change can be considered the most precise quantification method, provided an evidence that the set of the reactions and the products has been correctly identified. For the reduction of $MoO_3$, it was considered that the products could contain Mo and $MoO_2$ also on the basis of previous literature data [20,22]. Thus, only two possible reactions were taken into account:

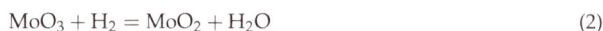

$$MoO_3 + 3H_2 = Mo + 3H_2O \tag{1}$$

$$MoO_3 + H_2 = MoO_2 + H_2O \tag{2}$$

In order to know the correct mass of the precursor, $MoO_3$ was baked in the vacuum oven to remove water traces before weighing.

## 3. Results and Discussion

### 3.1. MoO₃ Recovery from Mo-Rich Separation Module "Waste"

The starting material for the recovery procedure developed in the present study was coming from the separation module treatment of natural molybdenum foils. The module and the dissolution procedures are described in details by Martini et al. [12]. According to the EDS analysis of the dried Mo-rich "waste" fraction from the module, it contained mainly Mo, Na, and O, with a huge excess of Na in respect to Mo. According to the XRD analysis, the mixture contained sodium molybdate $Na_2MoO_4$·, disodium dimolybdate $Na_2Mo_2O_7$, sodium tetramolybdenum hexoxide $NaMo_4O_6$, and sodium peroxide $Na_2O_2$ (from partial conversion of sodium hydroxide NaOH).

Following the literature approach called "spent catalyst regeneration" (see Recipe MoO₃-1) a white-yellowish powder was obtained, containing Mo, Na, O, N, and Cl according to EDS analysis. Hence, we can suggest that $MoCl_6$, $Na_2MoO_4(H_2O)_x$, and $(NH_4)_2MoO_4(H_2O)_y$ were present on the basis of starting reagents and chemical reactions involved. Thus, the applied procedure appeared to be not efficient to transform all sodium molybdate from the mixture coming from the separation module into ammonium molybdate.

In order to minimize the amount of $MoCl_6$ impurity, hydrochloric acid was replaced by nitric acid (see Recipe MoO₃-2). According to EDS analysis, ~2% Na and ~11% Mo were still detected in the product. Thus, the second method also did not provide the full conversion of sodium molybdate into ammonium molybdate as expected.

The last method proposed here, described in Scheme (3), included a direct conversion to molybdenum oxide and molybdic acids, thus skipping the ammonium molybdate step (see Recipe MoO₃-3).

$$Na_2MoO_4 + 2HNO_3(excess) \xrightarrow{\text{80 °C, reflux}} 2NaNO_3 + H_2MoO_4$$

$$\Delta T$$

$$MoO_3\downarrow \quad + \quad H_2O\uparrow \tag{3}$$

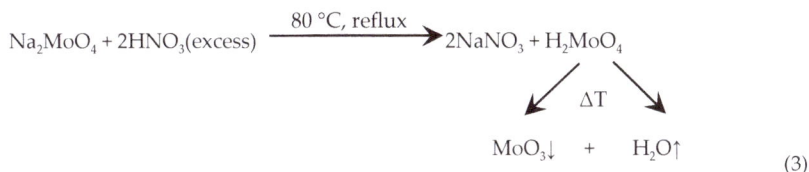

After precipitation, molybdic acids and sodium nitrate were separated on the basis of different water solubility [23]. Then, the solid was dried in a vacuum and analyzed by EDS and XRD. The structural analysis confirmed the presence of $MoO_3$ and $Na_2MoO_4$, while the EDS analysis led to quantify 0.3% Na and 15% Mo, thereby suggesting an enhanced separation efficiency with respect to previous experiments.

Furthermore, it has to be noted that Si impurity was detected in all the products coming from the separation module. It was supposed that Si impurity was produced by storing the product (Mo-rich waste from separation module) in glass vessels for several months' (simulating the way to operate with enriched material). In fact, sodium molybdate fraction has a high concentration of residual sodium hydroxide, thus, glass corrosion is likely to occur. It is worth to highlight that for the repeated cycles of recovery the absence of the impurities is a crucial factor. In particular, in the case of irradiation of recovered Mo with traces of Si at 10–25 MeV a set of undesired reaction channels can be opened. In order to avoid the described problem and unambiguously identify the source of Si the process of dissolution was reproduced starting from $^{nat}$Mo foil (see Recipe MoO$_3$-4), avoiding the storage in a glass vessel and realizing the recovery procedure immediately after the reproduced dissolution procedure. Two tests have been performed in parallel using in one case standard chemical glassware and in the other case Teflon labware. In both cases, Si impurity was not detected by EDS. This experiment unambiguously demonstrated, that silicon came from the glass vials corroded during the long term storage of the highly basic Mo-rich "waste" and that this contamination can be easily avoided.

It was found that the most efficient method to recover $MoO_3$ from the NaOH-rich waste of the $^{100}$Mo/$^{99m}$Tc-separation module is the one reported as Recipe MoO$_3$-4. Possible products are $Na_2MoO_4$, $NaNO_3$, and $MoO_3$. The same atomic content of Na and N was observed by EDS. Thus, $Na_2MoO_4$ should not be present any more in the mixture. In this case, the product contains $MoO_3$ with few $NaNO_3$ and, as expected, without any Si impurity. According to EDS analysis, Na impurity was present in ~3% at., (or recalculated ~2% weight $NaNO_3$) in $MoO_3$ product. This method allows to remove Na and leads to a $MoO_3$ yield of ~92%, which can be further improved by optimizing the precipitate washing conditions or using an ion exchange resin.

*3.2. MoO$_3$ Reduction to Mo Metallic*

The reduction of molybdenum (VI) oxide is realized in two steps:

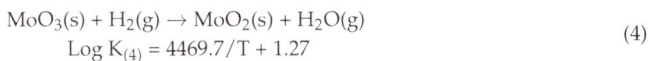

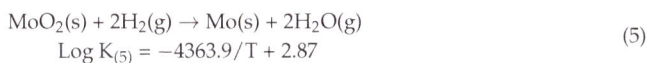

$$MoO_3(s) + H_2(g) \rightarrow MoO_2(s) + H_2O(g)$$
$$\text{Log } K_{(4)} = 4469.7/T + 1.27 \tag{4}$$

$$MoO_2(s) + 2H_2(g) \rightarrow Mo(s) + 2H_2O(g)$$
$$\text{Log } K_{(5)} = -4363.9/T + 2.87 \tag{5}$$

The first step of the conversion (Equation (4)) is exothermic and thermodynamically favorable at 550–600 °C [24,25]. The exothermic reaction can cause local overheating and $MoO_3$ volatilization. This aspect limits the velocity of heating. On the other hand, the second step (Equation (5)) is endothermic, and relatively high temperature and $H_2/H_2O$ ratio higher than two are required to achieve hydrogen reduction of $MoO_2$. According to the literature [24], the temperature necessary to carry on the second stage of the reduction can be varied in the range from 930 °C to 1000 °C depending upon the desired

Mo powder size. The lower the reduction temperature, the smaller the reduced Mo powder size. However, for low reduction temperatures, a longer reduction time is required in order to achieve a complete conversion [24]. The detailed description of the kinetics of the second step can be found in the work by Kim et al. [26].

The procedure for reduction under $H_2$ gas flow and the 3-step conversion starting from ammonium molybdate were described by Gupta [27] for nonenriched isotopes and by Gagnon et al. for enriched elements recovery [20]. Here, the reduction from $MoO_3$ to $MoO_2$ was considered to be completed at 500–750 °C keeping the heating rate at 2 °C/min, and the further reduction to metallic Mo was performed at 750–1100 °C with heating rate 5 °C/min and reduction time of 1 h at maximum temperature. An overall Mo recovery of 87% was reported [20].

The reduction thermal cycle optimized in the current work is presented in Figure 5. The heating velocity of 5 °C/min was used up to 450 °C, then the heating velocity was decreased to 2 °C/min until the temperature of 750 °C was reached. In this temperature range, the conversion to $MoO_2$ takes place. A lower heating ramp was used to minimize the sublimation of $MoO_3$, induced by local overheating caused by the exothermic reaction. The heating from 750 °C to 950 °C was realized with a 5 °C/min ramp rate. Then, the reduction from $MoO_2$ to Mo was performed at 950 °C, setting different time intervals for each experiment, as listed in Table 2. Subsequent cooldown was obtained through natural convection, simply by switching off the furnace. The product was exposed to air only after the system had reached the room temperature.

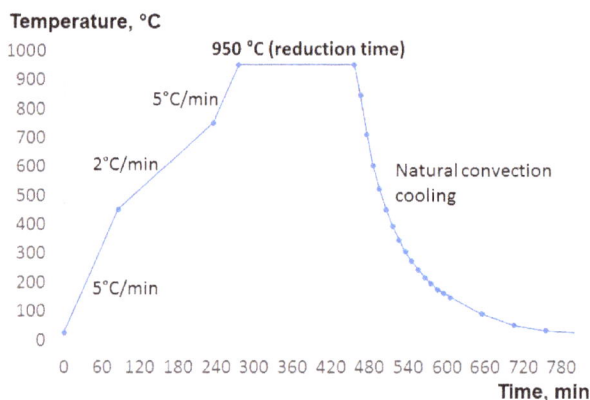

**Figure 5.** The optimized $MoO_3$ reduction thermal cycle.

**Table 2.** $MoO_3$ reduction experiments.

| Red. exp-t | $MoO_3$ Origin | $MoO_3$ Mass | Red. Time | Product Mass | Product Content, Mass % | | |
|---|---|---|---|---|---|---|---|
| | | | | | Mo | $MoO_2$ | Impurities |
| R-1 | Alfa Aesar | 587 mg | 1 h | 500 mg | ~12% | ~88% | C (<1%) |
| R-2 | Alfa Aesar | 389 mg | 6 h | 261 mg | ~98% | <2% | C (<1%) |
| R-3.1 | Alfa Aesar | 1 g | 1.5 h | 864 mg | ~15% | ~85% | C (<1%) |
| R-3.2 | Product of R-3.1 | 864 mg | 2 h | 687 mg | >90% | >2% | C (<2%) |
| R-4 | Product of rec. $MoO_3$-2 | 61 mg | 2 h | 37 mg | ~90% | - | $SiO_2$ (~9%) + $Na_2SiO_3$ (<1%) |
| R-5 | Product of rec. $MoO_3$-3 | 49 mg | 2 h | 34 mg | >95% | - | $SiO_2$ (<2%) |

### 3.2.1. Reduction of Commercial $MoO_3$

In order to develop the reduction method, commercially available $MoO_3$ powders from Sigma Aldrich were used. The grain size of the powders was <5 μm.

The parameters and the results of the reduction experiments are listed in Table 2. The first reduction experiment R-1, starting from ~0.6 g of commercial $MoO_3$, was realized with just 1 h reduction time at 950 °C. According to the results of the analyses (XRD, EDS, and quantification based on the mass of the product), this time was not sufficient to convert all MoO$_3$ into metallic Mo. Nevertheless, no MoO$_3$ was detected in the final product by XRD, thus indicating that the first step of the process (Equation (4)) was completed.

Increasing the reduction time to 6 h in the following experiment R-2 allowed us to obtain more than 95% of Mo metallic starting from ~0.4 g of commercial MoO$_3$ (Table 2). According to EDS analysis, still a few % of MoO$_2$ remained unreduced. The yield of the reaction is ~98%. It should be said, that it was possible to distinguish even visually the products R-1 rich of hygroscopic MoO$_2$ (see Figure 6a) and the R-2 product containing almost only metallic Mo powders (Figure 6b). The XRD spectra, normalized on the intensity of the most intense peaks, obtained from the reduction products are shown in Figure 7.

(a)                              (b)

**Figure 6.** MoO$_3$ reduction products: R-1 hygroscopic powder rich of MoO$_2$ (a) and R-2 almost only Mo metallic (b).

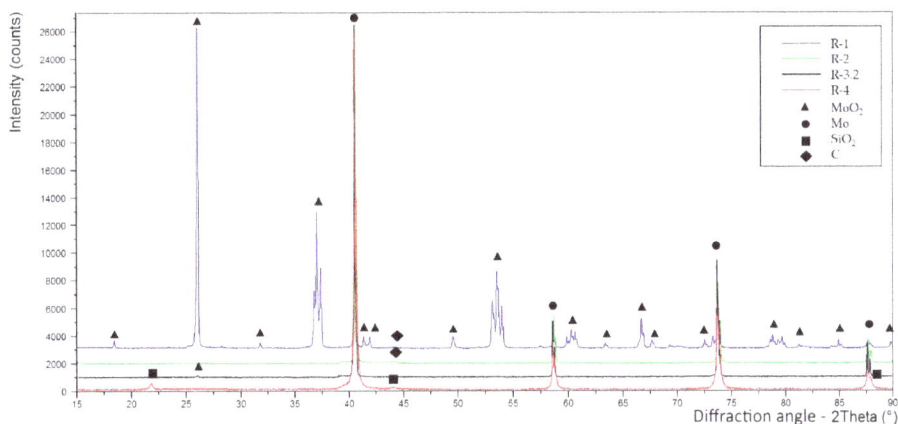

**Figure 7.** X-ray diffraction (XRD) spectra of the products of different reduction experiments.

The next experiment, with a higher amount of starting MoO$_3$, was realized in two steps. The R-3.1 during 1.5 h of reduction allowed to get only ~10% of Mo metallic (see Table 2). In order to convert more than 90% of molybdenum oxide to Mo metallic (Table 2), the additional 2 h of reduction were required (R-3.2).

Some carbon-based impurities were detected in the product by XRD (see Figure 7), probably related to previous heating of a graphite crucible in the same furnace. In order to avoid such contamination in the future, the tungsten reactor was mechanically polished.

The reduced molybdenum powders are characterized by the grain size of about 1 μm and display nonspherical, slightly irregular shape. The SEM analysis of the products of the reduction R-2 and R-3 is shown in Figure 8.

**Figure 8.** SEM of the products of reduction R-2 (**a**) and R-3.2 (**b**).

3.2.2. Reduction of Recovered $MoO_3$

Several reduction experiments starting from $MoO_3$ recovered from the module Mo-rich waste were realized. It should be taken into account that the precursor ($MoO_3$ from the recovery procedure) was not as pure as the commercial $MoO_3$.

From the analysis of the $MoO_3$ precursor coming from the recovery procedure, it was evident that it contained the impurities of glass $SiO_2$ and $Na_2SiO_3$. From the reduction product analysis (Table 2), it can be seen that the glass contaminants remained, when $MoO_3$ was reduced to Mo metallic. $Na_2SiO_3$ was not detected by XRD, owing to its amorphous nature. The quantification on the basis of reaction stoichiometry calculations is not reliable because of the glass contaminants. Nevertheless, since no $MoO_2$ was detected by XRD (see Figure 7), the reduction of $MoO_3$ could be considered complete.

Similarly to what observed in the previous experiment, the $SiO_2$ impurity coming from the $MoO_3$ recovered from the separation module waste remained in the product (Table 2). The appearance of the product was very similar to the R-2 metallic powders (see Figure 9a). From Figure 9b, showing the EDS mapping of the reduction product, it can be clearly observed that oxygen is associated with Si to form $SiO_2$. Moreover, according to EDS elemental analysis, the ratio between the atomic composition of Si and O is 1:2, thus indicating that all oxygen is associated with Si, and there is no excess oxygen to form molybdenum oxide. Therefore, the complete transformation of $MoO_3$ was demonstrated by the absence of $MoO_2$ in the product. For both experiments R-4 and R-5, two hours at 950 °C was sufficient to complete the reduction of ~50 mg $MoO_3$ to Mo metallic.

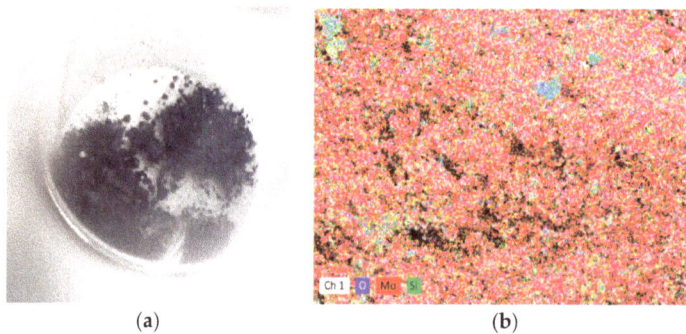

**Figure 9.** R-5 $MoO_3$ reduction product (**a**) and its energy-dispersive spectroscopy (EDS) map analysis (**b**).

The developed method allows for the recovery of Mo in metallic form, starting from the Mo-rich "waste" coming from the radiochemical separation module, in two steps with yields of approximately 92% and 98%, respectively, ultimately leading to the overall yield of more than 90%.

### 3.2.3. Recovered Mo Applicability for Further Cyclotron Target Preparation

Among the Mo cyclotron solid target preparation methods developed in TECHN-OSP project, the reduced Mo powders have been successfully tested only for electrostatic deposition (HIVIPP) technique. Mo target with ~3-μm thickness deposited by HIVIPP technique onto aluminium backing starting from reduced Mo powders coming from the experiment R-2 is shown in Figure 10a.

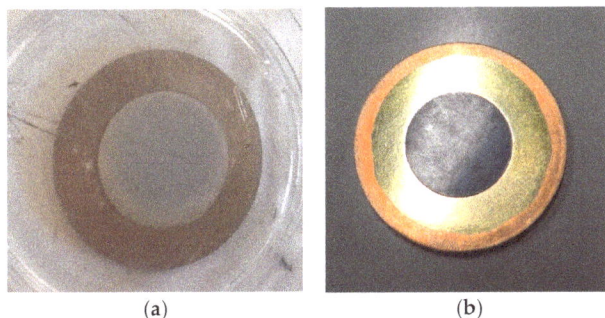

(a)                                    (b)

**Figure 10.** Mo targets prepared with high energy vibrational powders plating (HIVIPP) method starting from reduced powders (**a**) and spark plasma sintering (SPS)-sintered target from $^{100}$Mo of Isoflex (**b**).

From SEM analysis it was observed that Mo recovered powders (see Figure 8) had a very similar shape and size with respect to the commercial $^{100}$Mo-enriched powders (99.05% isotopic enrichment) of Isoflex (San Francisco, CA, USA), shown in Figure 11. This ensures the possibility to use recovered powders for the target preparation also using the SPS technique, as the $^{100}$Mo-enriched powders of Isoflex have been used (see Figure 10b).

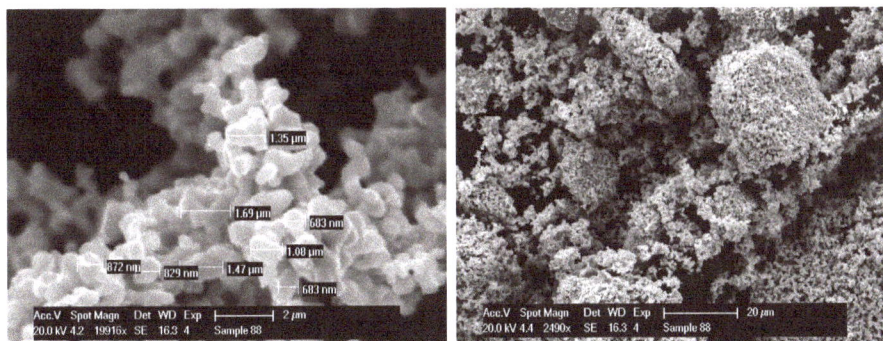

**Figure 11.** SEM analysis of the $^{100}$Mo commercial powders of Isoflex (USA).

Since all the Mo recovery tests presented in this work were realized with nonenriched Mo, the further study dedicated to a validation of the proposed methods using enriched Mo powders is essential and is currently in progress.

## 4. Conclusions

In order to get more attractive from the economic point of view the new, cyclotron-based, $^{99m}$Tc alternative production route, a procedure to close the loop and recover the costly $^{100}$Mo-enriched starting material from the separation module waste under metallic form was developed. The proposed method allowed to perform the recovery of MoO$_3$ with a 92% yield.

For the next recovery steps, a reliable method able to provide the reduction of batches of MoO$_3$ up to 1 g, either from the commercial supplier or using the "working" material coming from the separation module has been developed. Reduction yields up to 98% were achieved with a reduction time no longer than 4 h (+5 h heating). By increasing the reduction time, it may be possible to improve the transformation to metallic Mo or reduce higher amounts of MoO$_3$ in the same reactor. The overall yield of the two steps higher than 90% was achieved.

The recovered Mo powders have been successfully used for preparation of the targets by applying HIVIPP technique. Moreover, the powders display size and shape similar to the commercial $^{100}$Mo-enriched powders available from Isoflex, as evidenced by SEM analyses. Therefore, it can be inferred that the SPS technique can also be adopted to produce the Mo targets using recovered powder.

**Author Contributions:** The Conceptualization of the Mo reduction system was made by P.B. The Concept of the MoO$_3$ precipitation method by use of nitric acid was proposed by S.C. (Sara Carturan). The Methodology for MoO$_3$ recovery has been developed by H.S. and S.C. (Sara Carturan). The Methodology for MoO$_3$ reduction has been developed by C.R.A. and has been Validated by H.S. and S.C. (Sara Cisternino). Software for the reduction furnace automatic control was realized by C.R.A. Investigation of the content of the starting materials and all the products of recovery experiments was realized by H.S. Resources of the lab for MoO$_3$ recovery experiments have been provided by S.C (Sara Carturan) and J.E. Data Curation was the responsibility of H.S. and C.R.A. The $^{nat}$Mo-rich "waste" from the separation module, as the starting material for recovery procedure, has been provided by P.M. and A.B. Original Draft Preparation has been realized by H.S. Writing—Review & Editing, H.S., S.C. (Sara Cisternino), C.R.A., S.C. (Sara Carturan), J.E., A.B., and P.M. Work on visualization was realized by H.S.; J.E. as a responsible of TECHN-OSP project was in charge of work Supervision, Administration, and Funding Acquisition.

**Funding:** This research was realized in the framework of TECHN_ OSP project funded by CSN5 of the Istituto Nazionale di Fisica Nucleare, Italy for 2015-2017. National responsible: J. Esposito, INFN-LNL.

**Acknowledgments:** Special thanks should be given to A. Zanon and P. Bezzon for electronics and vacuum equipment support of the reduction apparatus. We are grateful to V. Palmieri responsible of the INFN-LNL Laboratories for Surface & Material Treatments for Nuclear Physics for the access to XRD and SEM. We also appreciate the help of the Mechanical workshop of INFN-LNL.

**Conflicts of Interest:** The authors declare no conflict of interest.

## References

1. SNMMI Addresses Mo-99 Isotope Shortage—SNMMI. Available online: http://www.snmmi.org/NewsPublications/NewsDetail.aspx?ItemNumber=30394 (accessed on 10 February 2019).
2. International Atomic Energy Agency. *Feasibility of Producing Molybdenum-99 on a Small Scale Using Fission of Low Enriched Uranium or Neutron Activation of Natural Molybdenum*; International Atomic Energy Agency: Vienna, Austria, 2015; ISBN 978-92-0-114713-4.
3. IAEA. *Cyclotron Based Production of Technetium-99m*; IAEA: Vienna, Austria, 2017; ISBN 978-92-0-102916-4.
4. Martini, P.; Boschi, A.; Cicoria, G.; Zagni, F.; Corazza, A.; Uccelli, L.; Pasquali, M.; Pupillo, G.; Marengo, M.; Loriggiola, M.; et al. In-house cyclotron production of high-purity Tc-99m and Tc-99m radiopharmaceuticals. *Appl. Radiat. Isot.* **2018**, *139*, 325–331. [CrossRef]
5. Gagnon, K.; Bénard, F.; Kovacs, M.; Ruth, T.J.; Schaffer, P.; Wilson, J.S.; McQuarrie, S.A. Cyclotron production of 99mTc: Experimental measurement of the 100Mo(p,x)99Mo, 99mTc and 99gTc excitation functions from 8 to 18 MeV. *Nucl. Med. Biol.* **2011**, *38*, 907–916. [CrossRef] [PubMed]
6. Benard, F.; Buckley, K.R.; Ruth, T.J.; Zeisler, S.K.; Klug, J.; Hanemaayer, V.; Vuckovic, M.; Hou, X.; Celler, A.; Appiah, J.-P.; et al. Implementation of Multi-Curie Production of 99mTc by Conventional Medical Cyclotrons. *J. Nucl. Med.* **2014**, *55*, 1017–1022. [CrossRef] [PubMed]

7.  Schaffer, P.; Bénard, F.; Bernstein, A.; Buckley, K.; Celler, A.; Cockburn, N.; Corsaut, J.; Dodd, M.; Economou, C.; Eriksson, T. Direct Production of 99m Tc via 100 Mo (p, 2n) on Small Medical Cyclotrons. *Phys. Procedia* **2015**, *66*, 383–395. [CrossRef]

8.  Sodium Pertechnetate (99mTc) Injection (Accelerator produced) Monograph No. 2891. January 2018: 2891. Volume 9.3, World Health Organization: Geneva, Switzerland, 2018; European Pharmacopoeia 9.3. 4801–4803.

9.  Andersson, J.D.; Thomas, B.; Selivanova, S.V.; Berthelette, E.; Wilson, J.S.; McEwan, A.J.B.; Gagnon, K. Robust high-yield ~1TBq production of cyclotron based sodium [99mTc]pertechnetate. *Nucl. Med. Biol.* **2018**, *60*, 63–70. [CrossRef] [PubMed]

10. Esposito, J.; Vecchi, G.; Pupillo, G.; Taibi, A.; Uccelli, L.; Boschi, A.; Gambaccini, M. Evaluation of Mo99 and Tc 99m Productions Based on a High-Performance Cyclotron. *Sci. Technol. Nucl. Ins.* **2013**, *2013*, 1–14.

11. Boschi, A.; Martini, P.; Pasquali, M.; Uccelli, L. Recent achievements in Tc-99m radiopharmaceutical direct production by medical cyclotrons. *Drug Dev. Ind. Pharm.* **2017**, *43*, 1402–1412. [CrossRef]

12. Martini, P.; Boschi, A.; Cicoria, G.; Uccelli, L.; Pasquali, M.; Duatti, A.; Pupillo, G.; Marengo, M.; Loriggiola, M.; Esposito, J. A solvent-extraction module for cyclotron production of high-purity technetium-99m. *Appl. Radiat. Isot.* **2016**, *118*, 302–307. [CrossRef]

13. Skliarova, H.; Cisternino, S.; Cicoria, G.; Marengo, M.; Palmieri, V. Innovative Target for Production of Technetium-99m by Biomedical Cyclotron. *Molecules* **2019**, *24*, 25. [CrossRef]

14. Palmieri, V.; Skliarova, H.; Cisternino, S.; Marengo, M.; Cicoria, G. Method for Obtaining a Solid Target for Radiopharmaceuticals Production. International patent application PCT/IB2018/056826, 7 September 2018. National Institute of Nuclear Physics, deposition reference P1183PC00.

15. Sugai, I. An application of a new type deposition method to nuclear target preparation. *NIM A* **1997**, *397*, 81–90. [CrossRef]

16. Uzunov, N.M.; Melendez-Alafort, L.; Bello, M.; Cicoria, G.; Zagni, F.; De Nardo, L.; Selva, A.; Mou, L.; Rossi-Alvarez, C.; Pupillo, G.; et al. Radioisotopic purity and imaging properties of cyclotron-produced $^{99m}$Tc using direct $^{100}$Mo($p, 2n$) reaction. *Phys. Med. Biol.* **2018**, *63*, 185021. [CrossRef] [PubMed]

17. Gumiela, M. Cyclotron production of 99m Tc. *Nucl. Med. Biol.* **2018**, *58*, 33–41. [CrossRef] [PubMed]

18. Park, K.H.; Reddy, B.R.; Mohapatra, D.; Nam, C.-W. Hydrometallurgical processing and recovery of molybdenum trioxide from spent catalyst. *Int. J. Miner. Process.* **2006**, *80*, 261–265. [CrossRef]

19. Kar, B.B.; Datta, P.; Misra, V.N. Spent catalyst: secondary source for molybdenum recovery. *Hydrometallurgy* **2004**, *72*, 87–92. [CrossRef]

20. Gagnon, K.; Wilson, J.S.; Holt, C.M.B.; Abrams, D.N.; McEwan, A.J.B.; Mitlin, D.; McQuarrie, S.A. Cyclotron production of 99mTc: Recycling of enriched 100Mo metal targets. *Appl. Radiat. Isot.* **2012**, *70*, 1685–1690. [CrossRef] [PubMed]

21. Statham, P.J. Limitations to Accuracy in Extracting Characteristic Line Intensities From X-Ray Spectra. *J. Res. Natl. Inst. Stand. Technol.* **2002**, *107*, 531–546. [CrossRef] [PubMed]

22. Arnoldy, P.; de Jonge, J.C.M.; Moulijn, J.A. Temperature-Programmed Reduction of MoO$_3$, and MoO$_2$. *J. Phys. Chem.* **1985**, *89*, 4517–4527. [CrossRef]

23. Ferris, L.M. Solubility of Molybdic Oxide and Its Hydrates in Nitric Acid, Nitric Acid-Ferric Nitrate, and Nitric Acid-Uranyl Nitrate Solutions. *J. Chem. Eng. Data* **1961**, *6*, 600–603. [CrossRef]

24. Pak, J.-J.; Jo, J.-O.; Park, C.-H.; Kang, J.-G.; Shin, D.-H. Recovery of Molybdenum from Spent Acid by Ammonia Gas Neutralization. *Mater. Trans.* **2008**, *49*, 202–207. [CrossRef]

25. Schulmeyer, W.V.; Ortner, H.M. Mechanisms of the hydrogen reduction of molybdenum oxides. *Int. J. Refract. Metals Hard Mater.* **2002**, *20*, 261–269. [CrossRef]

26. Kim, B.-S.; Kim, E.; Jeon, H.-S.; Lee, H.-I.; Lee, J.-C. Study on the Reduction of Molybdenum Dioxide by Hydrogen. *Mater. Trans.* **2008**, *49*, 2147–2152. [CrossRef]

27. Gupta, C.K. *Extractive Metallurgy of Molybdenum*; CRC Press: Boca Raton, FL, USA, 1992; ISBN 978-0-8493-4758-0.

*instruments*

MDPI

*Article*

# Development of New Target Stations for the South African Isotope Facility

Gideon F. Steyn *, Lyndon S. Anthony, Faiçal Azaiez, Shadley Baard, Robert A. Bark, A. Hugo Barnard, Philip Beukes, Johan I. Broodryk, J. Lowry Conradie, John C. Cornell, J. Garrett de Villiers, Stuart G. Dolley, Herman du Plessis, William D. Duckitt, Dirk T. Fourie, Mike E. Hogan, Ivan H. Kohler, Jacobus J. Lawrie, Chris Lussi, Joele P. Mira, K. Vuyo Mjali, Hendrik W. Mostert, Clive Naidoo, Fhumulani Nemulodi, David Saal, Nieldane P. Stodart, Reiner W. Thomae, Johan van Niekerk and Pieter A. van Schalkwyk

iThemba LABS, National Research Foundation, Old Faure Road, Somerset West 7129, South Africa; lyndon@tlabs.ac.za (L.S.A.); director@tlabs.ac.za (F.A.); sbaard@tlabs.ac.za (S.B.); bark@tlabs.ac.za (R.A.B.); hbarnard@tlabs.ac.za (A.H.B.); philip@tlabs.ac.za (P.B.); jibroodryk@tlabs.ac.za (J.I.B.); lowry@tlabs.ac.za (J.L.C.); dr.jc.cornell@gmail.com (J.C.C.); gdevil@tlabs.ac.za (J.G.d.V.); stuart@tlabs.ac.za (S.G.D.); herman@tlabs.ac.za (H.d.P.); wduckitt@tlabs.ac.za (W.D.D.); dirk@tlabs.ac.za (D.T.F.); hogan@tlabs.ac.za (M.E.H.); ivan@tlabs.ac.za (I.H.K.); lawrie@tlabs.ac.za (J.J.L.); chrislussi@tlabs.ac.za (C.L.); mira@tlabs.ac.za (J.P.M.); vmjali@tlabs.ac.za (K.V.M.); hendric@tlabs.ac.za (H.W.M.); clive@tlabs.ac.za (C.N.); fnemulodi@tlabs.ac.za (F.N.); dsaal@tlabs.ac.za (D.S.); nieldane@tlabs.ac.za (N.P.S.); rthomae@tlabs.ac.za (R.W.T.); johanvn@tlabs.ac.za (J.v.N.); tandem@tlabs.ac.za (P.A.v.S.)
* Correspondence: deon@tlabs.ac.za; Tel.: +27-21-843-1000

Received: 9 November 2018; Accepted: 6 December 2018; Published: 10 December 2018

**Abstract:** The development of new target stations for radioisotope production based on a dedicated 70 MeV commercial cyclotron (for protons) is described. Currently known as the South African Isotope Facility (SAIF), this initiative will free the existing separated-sector cyclotron (SSC) at iThemba LABS (near Cape Town) to mainly pursue research activities in nuclear physics and radiobiology. It is foreseen that the completed SAIF facility will realize a three-fold increase in radioisotope production capacity compared to the current programme based on the SSC.

**Keywords:** radionuclide production; target stations; targetry; 70 MeV cyclotron

## 1. Introduction

iThemba LABS is a multi-disciplinary research facility under the management of the National Research Foundation (NRF) of South Africa. The facility currently operates six accelerators at two campuses, located at Faure (near Cape Town) and Johannesburg. The main activities comprise research and training in the physical, biomedical and material sciences, provision of accelerator mass spectrometry (AMS) services, and the production of radioisotopes and radiopharmaceuticals for use in nuclear medicine. This paper discusses an initiative to achieve a three-fold increase in radioisotope production capacity.

The largest of the accelerators at iThemba LABS is a $K = 200$ separated sector cyclotron (SSC) [1] that has been in operation for more than 30 years. This machine is operated in conjunction with two solid-pole injector cyclotrons, one for high-intensity proton beams and one for heavy ions and polarized protons. The SSC is a shared facility. It provides accelerated ion beams for radioisotope production, nuclear physics research and, until recently, neutron and proton therapy. Nuclear physics experiments have mostly been conducted over weekends, the rest of the week scheduled for therapy (during the day) and production of both short-lived and long-lived radioisotopes (mostly at night).

A decision was recently taken that iThemba LABS would not continue with hadron therapy but would rather assist the medical community to pursue a dedicated proton therapy centre for South

Africa. The significant progress in recent years on dedicated proton therapy centres as well as the reduction in cost of these facilities (single treatment-room facilities) will benefit the country much more if established at one of the large public hospitals. In contrast, while the therapy facilities at iThemba LABS are still fully functional, they are becoming aged and can only provide limited services due to the restricted beam time available for therapy.

The sharing of beam time on the SSC is also limiting the other programmes. To increase the beam time for radioisotope production and nuclear physics research, the establishment of a new facility, currently known as the South African Isotope Facility (SAIF), has been proposed. The new facility will consist of two parts:

1. The Accelerator Centre for Exotic Isotopes (ACE Isotopes) will be a dedicated facility for radioisotope production. A commercial high-current 70 MeV $H^-$ cyclotron for this purpose will free the SSC and allow an increase in beam time for nuclear physics and related research. Recently, the South African Government approved the ACE Isotopes project and made available the first funding to finance the cyclotron.
2. The Accelerator Centre for Exotic Beams (ACE Beams) will be a radioactive ion beam (RIB) facility for nuclear physics research. The SSC will be used as a driver for an isotope separation on-line (ISOL) facility. A 66 MeV proton beam of up to 50 µA will be delivered by the SSC for producing radioactive beams from a target ion source. Phase 1 of this project will be a low-energy radioactive ion beam (LERIB) project without post acceleration. Phase 2 will be the post acceleration of the radioactive beams with a linear accelerator to energies between 4 and 5 MeV per nucleon. Some of the funding has already been secured for the development of LERIB and work on that project is in progress.

With the above "roadmap" for the future of iThemba LABS in place, a design study on new target stations for radionuclide production with a dedicated 70 MeV cyclotron commenced. This work is now in an advanced stage.

The main aspects of the new stations for batch targets (encapsulated materials and solids) are discussed below. The experience gained on the existing bombardment infrastructure largely influenced the new target station design. A brief historical overview is therefore given, followed by discussing the new station design and the changes necessary to the current facilities at iThemba LABS to accommodate them.

## 2. Materials and Methods

### 2.1. Existing Facilities

The routine production of radioisotopes with the SSC started in 1988 with the commissioning of the horizontal beam target station (HBTS, also called Elephant), mainly for short-lived isotopes such as $^{52}$Fe, $^{67}$Ga, $^{109}$Cd, $^{111}$In and $^{123}$I. Towards the late 1990s, production methods for the long-lived radioisotope $^{22}$Na were developed and commercialized. This was followed by targetry development for $^{68}$Ge and $^{82}$Sr. In 1996, a second target station was introduced for the bombardment of semi-permanent targets, including an enriched $^{18}$O-water target for $^{18}$F production, which was later transferred to a dedicated 11 MeV Siemens cyclotron. In 2006, the vertical beam target station (VBTS) was commissioned to exploit high-intensity proton beams delivered by the upgraded SSC. Target development for the VBTS focused on the relatively long-lived, high-value radioisotopes $^{22}$Na, $^{68}$Ge and $^{82}$Sr. Various tandem targets, e.g., Rb/Ga for $^{82}$Sr/$^{68}$Ge production and Mg/Ga for $^{22}$Na/$^{68}$Ge production, are routinely bombarded in the VBTS with 66 MeV proton beams with intensities up to 250 µA.

A beam splitter commissioned in 2007 allowed the simultaneous bombardment of four targets, i.e., beams on tandem targets in the HBTS and VBTS at the same time.

Thus, over the years several upgrades were undertaken to increase the production yield, especially for the long-lived radioisotopes, to meet an ever increasing market demand. The options for further growth on the SSC, however, have now been exhausted. Further growth can only be met by introducing a significant increase in allocated beam time for radioisotope production, which would come at a large expense to the other programmes, or by procuring a dedicated cyclotron for this purpose. This last option was chosen and will be achieved with the ACE Isotopes facility.

### 2.2. ACE Isotopes

The new radioisotope production facility will make use of three thick-walled concrete vaults previously occupied by the radiotherapy programme, as shown in Figure 1.

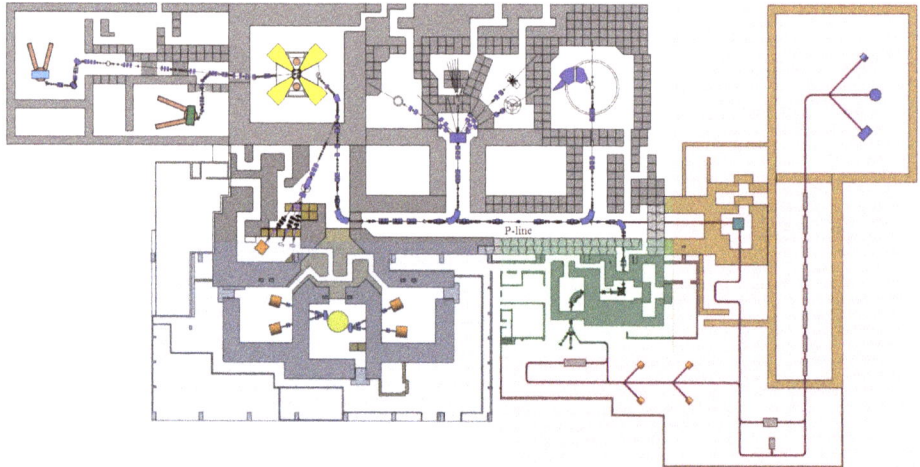

**Figure 1.** Layout of the cyclotron facility at iThemba LABS, showing ACE Isotopes (shaded blue), Phase 1 of ACE Beams, LERIB (green) and Phase 2 of ACE Beams (pink).

The layout of the ACE Isotopes facility is shown in Figure 2. The cyclotron will be located in the centre vault, with the bombardment of targets taking place in the two adjacent vaults. This will ensure flexibility since the production vaults are independent of each other, thus production can continue in one vault while maintenance is being performed in the other.

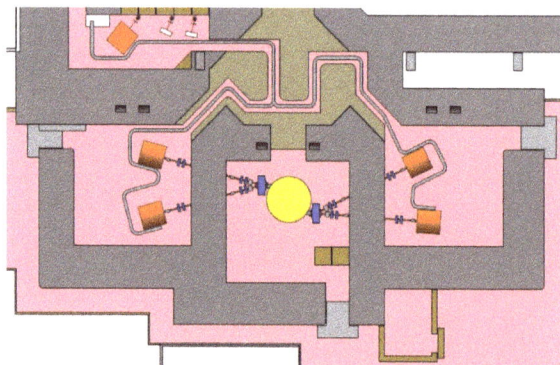

**Figure 2.** Layout of the ACE Isotopes facility, showing the cyclotron vault (middle), production vaults (left and right) and the expanded Telelift target transport system.

Each production vault will contain two target stations, classified as either "high intensity" (up to 350 µA) for production of long-lived radioisotopes or "medium intensity" (up to 100 µA) for production of short-lived radioisotopes. To maintain uninterrupted bombardment of batch targets, two high-intensity stations (one per production vault) and one medium-intensity station will be required. The fourth station will be dedicated to experimental development work and semi-permanent targets. The four beamlines all have the same design. An extracted beam will pass through a switching magnet and then continues through a straight section of beam pipe to a target station. Two sets of quadrupole doublets per beamline will be used for focussing. A "wobbler" system for sweeping the beam over the target surface will be located after the last quadrupole. Diagnostic systems will include profile grids for low-intensity beams and non-destructive capacitive beam position monitors for high-intensity beams.

### 2.3. New Target Stations

The transfer of the radioisotope production programme from the SSC to ACE Isotopes will take several years to complete. The existing target stations will continue to receive beam from the SSC while the new target stations are being built and commissioned.

Currently, the transfer of batch targets between the two existing target stations (HBTS and VBTS) and the processing hot-cell complex is by means of a Telelift rail system [2,3]. This system will be expanded to the two new production vaults. The existing PC-based transporter control system will also be modernized. Details on the targetry can be found in Ref. [2] and references therein.

The design of the new batch-target stations, shown in Figure 3, is now in an advanced stage. The four planned stations will be identical in all respects except for the entrance collimator aperture, which will be larger in the high-intensity stations. The local radiation shield will consist of complementary shielding materials, namely an inner iron shield, followed by borated paraffin wax as a middle layer, followed by an outer lead $\gamma$-ray shield. Monte Carlo radiation transport simulations provided the optimal layer thicknesses to achieve a dose attenuation of three orders of magnitude.

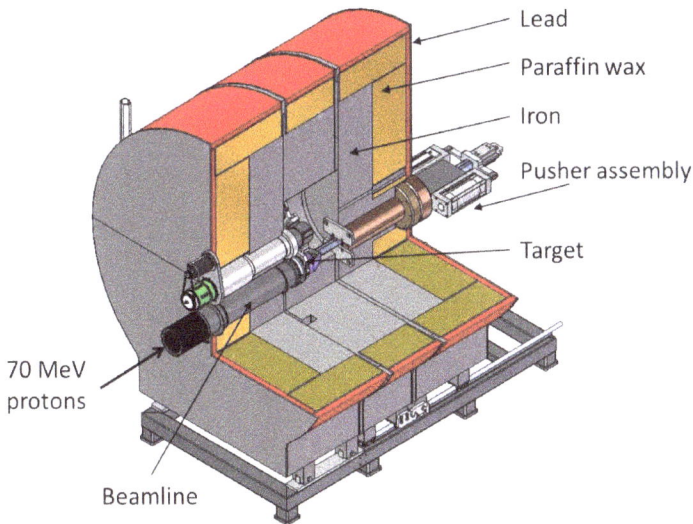

**Figure 3.** Cutaway view of the production station. The pusher assembly locates the target in the irradiation position and also connects the cooling-water lines. The beamline is 1.5 m above the floor.

The new target stations will all be fitted with a rotatable magazine that can hold three target holders. This will be a significant simplification compared to the current HBTS magazine, the latter of which can hold nine target holders. Figure 4 shows a cross-sectional view of a target station as seen in the opposite direction to the beam direction. The magazine is shown with one target in the bombardment position (the lowest position) with all three target positions filled. The structure mounted on the right is a pneumatically controlled robot arm for transferring target holders between a Telelift transporter and the station. A close-up view of a target magazine is shown in Figure 5. It can be rotated to put any of the target holders in the bombardment position or in the "transfer" position when loading or unloading a target holder from the station with the robot arm.

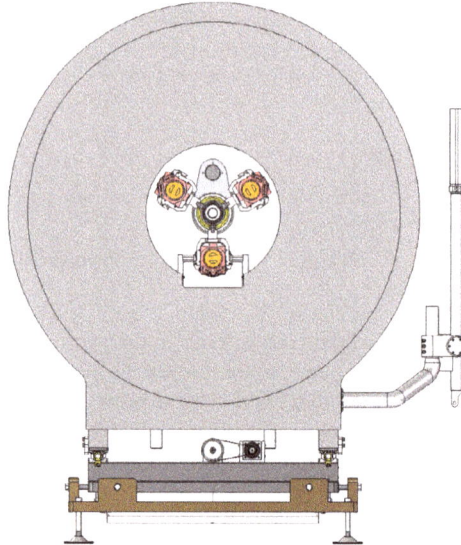

**Figure 4.** Simplified cross-sectional view of a target station. The robot arm is on the right. The beam axis is perpendicular to the page and points towards the reader.

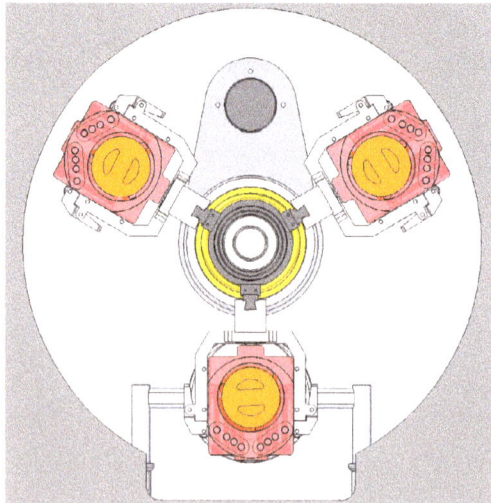

**Figure 5.** Close-up view of a target magazine with target holders loaded in all three positions. The target at the bottom is located on the beam axis (see also caption to Figure 4).

A new 70 MeV tandem Rb/Ga prototype target (see Figure 6) has recently been assembled and pressure-tested. It will be operated with a 20% higher cooling-water flow rate compared to the existing

VBTS targets. The new prototype target capsules have an outer diameter of 40 mm, similar to the present VBTS targets, however provision is made to increase the diameter of future capsules up to 52 mm. This will enable an increase in beam current from present nominal values of 250 µA to more than 300 µA. Figure 6 shows the encapsulated Rb and Ga targets mounted on a bayonet-mount plug, which in turn can be fitted onto a tandem target holder. The Rb metal is contained in a stainless steel capsule (grade 316) while the Ga target capsule is made from niobium [2].

**Figure 6.** The tandem Rb/Ga targets mounted on a bayonet-mount plug.

Figure 7 shows an open target holder while Figure 8 shows the closed target holder after the targets have been loaded. The eight holes at the bottom of the holder are cooling-water inlet and outlet ports.

**Figure 7.** An open tandem target holder with the bayonet-mount plug removed.

**Figure 8.** A closed tandem target holder with the bayonet-mount plug in position.

## 3. Results and Conclusion

Some development work towards Phase 1 of the ACE Beams project has commenced. The final approval for the ACE Isotopes project has recently been obtained from the South African Government and a funding model is in place. A dedicated 70 MeV cyclotron for radioisotope production will free the SSC to focus exclusively on scientific research, including a new RIB programme. The completed SAIF facility will be positioned to realize a three-fold increase in radioisotope production capacity.

**Acknowledgments:** This work was financially supported by the National Research Foundation (Pretoria, NRF Grant No. 85507).

**Conflicts of Interest:** The authors declare no conflict of interest.

## References

1. Botha, A.H.; Jungwirth, H.N.; Kritzinger, J.J.; Reitmann, D.; Schneider, S. Commissioning of the NAC separated-sector cyclotron. In Proceedings of the 11th International Conference on Cyclotrons and their Applications, Tokyo, Japan, 13–17 October 1986; Sekiguchi, M., Yano, Y., Hatanaka, K., Eds.; 1986; pp. 9–16. Available online: www.jacow.org/ (accessed on 8 November 2018).
2. Steyn, G.F.; Vermeulen, C.; Botha, A.H.; Conradie, J.L.; Crafford, J.P.A.; Delsink, J.L.G.; Dietrich, J.; du Plessis, H.; Fourie, D.T.; Kormány, Z.; et al. A vertical-beam target station and high-power targetry for the cyclotron production of radionuclides with medium energy protons. *Nucl. Instrum. Methods A* **2013**, *727*, 131–144. [CrossRef]
3. Telelift ETV System. Telelift GmbH, Germany. Available online: https://en.telelift-logistic.com/ (accessed on 8 November 2018).

![instruments logo] *instruments*

![MDPI logo]

*Article*

# First Steps at the Cyclotron of Orléans in the Radiochemistry of Radiometals: $^{52}$Mn and $^{165}$Er

**Justine Vaudon [1], Louis Frealle [1], Geoffrey Audiger [1], Elodie Dutilly [1], Mathieu Gervais [1], Emmanuel Sursin [1], Charlotte Ruggeri [1], Florian Duval [1,†], Marie-Laure Bouchetou [1], Aude Bombard [2] and Isidro Da Silva [1,*]**

[1]  CEMHTI (Conditions Extrêmes et Matériaux: Haute Température et Irradiation), CNRS, UPR3079, University of Orléans, F-45071 Orléans, France; justine.vaudon@insa-rouen.fr (J.V.); louis.frealle@cnrs-orleans.fr (L.F.); geoffrey.audiger@hotmail.fr (G.A.); edutilly@gmail.com (E.D.); mathieu.gervais@insa-rennes.fr (M.G.); sursin.emmanuel@laposte.net (E.S.); ruggeri.charlotte01@gmail.com (C.R.); florian.duval@cnrs-orleans.fr (F.D.); marie-laure.bouchetou@univ-orleans.fr (M.-L.B.)

[2]  TRISKEM, 3, rue des champs Géons ZAC de l'Eperon 35170 Bruz, France; abombard@triskem.fr

*  Correspondence: isidro.dasilva@cnrs-orleans.fr; Tel.: +33-238-255-427

†  Current address: ISTO, UMR7327, CNRS, University of Orléans, F-45071 Orléans, France.

Received: 17 April 2018; Accepted: 10 August 2018; Published: 16 August 2018

**Abstract:** This work describes the first real developments in radiochemistry around exotic radionuclides at the cyclotron of Orléans focusing on the radiochemistry of two radiometals $^{165}$Er and $^{52}$Mn. For these developments, targets were irradiated during 0.5–2 h at a maximum current of 2 μA. All activities have been determined by radiotracer method. The production of $^{165}$Er from a natural Ho target that was irradiated is described. Higher activities of $^{165}$Er were obtained via deuteron irradiation then proton with lower ratio $^{165}$Er/$^{166}$Ho (400/1 to 8/1). By using LN2 resin, the separation of adjacent lanthanides was made on various concentrations of HNO$_3$ (0.3 to 5 M). Weight coefficients (Dw) were defined in a batch test. Then, the first tests of separation on a semi-automated system were made: the ratio $^{166+nat}$Ho/$^{165}$Er in an isolated fraction was significantly reduced (1294 ± 1183 (*n* = 3)) but the reliability and reproducibility of the system must be improved. Then, a new Cr powder-based target for $^{52}$Mn production was designed. Its physical aspects such as mechanics, thermal resistance and porosity have been studied. Dw for various compositions of eluent Ethanol/HCl were evaluated by reducing contact time (1 h) comparative to the literature. A first evaluation of semi-automated separation Cr/Mn has been made.

**Keywords:** $^{165}$Er; $^{52}$Mn; irradiation; separation; resin; radiometals; heavy Rare Earth Elements (hREE)

---

## 1. Introduction

The cyclotron of Orléans [1–4] is a research accelerator allowing production of many beam types such as the proton, deuteron and alpha at variable energy from 10 to 45 MeV (Figure 1). Most of the applied uses of the beams focuses on materials studies and characterization. However, its properties give access to numerous radionuclides with potential applications especially for medical imaging. For example, in the period 2005–2013, $^{211}$At was produced once a week (300 MBq) (irradiation of target only, no radiochemistry made at Orléans) until the cyclotron ARRONAX (Nantes) was built and became operational to produce regular batch of this radionuclide for alpha-immunotherapy [5–8]. In a general context of the development of radiometals, such possibilities appear to be very interesting for accessing exotic radionuclides. The production of a radionuclide needs not only to perform a nuclear reaction but also to separate the radionuclide of interest from a bulk target. For this purpose, the cyclotron of Orléans is of interest in producing radionuclides for analytical applications allowing for the validation of the separation by the radiotracer method [9] and a method as sensitive as Inductively Coupled Plasma–Mass Spectrometry (ICP–MS). In this context, it was necessary to initiate tools and methodology in radiochemistry to develop further the accelerator in the area of medical radionuclides. However, two constraints relative to the choice of radionuclides were met at the cyclotron of Orléans. The first constraint was to produce gamma emission and the second was that the half-life of the radionuclide produced must be below 100 days (due to nuclear waste management considerations). Hence for the following work, we used $^{165}$Er/Er, $^{166}$Ho/Ho, $^{51}$Cr/Cr and $^{52}$Mn/Mn. Other tools to improve radiochemical separation based on resins, methodology, and target design must be developed. Two cases hereafter illustrate the method's development for the production and subsequent separation of $^{165}$Er and $^{52}$Mn.

**Figure 1.** The cyclotron of Orléans.

A task of the radiochemist is to define the conditions of separation of two elements: including resin, eluent (simple and compatible with the radiolabeling process) and the methodology (radiotracer method, irradiation conditions). These aspects have been developed during the production of $^{165}$Er. Difficulties were the separation of two adjacent lanthanides in disproportionate quantities and a separation compatible with the radiolabeling of molecules. $^{165}$Er decays with electron capture without gamma-ray emission, producing only X-rays (47–55 keV with a cumulative intensity of 73.7%) [10]. It is an ideal radionuclide for Auger electron therapy [11] and can be used for bimodality Magnetic Resonance Imaging / Single Photon Emission Computed Tomography (MRI/SPECT) [12]. It is accessible by irradiation of a natural Ho target by a proton or deuteron beam [13,14]. Its half-life of $T_{1/2}$ = 10.36 h is compatible with radiolabeling of small molecules. To access $^{165}$Er, irradiation with proton and deuteron beam has been used. The conditions of separation of Er/Ho have been defined by determination of weight coefficients (Dw). Tests on a small column (2 mL) have been performed to validated elution parameters. First tests on semi-automation of the separation have been made to allow manipulation of more than 100 MBq of $^{165}$Er. The ratio $^{166+nat}$Ho/$^{165}$Er has been determined to evaluate the efficiency of separation.

The access to the desired radionuclide might in some cases be challenging due to the irradiation path rather than to the radiochemical separation. Design of the target then becomes the key point. This is the case for $^{52}$Mn. It has a low average energy in β + and its range in biological tissues (244.6 keV, 0.63 mm) [15] is comparable to $^{18}$F (250 keV, 0.62 mm) [16,17]. These values are better than those of $^{51}$Mn (970.2 kev, 4.275 mm) in terms of Positron Emission Tomography (PET) resolution (ratio of 6). In addition, the short half-life of $^{51}$Mn ($T_{1/2}$ = 46.2 m.) requires working with high activities to optimize radiochemical separations. Finally, the need to use enriched targets in $^{50}$Cr or $^{54}$Fe in case of $^{51}$Mn is a real obstacle. In first approach, to the implementation of radiolabeling studies, $^{52}$Mn ($T_{1/2}$ = 5.59 days) has been suggested for the diagnosis and treatment of blood diseases [18], as a cationic perfusion tracer [19], and for the study of the involvement of manganese in the pathophysiology of degenerative neurological diseases [20]. To perform the first irradiations to determine Dw, according to the literature, irradiation of a target of Cr is necessary. Foil of Cr does not exist with convenient dimensions allowing direct irradiation to obtain $^{52}$Mn. Therefore, we have designed a target of Cr using powder of chromium.

## 2. Materials and Methods

### 2.1. Materials

All irradiations were made at the cyclotron of Orléans (a CGR-MeV 680 Type [21–25], (Figure 1) in a targetry named "hatch tip" (Figure 2) for the experiments presented in this work.

**Figure 2.** Targetry "hatch tip" low current (<2 μA) for development.

Ho and Cr elements were used to product various radionuclides. Commercial (Alfa Aesar) high purity foil of Ho (purity: 99.9%; thickness: 127 and 300 μm) was used. As commercial high purity chromium foil is not available with a thickness less than 500 μm without polymer support, chromium sample were made by sintering of high purity chromium powder (99%, 100 mesh metal basis density = 2.0–3.0 g/cm$^2$). The chromium powder was pressed to obtain 13 mm (outer diameter) pellets.

Nitric acid (67–69%; SCP Science Plasma Pure; Fisher Scientific), hydrochloric acid (32–35%; SCP Science; Fisher Scientific), alcohol ethanolic (Fisher Chemical), AG1x8$^®$ resin anion exchange (200–400 mesh) (Bio-Rad Laboratories) and LN2 resin (Triskem). A small column in 2 mL polypropylene (PP) including frits (pore size = 20 μm) (PE: polyethylene) was used with internal diameter: 7 mm and a filling height of resin > 39 mm, reference AC-142-TK from Triskem. All water used for the dilutions had a resistivity of 18 MΩ and was prepared using Purelab (Veolia) (Ultrapure Water Purification System).

Activated samples were diluted and put in a tube in a determinated geometry. Activity measurements were assessed by γ-ray spectrometry with an HPGe (High-Purity Germanium) detector. The HPGe detector was calibrated in energy and efficiency for different geometries with certified standard radioactive sources (Cerca, France). For activity measurement, γ-ray spectrum analysis software package, Genie 2000 (Canberra, France) was used. The concentrations of Er Ho, Cr and Mn were estimated following peaks centered at 53 keV for $^{165}$Er, 80 keV for $^{166}$Ho, 320 kev for $^{51}$Cr and 935 for $^{52}$Mn (744 and 1434 can be used too). All activities were determined at the end of beam (EOB).

For polishing the chromium target, Model ES 300GTL has been used. The magnetic putty (1, 3 and 6 μm) carpet for pre-polishing (graduation 120, 320, 600 and 1200) and carpet for polishing (1, 3 and 6 μm) came from ESCIL (France).

## 2.2. Methods for $^{165}Er$ (an Emitter Auger Electron and Bimodality)

### 2.2.1. Irradiation

$^{165}Er$ has been the subject of some publications concerning various methods of production from a cyclotron or a neutron reactor. The neutron pathway leads to a very large range of activities but the result is "carrier added" (i.e., the presence of other isotopes of erbium). However, a ligand radiolabeling application has to be a "no carrier added" pathway. Therefore, irradiation with a cyclotron, starting from a different element, can be more interesting.

Irradiation of an erbium target (natural or enriched) by a proton beam will be according to the reactions: $^{nat}Er$ (p, xn) $^{165}Tm \rightarrow$ (by decay) $^{165}Er$ or $^{nat}Er$ (d, xn) $^{165}Tm \rightarrow$ $^{165}Er$ [25]. The irradiation of this natural target will lead to $^{165}Tm$ (30.06 h) but also to $^{167}Tm$ (T $\frac{1}{2}$ = 9.25 days) and $^{168}Tm$ (T $\frac{1}{2}$ = 93.1 days). Indeed, the percentage (%) of every isotope on natural erbium is 0.139, 1.601, 33.503, 22.869, 26.978 and 14.910 respectively to isotope 162, 164, 166, 167, 168 and 170 of erbium.

To overcome this, the irradiation of a $^{166}Er$ enriched target guarantees a single nuclear reaction such as: $^{166}Er$ (p, 2n) $^{165}Tm$ which is quite challenging as it is not an easy and cheap way to obtain $^{165}Er$ for proof of concept.

The "easiest" way to access to $^{165}Er$ is the irradiation of a natural holmium target by proton beam according to the nuclear reaction $^{165}Ho$ (p, n) $^{165}Er$ at 16 MeV or deuteron beam by $^{165}Ho$ (d, 2n) $^{165}Er$ at 17.5 MeV. The proton beam method can be exploited in a commercial cyclotron (less than 20 MeV). At the cyclotron of Orléans, two ways of production have been explored. All cross sections were known and obtained by irradiation or simulation from calculation codes (ALICE-IPPE) [12,13,26,27].

The ratio $^{165}Er/^{166}Ho$ produced by proton and deuteron beam is respectively of around 400/1 and 8/1. The origin of the production of $^{166}Ho$ by proton irradiation is not clearly identified, but neutron activation seems to be the source.

### 2.2.2. Separation $^{165}Er/^{166}Ho$

The major difficulty in the development of $^{165}Er$ is its extraction from the holmium target. The selectivity factor for the couple Ho/Er (SF: Separation Factor around 1.5) is amongst the lowest in the lanthanides. In all publications relative to the irradiation of a Ho target, only one evokes a treatment of this target by a separation of the Ho/Er [11] pair. The complexing agent, α-HIBA (alpha-HydroxyIsoButyric Acid) is widely used in the separation of lanthanides [28]. However, these separations are often made at the analytical level with the use of an online chromatographic system (High-Performance Liquid Chromatography (HPLC)) on balanced mixtures or traces (some ng or pg) for each lanthanide. To separate a few ng of $^{165}Er$ from few mg of a target of about 200 mg of natural holmium becomes more complex. In addition, the separation conditions must allow the use of $^{165}Er$ in radiolabeling of ligands aiming vectorised therapy with Auger electron or used in radiotracer for biomodality MRI/SPECT.

HDEHP (di(2-ethylhexyl) phosphoric acid), HEHEHP (mono-2-ethylhexyl ester of phosphonic acid) and (H[DTMPeP]) bis(2,4,4-trimethylpentyl) phosphinic acid are the main acidic organophosphorous extractants [29] used in Rare Earth Element (REE) separation processes and have also been commercialized (for Triskem commercial name: LN, LN2 and LN3). HDEHP has been widely used for primary separation because the distribution coefficients of the REEs as a group differ markedly from impurities in leach groups (e.g., in spent fuel actinides and REEs) [30]. For this project, LN2 resin (Triskem) based on (HEHEHP) was used. This reagent was chosen because REEs can be stripped at lower acidities than from HDEHP [31]. They were used for the separation of adjacent lanthanides including the separation of a microcomponent from a macrocomponent [32–34]. Weight coefficients Dw have been determined for various concentrations of $HNO_3$. Additional evaluation of these coefficients on LN2 resin was performed with higher $^{166}Ho$ activity using deuteron irradiation, which should allow more sensitivity. The results presented here are part of a more complete study about Ho/Er with comparison LN, LN2 and LN3 (in the case of LN3, it is necessary to use a low concentration of $HNO_3$). The main drawbacks of HDEHP (LN) is a low adjacent separation factor, relatively low acidity for REE stripping, as well as low hydrolytic stability [35]. Phosphonic and phosphinic acids are more hydrolytically stable and effective extractants when compare with HDEHP. Moreover, they have demonstrated a high selectivity for the "heavy"REE (hREEs) compared to "light"REE (lREEs) [29].

For determination of distribution coefficient Dw, a target of natural Ho of 0.3 mm of thickness (196 mg) was irradiated during 2 h at 2 μA at 17.5 Mev deuterons ($A_{EOB}$: Activity (EOB)): around 700 MBq of $^{165}$Er/90 MBq of $^{166}$Ho were produced then dissolved in 1 mL of 5 M $HNO_3$ solution (no cut target for batch tests) and after that the complete dissolution of the target, adjusted to 5.5 mL with 18 MΩ water. This solution of 0.9 M $HNO_3$ was the crude solution (concentration around 35 mg/mL of holmium and 13 MBq of $^{165}$Er/1.6 MBq of $^{166}$Ho). Around 100 mg of LN2 resin was weighed for each test, which was always done in triplicate. In every tube, LN2 resin was mixed with 100 μL of crude solution and volume was adjusted to 1.5 mL with different volumes of $HNO_3$ and water 18 MΩ to obtain the final concentration of $HNO_3$.

For example, at 0.5 M $HNO_3$, 650 μL of 1 M $HNO_3$ and 750 μL of 18 MΩ water were added to 100 μL of crude solution of $^{165}$Er/$^{166}$Ho (initial concentration 0.9 M $HNO_3$). Every tube was shaken for 30 min [36,37] then allowed to rest and centrifuged for 20 min. 500 μL of each solution were extracted and diluted to 2 mL by adding 1.5 mL 18 MΩ water (geometry called as "Tube 2G"). Then samples were measured at spectrometry γ in this defined geometry with acceptable dead time (<5%). For several tubes, it was necessary to allow for one to two days of decay to perform a measurement.

Capacity of LN2 resin adsorption is 0.16 mmol/mL: maximum of 26.4 mg of holmium can be used for 0.37 g of resin (1 mL). However, Triskem recommends a surety factor of 50% on the capacity of adsorption. For that, in all batch tests 100 μL of crude solution contained around 3.5 mg of natural Ho. In a concentration of 0.4 M $HNO_3$, two tests (denote 0.4 100 and 0.4 200) with 100% and 200% capacity of resin had added, respectively with 200 and 400 μL of crude solution to evaluate this factor on values Dw and SF.

Weight distribution $D_w$ (mL/g) and separation factor ($SF_{Er/Ho}$) were calculated using the following equation:

$$D_w = \frac{A_0 - A_e}{A_e} \times \frac{V}{m} \tag{1}$$

Here $A_0$ and $A_e$ are the liquid phase metal activities (Bq) before and after equilibrium, $m$ is the weight of the LN2 resin in grams and V is the volume of liquid phase in mL.

$$SF_{Er/Ho} = \frac{D_w(Er)}{D_w(Ho)} \tag{2}$$

Then tests on a 2 mL column with around 0.37 g of LN2 resin (1 mL) were made first by the gravimetry method then with variation of flow (1 mL/min and 3 mL/min) at atmospheric pressure using a peristaltic pump in the outlet of the column. A gradient of concentration on $HNO_3$ was used: 0.4 M 0.8 M, 1 M and 2 M or/and 5 M if necessary. Tests at 1 bar of Nitrogen (maximum pressure of solenoid valves from Bio-Chem (used in semi-automated system) is 1.4 bars) were made. Around 4 g of LN2 (for a maximum of mass target of 140 mg, around 11 mL) 50–90 mg of the target were used in semi-automated tests. Four tests on a homemade automation system (drive with the homemade software ACCRA (Automatisme & Contrôle Commande Radiochimie)) were made using the remote control sketch below (Figure 3) from the dissolution step of the target until the final evaporation step of $HNO_3$.

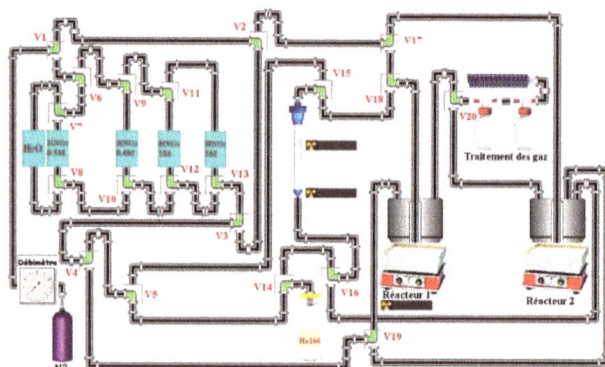

**Figure 3.** Remote control of Er/Ho separation.

In a semi-automated system 50 µL of solution were collected manually to determine the extraction yield of radionuclide in each test after adding 2 mL of 0.3 M $HNO_3$ in the reactor. The reactor was weighed before and after addition of 0.3 M $HNO_3$ (no weighing for test n°3: no extraction yield). This was the activity reference for $^{165}$Er and $^{166}$Ho ($A_{ref}$: deposited activity). With these values, the initial activity of $^{165}$Er and $^{166}$Ho in the cut target were determined. A relationship activity-weight was established by the ratio $^{165}$Ho/$^{166}$Ho: activity of $^{166}$Ho in Bq by mg of $^{165}$Ho (cut target). the initial ratio Ho/$^{165}$Er (for Ho, $^{165}$Ho was considered because the mass of $^{166}$Ho was insignificant) on the target (cut target mass) and the ratio in an isolated fraction of $^{165}$Er was calculated. A Ho decontamination factor was determined according to the following formula:

$$F_{Ho\ (cont.)} = \frac{A^{166}_{0(Ho)} / Act^{166}_{f(Ho)}}{A^{165}_{0(Er)} / A^{165}_{f(Er)}}$$

$F_{Ho\ (cont.)}$: Contamination factor in Ho

$A^{166}_{0(Ho)}$: Activity $^{166}$Ho in cut target

$A^{166}_{f(Ho)}$: Activity $^{166}$Ho in isolated fraction of $^{165}$Er

$A^{165}_{0(Er)}$: Activity $^{165}$Er in cut target

$A^{165}_{f(Er)}$: Activity $^{165}$Er in isolated fraction of $^{165}$Er

All dilution factors were determined by weighing aliquots of fraction. A geometry "Tube 2G" was used to determine the activities of the samples. Extraction yield (%) was defined as:

$$Y\ (\%) = \frac{\sum Activities\ all\ tubes}{A_{ref}}$$

$\sum Activities\ all\ tubes$: sum of all activities in fraction tubes for one radionuclide ($^{166}$Ho or $^{165}$Er)
$A_{ref}$: deposited activity on column.

The same column was used and between different tests (4), column is conditioned with 30 mL of 5 M $HNO_3$ then 18 MΩ water until pH was neutral. Then, 30 mL of 0.3 M $HNO_3$ was used to conditioning column. Manually target was cut and around 50% of target was dissolved (more than 90% of activity of target). All weights listed for the target were referenced to cut the target (not all the irradiated mass of the target). The height of the column (glass) was 14 cm and the inner diameter 1 cm. The LN2 resin was shaken for 30 min with 18 MΩ water then put on the column. It was conditioned with 20 mL of 18 MΩ water then 30 mL of 0.3 M $HNO_3$ before depositing the sample. Then, the cut target was dissolved in 1 mL of 5 M $HNO_3$ in a reactor of 3 mL. Nitric acid was evaporated until it was in a pink gel formation. Then 2 mL of 0.3 M $HNO_3$ was added to dissolve the pink residue and transfer it to the column under flow of nitrogen (20 mL/min). The solution on top of column was adsorbed under the flow of nitrogen (100 mL/min). When the solution on top of the column was adsorbed, 0.4 M $HNO_3$ (100–120 mL) was used to elute Ho at flow of 1–2 drops/s (1 mL/min) under 1 bar of Nitrogen (at flow of 100 mL/min). Then 1 M $HNO_3$ was used (around 40 mL except for sample 3 (22 mL)) and finally, in some cases, 2 M (15 mL for sample 3) and 5 M $HNO_3$ (sample 1 and 3) more to elute residue activity on the column. The radiotracer method was used for this separation using [166]Ho for Ho and [165]Er for Er.

### 2.3. Methods for [52]Mn (Radioimmunotherapy and Bimodality)

To obtain this radionuclide from a cyclotron, the irradiation of chromium by a proton beam is the most used route, by the reaction [52]Cr (p,n)[52]Mn to 16–17 MeV [38], because it is also the most accessible by commercial cyclotrons. Then, the separation of [52]Mn from a chromium target was carried out by liquid–liquid extraction [39], but the most interesting results concern the use of an ion exchange resin [40,41]. This route seems more suitable for subsequent automation of the separation step. The choice of the best way to access [52]Mn must be focused in part on the simplicity of the separation process. Indeed, because of its three high gamma rays with an emission of more than 90%, the radiation safety for operators is a paramount criterion. Here, a target design (Cr pellet) is described in detail. Various studies have been made to validate the specifications of a pellet of chromium for irradiation.

As chromium is a compound that oxidizes very easily with air, the sintering of Cr powder has to be done in an inert or reducing atmosphere in an electrical furnace (Nabertherm with controller P 320). Therefore, we used an $H_2/N_2$ mixture (5% hydrogen). In addition, to ensure that the chromium does not react with the residual oxygen, we surrounded it with crushed graphite. We placed the whole in an alumina crucible (Figure 4):

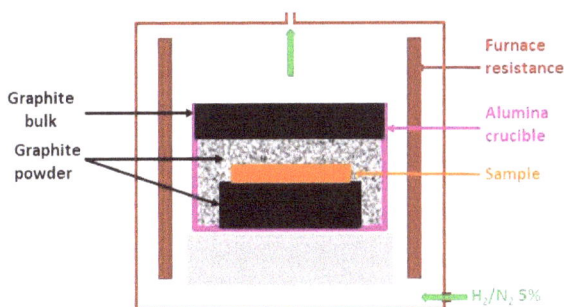

**Figure 4.** Set-up of sintering.

Finally, we proceeded to the heat treatment with the following thermal cycle:

- Temperature rise rate: 5 °C·min$^{-1}$
- Isothermal run time, 30 min
- Cooling speed: 5 °C·min

The same program was used in vacuum conditions. The vacuum was $10^{-7}$ mBar and the furnace is a homemade system where the measurement of temperature is determined by pyrometry. The sample was put on a block next to a block of molybdenum. By measuring the emissivity of molybdenum, the temperature of the sample was determined.

After sintering, Cr pellets were tested in bending conditions (to determine stress resistance of pellet). The device used was unconventional and inspired by the "Ball on ring" test (Figure 5). A moving speed (0.5 mm/min) of a ball of alumina (9.32 mm diameter) was applied. The measurement stops when the target broke. To compare mechanical properties in bending conditions of each Cr pellet, we compared the load required to break the pellet.

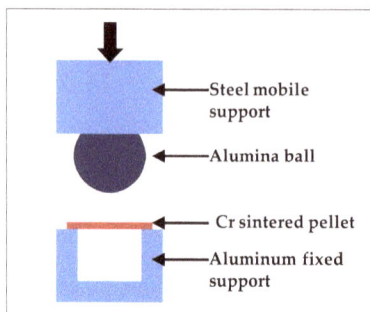

**Figure 5.** Bending test.

Using a target of Cr designed (800 mg pressed at 8.3 tons and sintered 1 h at 800 °C), a first determination of Dw for separation Cr/Mn was done. The radiotracer method was used to follow two elements $^{51}$Cr for Cr and $^{52}$Mn for Mn. To obtain a significant activity of $^{51}$Cr, irradiation of the Cr pellet was performed at 23 MeV. In these experimental conditions, nuclear reactions $^{nat}$Cr (p,n)$^{52}$Mn and $^{nat}$Cr(p,x)$^{51}$Cr occurred.

For the resin exchange anions, the AG1x8® resin was used and different concentrations of HCl (1, 4, 8 and 12 M) and various percentages of Ethanol (0 and 93 to 98% by step of 1%) in eluted solution (ethanol/HCl) were tested. Contact time was 1 h instead of several hours reference studies [14,40].

Briefly, a Cr pellet of around 800 µm of thickness (800 mg) was irradiated during 0.5 h at 0.2 µA(EOI: around 2.9 MBq of $^{52}$Mn/1.45 MBq of $^{51}$Cr). The target was dissolved in 5 mL of 12 M HCl and evaporated to dryness. Then, we dissolved the residue with 2 mL 2 M HCl. This was the crude solution of $^{52}$Mn/$^{51}$Cr. Around 80 mg of AG1x8® resin were weighed for each test (all tests were triplicated). In every tube, AG1x8® resin was mixed with 970 µL of elution solution. Then 30 µL of crude solution was added. Every tube was shaken for 1 h then allowed to rest, and finally centrifuged for 10 min. 500 µL of each solution were extracted and diluted to 2 mL by adding 1.5 mL 18 MΩ water. Then samples were measured at spectrometry γ in this geometry called "Tube 2G" with acceptable dead time. For several tubes, it was necessary to allow for one to two days of decay to perform a measurement. All calculations were made according to Equations (1) and (2) in Section 2.2.2. Separation $^{165}$Er/$^{166}$Ho.

For a test of the semi-automation of Cr/Mn separation, a 1.2 mL of dissolved chromium solution (eluent 1: Ethanol/HCl 12 M (97/3)) was used. No calculations of a decontamination Cr factor were done. This aliquot which contained 35,000 Bq of $^{51}$Cr and 3100 Bq of $^{52}$Mn, was deposited on column 1. Three columns with 300, 200 and 100 mg of AG1x8® resin were used, respectively, as columns 1, 2 and 3. The sample was deposited on column (all have been conditioned with 10 mL of water 18 MΩ, then 10 mL ethanol and finally 10 mL of eluent 1). Column 1 was eluted first with 8 mL of eluent 1 then rinsed with 10 mL of eluent 1 divided into 3 fractions of 3 mL. Finally 1 mL of HCl 12 M was used to recover $^{52}$Mn. An aliquot of this fraction was used to quantify $^{51}$Cr and $^{52}$Mn. 30 mL of pure ethanol were added and mixture was shaken 10 min then deposited on column 2. After that, 3 mL of eluent was added and 1 mL of HCl 12 M desorbed $^{52}$Mn. After collecting the fraction for measurements, 30 mL of pure ethanol was added and the batch shaken for 10 min. The batch was adsorbed on column 3 and eluted with 3 mL of eluent 1. Finally, $^{52}$Mn was desorbed with 1.5 mL of 12 M HCl.

For optimizing this process and reducing the quantity of chromium, pre-polishing and polishing on the chromium target was made like this:

■ Pre-polishing is necessary to reduce dramatically thickness of pellet:

- 30 min with carpet of 120 at 100 rpm;
- 8–10 min with new carpet of 120 at 100 rpm.

Then, the thickness was checked because on following operations, every carpet decreased in thickness by about 100 μm around:

- 5 min with carpet of 320 at 100 rpm;
- 10 min with carpet of 600 at 100 rpm;
- 20 min with carpet of 1200 at 100 rpm;

■ Polishing:

- 30 min with carpet of 6 μm at 100 rpm;
- 45 min with carpet of 3 μm at 100 rpm;
- 45 min with carpet of 1 μm at 100 rpm.

When illustrations showed polished samples, only the polishing step was taken. Irradiation tests and separations were performed on unpolished targets.

## 3. Results

### 3.1. $^{165}Er$

Concentrations of 0.3 M, 0.4 M, 0.5 M, 0.8 M, 1 M, 2 M and 5 M $HNO_3$ were evaluated for the separation at 50% of capacity adsorption of resin. They were also evaluated for concentration 0.4 M 100% (point 0.4 100) and 200% (0.4 200) of capacity (Figure 6).

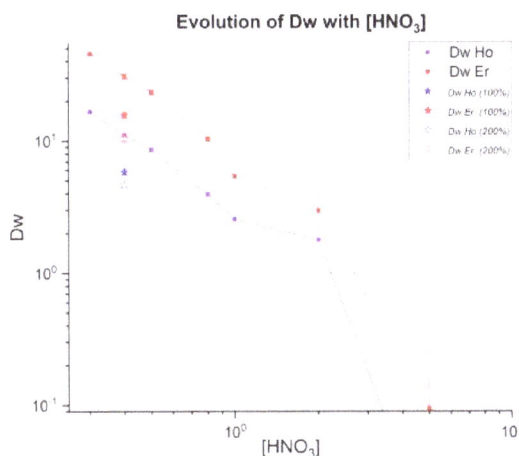

**Figure 6.** $D_w$ de Er/Ho with LN2.

Weight distribution $D_w$ (mL/g) and separation factor $SF_{Er/Ho}$ was determined (Table 1):

Table 1. $D_w$ and SF of Er and Ho in $HNO_3$.

| $HNO_3$ | $D_w$ (Er) | $D_w$ (Ho) | SF $_{Er/Ho}$ |
|---|---|---|---|
| 0.3 M | 45.7 ± 1.3 | 16.7 ± 0.4 | 2.74 |
| 0.4 M | 30.8 ± 2.0 | 11.2 ± 0.5 | 2.75 |
| 0.4 M 100 | 15.4 ± 1.2 | 5.8 ± 0.4 | 2.70 |
| 0.4 M 200 | 10.5 ± 1.1 | 4.7 ± 0.2 | 2.25 |
| 0.5 M | 23.4 ± 1.5 | 8.6 ± 0.4 | 2.71 |
| 0.8 M | 10.4 ± 0.2 | 3.9 ± 0.1 | 2.65 |
| 1 M | 5.4 ± 0.2 | 2.6 ± 0.1 | 2.11 |
| 2 M | 3.0 ± 1.0 | 1.8 ± 0.4 | 1.67 |
| 5 M | 0.1 ± 0.4 | −0.4 ± 0.3 | −0.21 |

There was no difference between $SF_{Er/Ho}$ at 0.3 and 0.4 M $HNO_3$ but Dw were high in the case of 0.3 M $HNO_3$: concentration for dissolution of target. Then, to start elution 0.4 M $HNO_3$ can considered a better choice than 0.3 M $HNO_3$. The separation factor was influenced by % of capacity resin involved in separation: at 200% of capacity, this factor SF decreased significantly and Dw was weaker. At this ratio, selectivity was low and a significant part of the erbium could be eluted with holmium. That is why some tests of elution on a 2 mL column were done with the following conditions: target dissolved in 0.3 M $HNO_3$, starting elution of batch with 100–120 mL 0.4 M or gradient with 0.4 M/0.8 M $HNO_3$ before desorbing holmium with 1 M $HNO_3$ and then erbium. If 1 M $HNO_3$ was used directly, no separation was done, but using 0.4 M $HNO_3$ generated a difference of elution between Ho and Er more important than the other eluent.

On a 2 mL column, by gravimetry, time elution was more than 1 h with 370 mg of resin LN2. There was no influence of flow between 1 mL/min and 3 mL/min in separation only in separation time. At 1 bar of nitrogen, the flow was too quick and was reduced to 1 mL/min in setting pressure of nitrogen to 0.7 bars.

With these previous tests, a semi-automated system of separation of Ho/Er was evaluated on a column of 4 g of LN2 resin (Table 2).

Table 2. Elution of 4 samples on semi-automated system separation of Er/Ho.

| | Volume of Eluent (mL) | | | | Repartition of Ho (%) | | | | Repartition of Er (%) | | | |
|---|---|---|---|---|---|---|---|---|---|---|---|---|
| Sample | 1 | 2 | 3 | 4 | 1 | 2 | 3 | 4 | 1 | 2 | 3 | 4 |
| $HNO_3$ 0.4 M | 102 | 104 | 123 | 110 | 87.9 | 82.4 | 90.9 | 89.6 | 1.5 | 0.5 | 0.0 | 1.6 |
| $HNO_3$ 1 M | 45 | 38 | 22 | 38 | 12.1 | 17.6 | 9.0 | 10.4 | 98.5 | 99.6 | 24.3 | 98.4 |
| $HNO_3$ 2 M | N.u. | N.u. | 15 | N.u. | - | - | 0.1 | - | | | 75.7 | |
| $HNO_3$ 5 M | 45 | N.u. | 5 | N.u. | 0.0 | | 0.0 | | 0.0 | | 0.0 | |

N.u.: Not used.

The isolated fraction in all samples (Table 3) contained more than 60% (72.3 ± 5.1 % ($n = 4$)) of [165]Er activity.

**Table 3.** % of $^{165}$Er and $^{166}$Ho in isolated fraction against all fractions.

| Sample | HNO$_3$ | | Isolated Fraction of $^{165}$Er | |
|---|---|---|---|---|
| | Volume | Concentration | % of Rn | |
| | (mL) | (M) | $^{166}$Ho | $^{165}$Er |
| 1 | 11 | 1 | 0.07 | 72.7 |
| 2 | 10 | 1 | 0.00 | 63.9 |
| 3 | 15 | 2 | 0.08 | 75.7 |
| 4 | 4 | 1 | 0.37 | 76.8 |
| mean | | | 0.13 | 72.3 |
| σ | | | 0.14 | 5.1 |

% of Rn: percentage of radionuclide in isolated fraction against all fractions.

The separation profile is illustrated in Figure 7 with the sample 3.

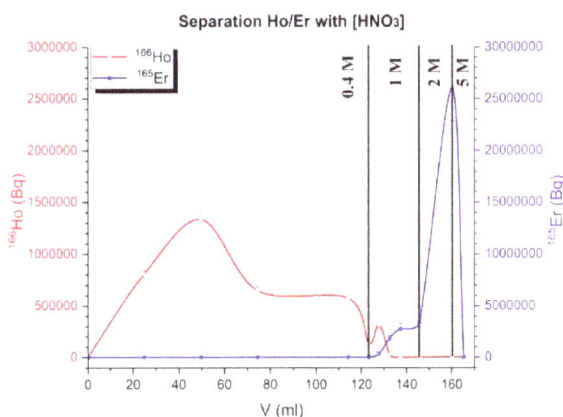

**Figure 7.** Profile of elution $^{165}$Er/$^{166}$Ho of sample 3.

The extraction yields of erbium and holmium were higher than 90% (except sample 4). A Ho decontamination factor was calculated on sample 1, 2 and 4 (not possible for 3 because no reference activity $A_{ref}$ (deposited activity on top of column)). The semi-automated separation system was significantly reduced quantity of holmium on isolated fraction of $^{165}$Er by more than a factor $1294 \pm 1183$ ($n = 3$) (Table 4).

**Table 4.** Analysis of isolated fraction from 4 tests on semi-automated separation system.

| Test | Extraction Yield | | Cut Target | | | | Isolated Fraction of $^{165}$Er | | | $F_{Ho (cont.)}$ |
|---|---|---|---|---|---|---|---|---|---|---|
| | Y (%) | | Mass (mg) | Activity (Bq) | | Ratio $^{165}$Er/$^{166}$Ho | Activity (Bq) | | Ratio $^{165}$Er/$^{166}$Ho | |
| | $^{166}$Ho | $^{165}$Er | $^{165}$Ho | $^{166}$Ho | $^{165}$Er | | $^{166}$Ho | $^{165}$Er | | |
| 1 | 92 | 98 | 88.4 | $9.5 \times 10^6$ | $8.0 \times 10^7$ | 8.4 | $5.7 \times 10^3$ | $5.4 \times 10^7$ | 9397 | 1123 |
| 2 | 95 | 98 | 57.4 | $6.3 \times 10^6$ | $5.2 \times 10^7$ | 8.3 | $1.4 \times 10^3$ | $3.0 \times 10^7$ | 21,070 | 2553 |
| 3 | ND | ND | 53.8 | ND | ND | ND | $3.1 \times 10^3$ | $2.6 \times 10^7$ | 8356 | ND |
| 4 | 86 | 86 | 74.7 | $8.3 \times 10^6$ | $7.2 \times 10^7$ | 8.6 | $1.7 \times 10^3$ | $3.0 \times 10^7$ | 1773 | 206 |
| Mean | 91 | 94 | | | | 8.4 | | | 10,149 | 1294 |
| σ ($n = 3$) | 5 | 7 | | | | 0.2 | | | 6950 * | 1183 |

* ($n = 4$).

## 3.2. $^{52}$Mn

### 3.2.1. Design Target Conception

In order to obtain $^{52}$Mn, a natural chromium target was irradiated by a proton beam. Once irradiated, the target was then chemically treated to isolate $^{52}$Mn from Cr. However, the preparation of this target requires special attention. No commercial Cr foil exists with a thickness less than 500 μm without polymer support. However, this element is commercially available in the form of powder. Because of the risks of cyclotron contamination, powdery targets are difficult to irradiate. A Cr pellet was therefore designed taking into account the following criteria: the risks of contamination during irradiation, dissolution of the target, limitation of the amount of chromium not activated in radiochemical separation (possibly through lower thickness). For these purposes, the physical aspects of the target such as mechanics, thermal resistance and porosity (closed and opened porosity) were studied.

### 3.2.2. Compaction

A study about optimization of parameters (force, mass and time of compaction of pellets) was made to obtain a pellet of 13 mm diameter with good visual mechanical resistance. First, a force of 8.3 tons was applied on the powder during 5 min. Several quantities of powder were tested from 200 to 1200 mg (Table 5). The aim was to obtain a complete pellet with a lower mass.

**Table 5.** Various mass of pellet of Cr.

| Powder (mg) | Pellet | Description |
|:---:|:---:|:---:|
| 200 | | pellet incomplete |
| 400 | | pellet incomplete |
| 600 | | pellet incomplete |
| 800 | | pellet perfect |
| 1200 | | pellet complete but too thick |

Tests were carried out to reduce this mass (therefore the thickness) by varying time and compaction force. Results obtained are presented on the following Table 6.

Table 6. different masses and thicknesses of the pellets.

| Force Tons (Bars) | Mass Powder (mg) | Time Compaction (min) | Thickness (mm) | Visual Aspect |
|---|---|---|---|---|
| 4 (150) | 649.6 | 5 | 1.08 | Bad |
| | 649.5 | 10 | 1.01 | Bad |
| | 649.8 | 60 | 1.04 | Bad |
| | 699.7 | 5 | 1.09 | Bad |
| 8.3 (300) | 599.7 (3) * | 5 | 0.89 (3) | Bad |
| | 600.2 | 30 | 0.84 | Good |
| | 650.6 (3) * | 2 | 0.88 | Good (2) ** |
| | 650.8 (2) * | 5 | 0.93 | Good |
| | 650.2 (4) * | 30 | 0.9 | Good |
| | 650 (2) * | 60 | 0.89 | Bad |
| | 699.6 (3) * | 2 | 0.96 | Good (2) ** |
| | 699.4 (3) * | 5 | 0.95 | Good |
| | 700.2 | 60 | 1.04 | Bad |

* () number of tests; ** () number of tests with described visual aspect.

Eleven pellets were made with better conditions of compaction, of 650 mg 30 min at 8.3 tons (300 bars) with a good visual aspect and these were added to 4 obtained in first tests (Table 6). Thus a variability, was made among 15 pellets:

- weight average 649.6 ± 0.6 mg (15),
- thickness = 0.87 ± 0.02 mm (13) (2 broken pellets by manipulating).

### 3.2.3. Sintering

Despite compaction, the pellets remained fragile. This was a problem for the irradiation as it means that in contact with a beam of high intensity, the pellet was likely to crumble or break. In order to improve the mechanical strength of the pellets, a study of sintering heat treatment was carried out on the pellet at various temperatures and different specific conditions of sintering (in air, vacuum and inert atmosphere).

The first tests were operated under atmosphere controls in oxygen (500 °C to 1400 °C) (Figure 8). A bending test has been performed to check the flexural strength of the pellet. A load was applied, via a ball, to the center of the sample that would be biased in bending [42,43]. The force applied was recorded as a function of the vertical displacement of the ball. The tests were carried out on pellets of 800 mg of compaction at 8.3 tons during 5 min, sintered at 500 °C, 800 °C and 1000 °C for 1 h. Each test was repeated three times. Recording was stopped when the ball was cracked.

Figure 8. Micrography of polished samples before and after sintering at (800 °C).

Subsequently, the average maximum stress resistance was determined (Figure 9).

We could observe that stress resistance increases with sintering temperature; however, above 800 °C no differences appeared. This higher stress resistance could be explained by a decrease of porosity as the temperature of sintering increases. An optimization of conditions of sintering was made at 600 and 800 °C with 650 mg and 800 mg pellets (Table 7).

According to these results, a temperature of 800 °C was necessary to obtain a good resistance structure. The presence of oxygen or a $H_2/N_2$ atmosphere during the sintering have no influence on the mechanical properties. It only influences the chemical composition of the pellet: in the presence of oxygen, a high amount of chromium oxide appears during the heat treatment.

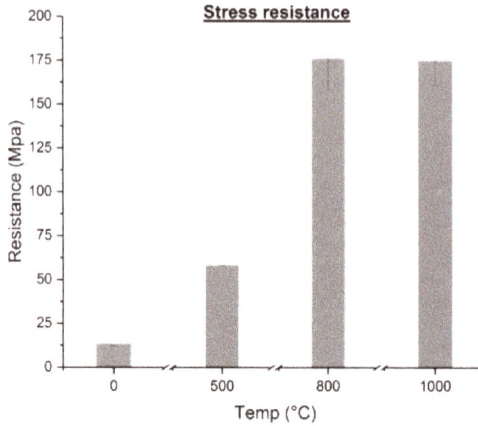

**Figure 9.** Comparison of maximum stress of various sintered target.

**Table 7.** Mechanical resistance for different masses and thicknesses of the pellets.

| Mass (mg) | Time Compaction (min) | Sintering Conditions | T (°C) | Resistance Rupture (MPa) | Standard Deviation (MPa) |
|---|---|---|---|---|---|
| 650 | 30 | V. | 600 | 13 | 1 |
| | | | 800 | 16 | 1 |
| | | N.C.A. | 600 | 105 | 13 |
| | | | 800 | 221 | 29 |
| | | C.A. | 600 | 123 | 14 |
| | | | 800 | 204 | 29 |
| 800 | 5 | C.A. | 800 | 211 | 28 |
| | 30 | N.C.A. | 600 | 100 | 7 |
| | | | 800 | 221 | 15 |
| | | C.A. | 600 | 103 | 13 |
| | | | 800 | 205 | 13 |
| 650 | 30 | No sintering | / | 12 | 2 |

V.: vacuum/C. A.: Controlled Atmosphere/C. A.N.: Not Controlled Atmosphere.

### 3.2.4. Dissolution of Target

Tests of dissolution were then made on various pellets (variation of temperature of sintering) with 12 M HCl and heating at 200 °C on a block. These results were obtained with a pellet of 800 mg 5 min of compaction at 8.3 tons.

Dissolution tests have been shown to have difficulties solving the target sintered up to 1000 °C, but it was not necessary to complete the dissolution of pellet to extract all the $^{52}$Mn [41]. However, the results reported in Table 8 would be very useful for optimizing target dissolution.

**Table 8.** Dissolution of target.

| Sintering T (°C) | 1300 | 1200 | 1000 | 800 | 650 | 600 | 600 | 550 | 500 |
|---|---|---|---|---|---|---|---|---|---|
| Volume 12 M HCl (mL) | 5 + 5 | 5 + 5 | 10 | 10 | 10 | 10 | 20 | 10 | 10 |
| T(°C) heating | 100 | 200 | 200 | 200 | 200 | 200 | 200 | 200 | 200 |
| Dissolution | P. | P. | P. | P. | P. | P. | C. (20 min) | C. (10 min) | C. (10 min) |
| % Weigh dissolved in 30 min | 65 | 75 | 76 | 80 | 80 | 80 | 100 | 100 | 100 |

P.: Partial/C.: Complete.

Although sintering at 600 °C could be suitable for the dissolution of the pellet, the breaking strength remains an essential criterion in the choice of the preparation. The sintering at 800 °C was selected for the design of this target. For the development of the separation method, this target was used (however, the thickness was still substantial at 900 μm for a weight of 650 mg).

The dissolution of the Cr pellet by hydrochloric acid solution starts at the grain boundaries and increases the porosity. Scanning Electron Microscope (SEM) observations confirm that grain boundaries are attacked and material becomes more porous after the dissolution step (Figure 10).

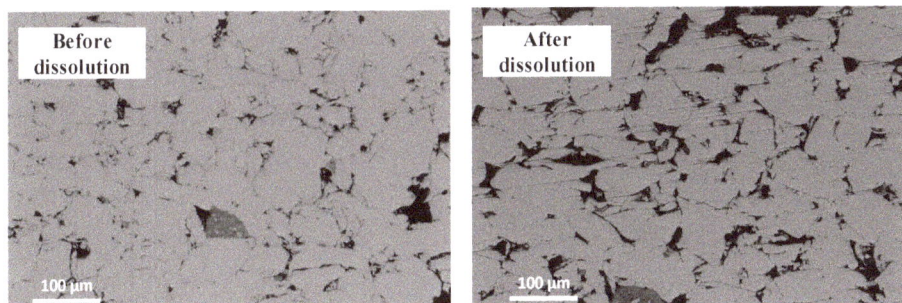

**Figure 10.** Scanning electron microscope (SEM) micrograph (mode BSE (back-scatter electron). of sintered pellet at 800 °C, before and after dissolution.

For temperature higher than 600 °C, the densification is very high. The residual porosity is too low, the reactive surface is, therefore, too small and the acid cannot penetrate the core of the material.

Chromium is a compound that oxidizes very easily: a small amount of chromium oxide crystallizes during the sintering step or during dissolution when in contact with the acid. Chromium oxide is a stable compound that is impossible to dissolve. During dissolution, the acid solution is charged with chromium ions. At 600 °C, if 10 mL of hydrochloric acid is set from the beginning, and the pellet does not dissolve. In contrast, an initial addition of 20 mL allows complete dissolution.

During sintering, the pellet is in the presence of carbon to avoid oxidation. It is possible that chromium carbide is formed and can then limit the dissolution (Figure 11) [44].

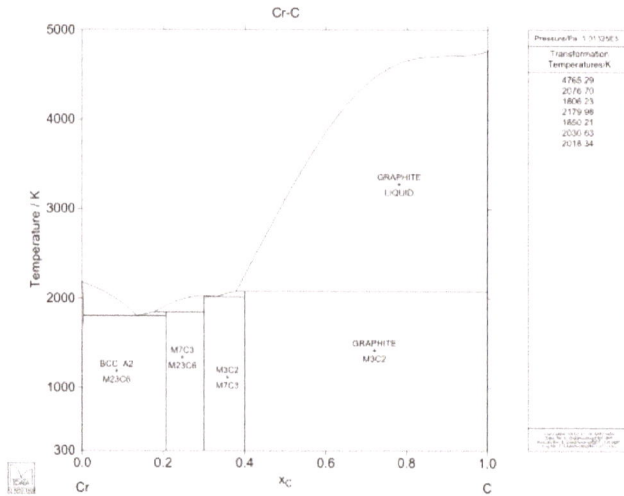

**Figure 11.** Cr-C phase diagram.

Nevertheless, it should be noted that the higher the density, the more difficult it is to attack with the acid during dissolution because the grains are very close. Indeed, a pellet sintered at 650 or 800 °C does not dissolve completely while a sintering at 600 °C dissolves completely.

3.2.5. Porosity of Target

To validate these explications, more investigations were undertaken. First, average density of the Cr pellet was determined at different conditions of sintering (by weighing pellets and measurement of thickness and diameter) (Table 9):

**Table 9.** Evolution of pellet density relative to sintering temperature.

| Temperature (°C) | 500 | 550 | 600 | 650 | 800 |
|---|---|---|---|---|---|
| Density (%) | 76 | 80 | 75 | 77 | 82 |

(% of theoretical density of Cr for the same volume).

As the sintering temperature increases, the density increases. Before sintering, the density of the 800 mg pellet was 73%. After sintering, the density increased. A heat treatment was used to densify the material (Figure 12). It is clear that the porosity decreased. Grains were close together or even welded.

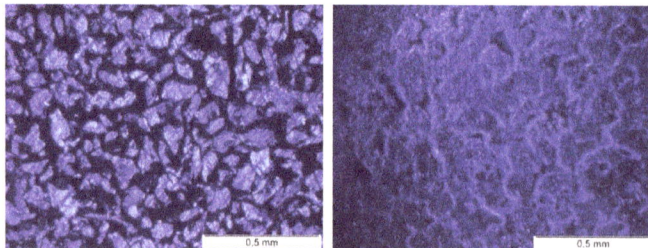

**Figure 12.** Micrograph of non-sintered pellet at left and sintered pellet at 1300 °C at right.

The porosity of the pellets was then defined: this was opened and closed porosity. The closed porosity corresponds to a pore which is isolated from the others and which does not communicate with the outside. On the contrary, we observed opened porosity when the pores communicated with the outside. The results obtained are summarized in the following Figure 13.

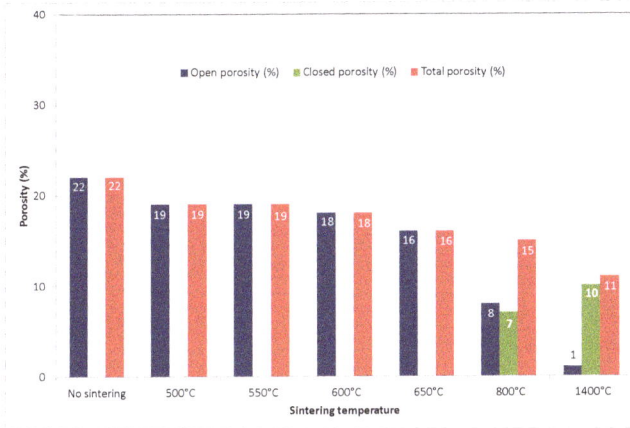

**Figure 13.** Evolution of the porosity of the Cr pellet with the sintering temperature.

These results show that the non-sintered chromium pellets is very porous. The higher the sintering temperature, the lower the porosity obtained. Therefore, when the porosity was very low, the attack of the acid for dissolution was difficult because the acid could not penetrate the core of the pellet and attack grain boundaries and grains. This may also explain why a pellet sintered at 600 °C dissolved easily while sintering at 800 °C did not completely dissolve.

To understand the ability to dissolve as a function of the sintering temperature, the evolution of the microstructure as a function of temperature was followed. For this, observations in optical microscopy and SEM were performed. Some micrographs are presented in the following Figures 14 and 15.

The heat treatment of the Cr pellet in $N_2/H_2$ atmosphere and in a crucible containing graphite powder induced the formation of small areas of chromium oxide (because of residual oxygen adsorbed in the porosity) and the formation of a chromium carbide, when the temperature was higher, for example 1400 °C: chromium carbide was formed from the chromium of the pellet and the carbon powder. During the dissolution step, these two compounds will not facilitate the attack of the pellet with hydrochloric acid.

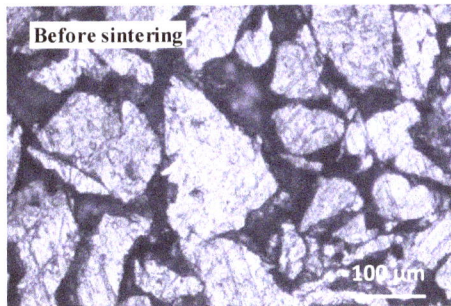

**Figure 14.** Micrography of a 800 mg Cr pellet before sintering (optical microscope).

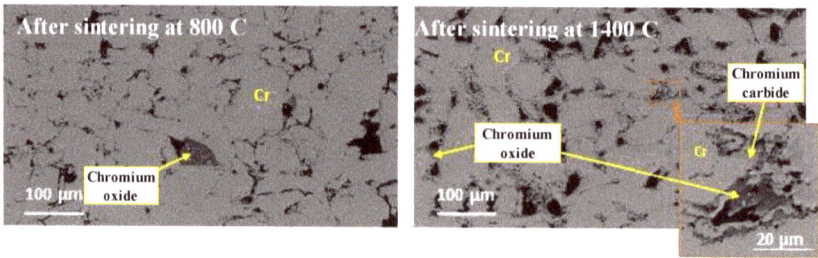

**Figure 15.** Micrography of a 800 mg Cr pellet after sintering and polished (SEM).

X-ray diffraction (XRD) analysis was performed on a pellet sintered at 800 °C (Figure 16). These results highlights the presence of chromium metal in the majority phase and the presence of traces of eskolaïte $Cr_2O_3$. This result is consistent with observations and analysis made by SEM.

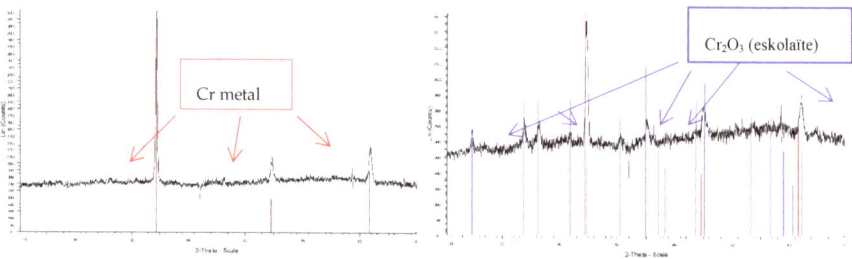

**Figure 16.** XRD diffractograms of pellet before and after sintering at 800 °C.

3.2.6. $D_w$ in Separation of Cr/Mn

Figure 17 presents these results. With an eluent without ethanol, no separation between Cr and Mn was observed. An ethanolic eluent (93 to 98%) modified impressively this separation. Cr was not retained on AG1x8® resin whereas Mn was and $D_w$ was bigger as HCl concentration of the eluent solution increased.

These results were in agreement with literature [40] but, in our case, less activity was undertaken for nuclear safety considerations and less contact time (1 h). Agreements of $D_w$ with published results have led us to make a first test of the semi-automation of the Cr/Mn separation.

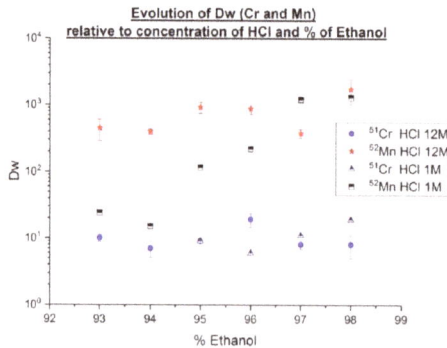

**Figure 17.** $D_w$ of Mn and Cr.

### 3.2.7. Semi-Automated Separation of Cr/Mn

It is necessary to modify the remote-control sketch of $^{165}$Er for use in the extraction of $^{52}$Mn, and this requires three columns at least to obtain a suitable separation (Figure 18).

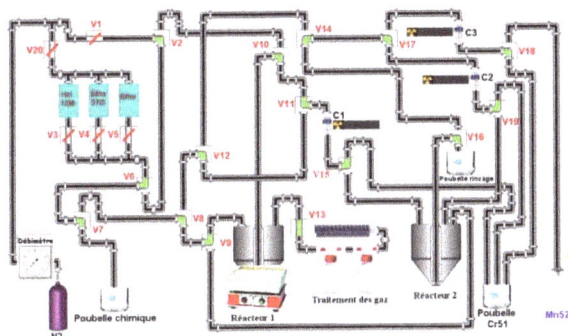

**Figure 18.** Remote control of Mn/Cr separation.

Results for the first test, with weak activity, on a semi-automation of separation Mn/Cr are listed in Table 10.

**Table 10.** Extraction yield in semi-automation separation of Mn/Cr.

|  |  | Volume | Y (%) | |
|---|---|---|---|---|
|  |  | (mL) | $^{51}$Cr | $^{52}$Mn |
| Column 1 | Sample | 8.2 | 91.2 | 0.0 |
|  | Eluent F1 | 3.7 | 2.8 | 0.0 |
|  | Eluent F2 | 3.4 | 0.6 | 0.0 |
|  | Eluent F3 | 3.3 | 0.3 | 0.0 |
|  | HCl 12 M | 1.3 | 0.0 | 76.5 |
| Column 2 | Sample | 29.6 | 0.0 | 0.0 |
|  | Eluent F1 | 3.4 | 0.0 | 0.0 |
|  | HCl 12 M | 1.0 | 0.0 | 81.7 |
| Column 3 | Sample | 21.8 | 0.0 | 0.0 |
|  | Eluent F1 | 3.5 | 0.0 | 0.0 |
|  | HCl 12 M | 1.5 | 0.0 | 83.8 |
| Global Yield (%) |  |  | 94.9 | 46.7 * |

Chromium was recovered in the first batch: it was not so much adsorbed on column 1 as was $^{52}$Mn. Global recovery of $^{52}$Mn was less than 50% (* calculated on the final batch of $^{52}$Mn: initial activity/final activity in batch).

### 3.2.8. Reduction of Pellet Thickness

To optimize the target design of a Cr pellet, a reduction of the target thickness and, by this way, a limitation of non-irradiated Cr mass, polishing is a useful solution.

This operation reduced the amount of chromium handled during the extraction process. Thickness was reduced by one third (from 900 μm to 350 μm) and the oxide layer on surface of target was removed.

## 4. Discussion

The cyclotron of Orléans is a multipurpose research accelerator (proton, deuteron and alpha at various energy from 10 to 45 MeV). With these specifications, more radionuclides can be accessed than in a commercial cyclotron. In this work, we demonstrate its capabilities for development of exotic radionuclides like [165]Er and [52]Mn. It is a wonderful tool for a radiochemist to product radionuclides aimed at imaging as well as analytical applications too (radiotracer method). But, with energy of 30 MeV, most of SPECT isotopes are produced by reactions (p, 2n) or (p, 3n) [45]. Even when one has the choice between (p, n) and (p, 2n), e.g., for [67]Ga one favors (p, 2n) reaction for the yield (thicker targets, less warming by μA). For these two radionuclides produced at the cyclotron of Orléans, the first tool developments for separative radiochemistry have been made.

In the case of [165]Er, separation Er/Ho needed to be better understood and a real improvement could also be made to facilitate semi-automation.

For a limited volume of isolated [165]Er fraction with elution of 1 M $HNO_3$, a NaI detector with two windows at 53 keV and 80 keV could be a solution. On the other hand, more detailed developments on geometric parameters on the elution curve are necessary because LN2 resin has a limited capacity for adsorption.

Semi-automation proved necessary with increased activity of [165]Er by the deuteron method. A more sensitive evaluation of the quality of a [165]Er batch and its ratio [166+165(nat)]Ho/Er must be performed more carefully regarding the quality of the irradiated target and the by-products generated. Additional analysis by ICP–MS is necessary to complete these validations. At least, this protocol of separation Ho/Er can already be adapted as it answers issues of radiolanthanides extraction where the separation of micro- from macrocomponents is crucial.

This work made it possible to obtain a Cr target for the production of [52]Mn for irradiation and method development tests: 650 mg of powder sintered at 800 °C for 1 h (thickness 900 μm).

In a detailed analysis of results of compaction, a test at 4 tons on 650 mg pellet gave a bad visual aspect and greater thickness of 1000 μm than those made at 8.3 tons (900 μm): density must be lower which is not interesting for an irradiation (mechanical firmness). A surprising result is compaction of 60 min at 8.3 tons on 650 mg pellet that gave a bad visual aspect. This can be explained by the fact that its time in pressed state can generated micro crack. Then, when the pellet is relieved of stress, the pellet cracked.

Regarding the conditions of sintering, at vacuum no difference with a pellet not-sintered has seen. This can be perhaps explain by system of sintering: temperature has defined by emissivity of molybdenum where pellets are put. A second explanation might be that more vacuum is, less tension surface on grain is. With a reduction of tension surface, grain boundaries are more difficult to form.

Between 600 °C and 800 °C, resistance rupture is twice as big. There is no relation between weight and resistance rupture. The optimum temperature of state transition is between 600 °C and 800 °C. Resistance rupture is less altered by the weight of the pellet than the temperature.

Breaking measurements must be carried out on the polished targets of 300–400 μm thickness as well as dissolution tests.

## 5. Conclusions

The production of [165]Er from a natural Ho target was obtained via deuteron irradiation. This form of irradiation generated more activity than the proton method, with much activity of [166]Ho (a radiotracer for Ho). LN2 resin (from Triskem) was used to separate Er and Ho with nitric acid solution. Selectivity between two radiolanthanides was weak (SF Er/Ho = 2.75 at 0.4 M $HNO_3$) but an isolated fraction with 72.3 ± 5.1% ($n = 4$) of initial [165]Er activity can be achieved by an appropriate gradient of elution with 0.4 M (120 mL) and 1 M $HNO_3$ (40 mL). In this fraction, the mass of holmium was reduced by a factor 1294 ± 1183 ($n = 3$). Regarding [52]Mn production, a design of the target has been established: a Cr pellet of 800 mg compressed at 8.3 tons (300 bars) and sintered at 800 °C. Elution on AG1x8® resin with a mixture of ethanol/HCl (12 M) (97/3 ($v/v$) gave good results for the separation of [51]Cr from [52]Mn. A test of semi-automation separation using three columns was made with a diluted solution of [51]Cr and [52]Mn. A global extraction of 47% on [52]Mn was obtained.

For these two radiometals, [165]Er and [52]Mn, results were encouraging for the following experiments: some improvements are underway concerning the irradiation and the targeting. For the separation, a methodology about more detailed developments on geometric parameters on the elution curve must have been set. The first results of the Er/Ho and Mn/Cr separation on a semi-automated system demonstrated the versatility of the homemade system. However, these tests have been shown to require more reliability and reproducibility in the separation step with a better follow-up on line using adapted sensors.

**Author Contributions:** E.D. and F.D. worked on production of radionuclides for the ratiotracer method (definition of beam) and developed the simple tool to predict a nuclear reaction for a radiochemist; J.V. and M.G. worked on the design of the Cr-target and performed experiences on chemical and physical specifications of Cr-pellet; M-L.B. contributed to the design of the Cr-target through knowledge in materials and methods of analysis; G.A. (spectrometry γ), E.S. ($^{52}$Mn) and C.R. ($^{165}$Er) performed the experiments for determination of Dw and tests of resin; L.F. designed the automation of the process and analyzed the data by spectrometry γ; A.B. contributed reagents/materials/analysis tools and the definition of Er/Ho separation by using resin LN2 (Triskem); I.D.S. conceived the radiotracer methods, defined conditions of experiments and participated in all steps of these various works, and wrote the paper; I.D.S. would like to thank G. Junot ($^{64}$Cu), K. Blanchard ($^{64}$Cu), D. Barré ($^{11}$C), A. Fournier ($^{64}$Cu), L. Colas ($^{11}$C), L. Léost ($^{64}$Cu), A. Sabourin ($^{11}$C), C. Dupressoir ($^{64}$Cu) and N. Mialon ($^{11}$C) for their work contributing to the setup of radiochemistry at the cyclotron of Orléans and, indirectly, to this article.

**Funding:** This research had financial support of Triskem Society and AAP—Défi 2017 "Instrumentation aux limites "project "CiSCoTe": Cible Solide irradiée Contrôlée en Température". It funded too by the French National Research Agency (grant ANR-13-JS07-0007).

**Conflicts of Interest:** The authors declare no conflict of interest.

## References

1. Zeglis, B.M.; Houghton, J.L.; Evans, M.J.; Viola-Villegas, N.; Lewis, J.S. Underscoring the Influence of Inorganic Chemistry on nuclear Imaging with Radiometals. *Inorg. Chem.* **2014**, *53*, 1880–1899. [CrossRef] [PubMed]

2. Brasse, D.; Nonat, A. Radiometals: Towards a new success story in nuclear imaging? *Dalton Trans.* **2015**, *44*, 4845–4858. [CrossRef] [PubMed]

3. Müller, C.; Van Der Meulen, N.P.; Benesová, M.; Schibli, R. Therapeutic Radiometals Beyond $^{177}$Lu and $^{90}$Y: Production and Application of Promising α-particle, β$^-$-Particle, and Auger Electron Emitters. *J. Nucl. Med.* **2017**, *58*, 91S–96S. [CrossRef] [PubMed]

4. Amoroso, A.J.; Fallis, I.A.; Pope, S.J.A. Chelating agents for radiolanthanides: Applications to imaging and therapy. *Coordin. Chem. Rev.* **2017**, *340*, 198–219. [CrossRef]

5. Guérard, F.; Rajerison, H.; Faivre-Chauvet, A.; Barbet, J.; Meyer, G.J.; Haddad, F.; Da Silva, I.; Gestin, J.-F. Radiolabeling proteins with stabilised hypervalent astatine-211: Feasability study and evaluation of the in vitro stability. *Eur. J. Nucl. Med. Mol. Imaging* **2010**, *37*, S361.

6. Rajerison, H.; Guérard, F.; Mougin-Degraef, M.; Bourgeois, M.; Da Silva, I.; Chérel, M.; Barbet, J.; Faivre-Chauvet, A.; Gestin, J.-F. Radioiodinated and astatinated NHC rhodium complexes: Synthesis. *Nucl. Med. Biol.* **2014**, *41*, e23–e29. [CrossRef] [PubMed]

7. Sergentu, D.C.; Teze, D.; Sabatié-Gogova, A.; Alliot, C.; Guo, N.; Bassal, F.; Da Silva, I.; Deniaud, D.; Maurice, R.; Champion, J.; et al. Advances on the Determination of the Astatine Pourbaix Diagram: Predomination of AtO(OH)2− over At− in Basic Conditions. *Chem. Eur.* **2016**, *22*, 2964–2971. [CrossRef] [PubMed]

8. Alliot, C.; Chérel, M.; Barbet, J.; Sauvage, T.; Montavon, G. Extraction of astatine-211 in diisopropylether (DIPE). *Radiochim. Acta* **2009**, *97*, 161–165. [CrossRef]

9. Meijs, W.E.; Herscheid, J.D.M.; Haisma, H.J.; Wijbrandts, R.; Van Langevelde, F.; Van Leuffen, P.J.P.; Mooy, R.; Pinedo, H.M. Production of Highly Pure no-carrier added $^{89}$Zr for the labelling of antibodies with a positron emitter. *Appl. Radiat. Isot.* **1994**, *45*, 1143–1147. [CrossRef]

10. National Nuclear Data Center. Available online: https://www.nndc.bnl.gov/chart/decaysearchdirect.jsp?nuc=165ER&unc=nds (accessed on 23 February 2018).

11. Beyer, G.J.; Zeisler, S.K.; Becker, D.W. The Auger-electron emitter $^{165}$Er: Excitation function of the $^{165}$Ho(p,n)$^{165}$Er process. *Radiochim. Acta* **2004**, *92*, 219–222. [CrossRef]

12. Malikidogo, P.K.; Da Silva, I.; Morfin, J.-F.; Lacerda, S.; Barentin, L.; Sauvage, T.; Sobilo, J.; Lerondel, S.; Toth, E.; Bonnet, C. A cocktail of 165Er$^{3+}$ and Gd$^{3+}$ complexes for quantitative detection of zinc by SPECT/MRI. *Chem. Commun.* **2018**. [CrossRef] [PubMed]

13. Tárkányi, F.; Hermanne, A.; Takács, S.; Ditrói, F.; Király, B.; Kovalev, S.F.; Ignatyuk, A.V. Experimental study of the $^{165}$Ho(d,2n) and $^{165}$Ho(d,p) nuclear reactions up to 20 MeV for production of the therapeutic radioisotopes $^{165}$Er and $^{166g}$Ho. *Nucl. Instrum. Meth. B* **2008**, *266*, 3529–3534. [CrossRef]

14. Tárkányi, F.; Hermanne, A.; Király, B.; Takács, S.; Ditrói, F.; Baba, M.; Ohtsuki, T.; Kovalev, S.F.; Ignatyuk, A.V. Study of activation cross-sections of deuteron induced reactions on erbium: Production of radioisotopes for practical applications. *Nucl. Instrum. Meth. B* **2007**, *259*, 829–835. [CrossRef]

15. Graves, S. Production and Applications of Long-Lived Positron Emitting Radionuclides. Ph.D. Thesis, University of Wisconsin-Madison, Madison, WI, USA, 6 January 2017.

16. Le Loirec, C. Track structure simulation for positron emitters of physical interest. Part II: The case of the radiometals. *Nucl. Instrum. Meth. A* **2007**, *582*, 654–664. [CrossRef]

17. Disselhorst, J.A.; Brom, M.; Laverman, P. Image-Quality Assessment for Several 570 Positron Emitters Using the NEMA NU 4-2008 Standards in the Siemens Inveon Small-Animal PET Scanner. *J. Nucl. Med.* **2010**, *51*, 610–617. [CrossRef] [PubMed]

18. Sastri, C.S. Production of Manganese-52 of high isotopic purity by $^{3}$He activation of Vanadium. *Int. J. Appl. Radiat. Isot.* **1981**, *32*, 246–247. [CrossRef]

19. Daube, M.E.; Nickles, R.J. Development of myocardial perfusion tracers for positron emission tomography. *Int. J. Nucl. Med. Biol.* **1985**, *12*, 303–314. [CrossRef]

20. Tolmachev, V.; Bruskin, A.; Lunqvist, H. *Neutron Deficient Nuclides for Positron Emission Tomography. IX. Preliminary Study of Production Routes of Positron Emitting Isotopes of Manganese*; Institute of Theoretical and Experimental Physics: Moscow, Russia, 1994; pp. 47–94.

21. Calandroni, D.; Dupuis, A.; Grémont, B.; Hurt, B.; Kervizic, J.; Meyrand, G.; Tran, D.T. Some characteristics from Orléans and Liège cyclotrons. In Proceedings of the Seventh International Conference on Cyclotrons and their Applications, Zurich, Switzerland, 19–22 August 1975; pp. 88–91.

22. Dupuis, A.; Kervizic, J.; Launé, B.; Meyrand, G.; Tran, D.T.; Tronc, D.; Goin, G. Cyclotron center region study and beam diagnostic at Orléans cyclotron. In Proceedings of the Seventh International Conference on Cyclotrons and their Applications, Zurich, Switzerland, 19–22 August 1975; pp. 275–278.

23. Goin, G. Status report on the CNRS Orleans' cyclotron. In Proceedings of the 9th International Conference on Cyclotrons and their Applications, Caen, France, 7–10 September 1981; pp. 133–135.

24. Goin, G.; Vernois, J.; Cimetiere, C.; Leger, J. Iodine 123 production by irradiation at the CNRS cyclotron in Orléans. In Proceedings of the 9th International Conference on Cyclotrons and theirs Applications, Caen, France, 7–10 September 1981; pp. 697–698.

25. Breteau, N.; Sabattier, R.; Gueulette, J.; Bajard, J.C. The Orléans neutrontherapy facility. In Proceedings of the 9th International Conference on Cyclotrons and theirs Applications, Caen, France, 7–10 September 1981; pp. 703–705.

26. Tárkányi, F.; Takács, S.; Hermanne, A.; Ditrói, F.; Király, B.; Baba, M.; Ohtsuki, T.; Kovalev, S.F.; Ignatyuk, A.V. Investigation of production of the therapeutic radioisotope $^{165}$Er by proton induced reactions on erbium in comparison with other production routes. *Appl. Radiat. Isot.* **2009**, *67*, 243–247. [CrossRef] [PubMed]

27. Zandi, N.; Sadeghi, M.; Afarideh, H. Evaluation of the cyclotron production of $^{165}$Er by different reactions. *J. Radioanal. Nucl. Chem.* **2013**, *295*, 923–928. [CrossRef]

28. Korkisch, J. *Rare Earth Elements in Handbook of Ion Exchange Resins: Their Application to Inorganic Analytical Chemistry*; CRC Press: Boca Raton, FL, USA, 1989; ISBN 0-8493-3191-9.

29. Nifant'ev, I.E.; Minyaev, M.E.; Tavtorkin, A.N.; Vinogradov, A.A.; Ivchenko, P.V. Branched alkylphosphinic and disubstituted phosphinic and phosphonic acids: Effective synthesis based on α-olefin dimers and applications in lanthanide extraction and separation. *RSC Adv.* **2017**, *7*, 24122–24128. [CrossRef]

30. Xie, F.; Zhang, T.A.; Dreisinger, D.; Doyle, F. A critical review on solvent extraction of rare earths from aqueous solutions. *Miner. Eng.* **2014**, *56*, 10–28. [CrossRef]

31. Reddy, M.L.P.; Rao, P.; Damodaran, A.D. Liquid-liquid extraction processes for the separation and purification of rare earths. *Miner. Process. Extract. Metall. Rev.* **1995**, *12*, 91–113. [CrossRef]

32. Horwit, E.P.; Bloomquist, C.A.A. Chemical separation for heavy elements searches in irradiated uranium targets. *J. Inorg. Nucl. Chem.* **1975**, *37*, 425–434. [CrossRef]

33.  Kazakov, A.G.; Aliev, R.A.; Bodrov, A.Y.; Priselbova, A.B.; Kamlykov, S.N. Separation of radioisotopes of terbium from a europium target irradiated by 27 MeV α-particles. *Radiochim. Acta* **2018**, *106*, 135–140. [CrossRef]

34.  Hidaka, H.; Yoneda, S. Sm and Gd isotopic shifts of Apollo 16 and 17 drill stem samples and their implications for neolith history. *Geochim. Cosmochim. Acta* **2007**, *71*, 1074–1086. [CrossRef]

35.  Monroy-Guzman, F.; Jaime Salinas, E. Separatin of Micro-Macrocomponent Systems: 149Pm—Nd, 161Tb-gd, 166Ho-Dy, and 177Lu-Yb by Extraction Chromatography. *J. Mex. Chem. Soc.* **2015**, *59*, 143–150.

36.  Horwitz, E.P.; McAlister, D.R.; Dietz, M.L. Extraction chromatography versus solvent extraction: How similar are they? *Sep. Sci. Technol.* **2006**, *41*, 2163–2182. [CrossRef]

37.  Shu, Q.; Khayambashi, A.; Zou, Q.; Wang, X.; Wei, Y.; He, L.; Tang, F. Studies on adsorption and separation characteristics of americium and lanthanides using a silica-based macroporous bi(2-ethylhexyl)phosphoric acid (HDEHP) adsorbent. *J. Radioanal. Nucl. Chem.* **2017**, *313*, 29–37. [CrossRef]

38.  Lake Wooten, A. Cross-sections for (p,x) reactions on natural chromium for the production of $^{52,52m}$Mn,$^{54}$Mn radioisotopes. *Appl. Radiat. Isot.* **2015**, *96*, 154–161. [CrossRef] [PubMed]

39.  Lahiri, S.; Nayak, D.; Korschinek, G. Separation of no-carrier-added $^{52}$Mn from bulk chromium: A simulation study for Accelerator Mass Spectrometry measurement of $^{53}$Mn. *Anal. Chem.* **2006**, *78*, 7517–7521. [CrossRef] [PubMed]

40.  Fonslet, J.; Tietze, S.; Jensen, A.I.; Graves, S.A.; Severin, G.W. Optimized procedures for manganese-52: Production, separation and radiolabelling. *Appl. Radiat. Isot.* **2017**, *121*, 38–43. [CrossRef] [PubMed]

41.  Graves, S.; Hernandez, R.; Fonslet, J.; England, C.G.; Valdovinos, H.F.; Ellison, P.A.; Barnhart, T.E.; Elema, D.R.; Theuer, C.P.; Cai, W.; et al. Novel preparation methods of 52Mn for immunoPET imaging. *Bioconjugate Chem.* **2015**, *26*, 2118–2124. [CrossRef] [PubMed]

42.  Westergaard, H.M. Stresses in Concrete Pavements Computed by theoretical analysis. *Public Roads* **1926**, *7*, 2–25.

43.  François, D.; Pineau, A.; Zaoui, A. *Comportement Mécanique des Matériaux*; Hermès: Paris, France, 1995; Chapter 4; pp. 402–437.

44.  National Physical Laboratory. Available online: http://resource.npl.co.uk/mtdata/phdiagrams/ccr.htm (accessed on 22 June 2018).

45.  Scientific Report of Nuclear Physics European Collaboration Committee (NuPECC) 2014 Nuclear Physics for Medicine. Available online: http://www.nupecc.org/pub/npmed2014_brochure.pdf (accessed on 22 June 2018).

![instruments logo] *instruments*

*Review*

# Cyclotron Production of Unconventional Radionuclides for PET Imaging: the Example of Titanium-45 and Its Applications

Pedro Costa [1,2,*], Luís F. Metello [1,3], Francisco Alves [4,5] and M. Duarte Naia [6,7]

[1]   Nuclear Medicine Department, School of Health, Polytechnic Institute of Porto, 4200-072 Porto, Portugal; lfm@ess.ipp.pt
[2]   Health and Environment Research Center (CISA), Polytechnic Institute of Porto, 4200-072 Porto, Portugal
[3]   Isótopos para Diagnóstico e Terapêutica (IsoPor), S.A., 4445-526 Ermesinde, Portugal
[4]   Institute of Nuclear Sciences Applied to Health (ICNAS), University of Coimbra, 3000-548 Coimbra, Portugal; franciscoalves@uc.pt
[5]   Coimbra Health School, Polytechnic Institute of Coimbra , 3046-854 Coimbra, Portugal
[6]   Physics Department, University of Trás-os-Montes e Alto Douro, 5000-801 Vila Real, Portugal; duarte@utad.pt
[7]   Centre for Mechanical Engineering, Materials and Processes (CEMMPRE), University of Coimbra, 3030-788 Coimbra, Portugal
*    Correspondence: psc@ess.ipp.pt

Received: 2 May 2018; Accepted: 31 May 2018; Published: 3 June 2018

**Abstract:** Positron emitting radionuclides are used to label different compounds, allowing the study of the major biological systems using PET (positron emission tomography) imaging. Although there are several radionuclides suited for PET imaging, routine clinical applications are still based on a restrict group constituted by $^{18}$F, $^{11}$C, and, more recently, $^{68}$Ga. However, with the enlarged availability of low-energy cyclotrons and technical improvements in radionuclide production, the use of unconventional radionuclides is progressively more common. Several examples of unconventional radionuclides for PET imaging are being suggested, and $^{45}$Ti could be suggested as a model, due to its interesting properties such as its abundant positron emission (85%), reduced positron energy ($\beta^+$ endpoint energy = 1040 keV), physical half-life of 3.09 h, and interesting chemical properties. This review aims to introduce the role of cyclotrons in the production of unconventional radionuclides for PET imaging while using $^{45}$Ti as an example to explore the potential biomedical applications of those radionuclides in PET imaging.

**Keywords:** cyclotron; PET; titanium-45; unconventional radionuclides

---

## 1. Introduction: Unconventional Radionuclides for PET Imaging

The incorporation and utilization of technological developments in clinical routine is becoming normal in modern practice of medicine. The application of imaging modalities is improving the quality of medical care and procedures that are available in medicine, which is an attitude that enhances the rapid implementation of modern paradigms of evidence-based medicine and science-based medicine. In fact, medical imaging occupies a primary position in clinical decision algorithms. In such context, non-invasive imaging modalities could allow for accurate diagnoses, increase precision in treatment choice/planning, and provide opportunities to follow the evolution of the clinical status of patients [1]. These modalities could also act over wide ranges of time and size scales involved in biological and pathological processes. Multiple imaging techniques are available, including X-ray imaging, computerized tomography (CT), magnetic resonance (MR) imaging, ultrasonography (US), optical imaging using fluorescent molecules (OF), and nuclear medicine (NM) using radioisotopes for different procedures such as planar imaging, single photon emission computed tomography (SPECT) and positron emission tomography (PET) [2].

Nuclear medicine has been an independent medical specialty since 1972, and is defined by the World Health Organization (WHO) as one that *"encompasses applications of radioactive materials in diagnosis, treatment, or in medical research, with the exception of the use of sealed radiation sources in radiotherapy"* [3]. In practice, nuclear medicine procedures are methods and techniques based in radionuclides. For diagnostic purposes, nuclear medicine uses primarily two types of radionuclides: (i) gamma-photon emitting nuclides (the ones that decay by isomeric transition or electron capture) for planar imaging and SPECT; and (ii) positron emitting nuclides for PET, which will be deeply discussed along this paper.

Radionuclides are used to label very small amounts of a pharmaceutical compound of interest, creating radiopharmaceuticals. According to the decay type of the nuclide component, the radiopharmaceutical can be optimized for a given procedure (SPECT versus PET). This industry of radiopharmaceuticals in general present a significant impact at the economical level. According to recent reports, the global radiopharmaceuticals market in 2017 was evaluated in $5.5 billion USD [4].

There are several factors that enhance the success of the application of a certain radionuclide. Considerations such as the physicochemical properties, availability, cost, and ease of use are some of the most relevant. For instance, according the literature, the usual clinical practice of PET imaging relies just on four radionuclides, namely: carbon-11 ($^{11}$C), nitrogen-13 ($^{13}$N), oxygen-15 ($^{15}$O), and fluorine-18 ($^{18}$F). All of them are characterized by relatively short physical half-lives [5]. The commercially available radiopharmaceuticals are almost invariably associated with $^{18}$F, due to the technical constraints in the production and distribution processes of the other short-lived nuclides. In fact, radiopharmaceuticals labeled with $^{18}$F are very interesting and commonly used mainly because of physical properties, such as a period of half-life that allows local (in situ) production and short-medium scale distribution, and the low $\beta^+$ energy endpoint, which guarantees a good image quality [6].

Recently, a global trend was observed of exploring several new radionuclides for PET imaging that were different than the most classical ones, in order to increase the range of applications of the technology, as well as diversify the radiolabeled compounds. This become possible due to new insights into the production processes, radiolabeling methodologies, and interest in clinical applications of different radionuclides [7]. Examples of these unconventional radionuclides for PET imaging are: Scandium-44 ($^{44}$Sc), Titanium-45 ($^{45}$Ti), Cobalt-55 ($^{55}$Co), Copper-60/61/64 ($^{60}$/$^{61}$/$^{64}$Cu), Bromium-76 ($^{76}$Br), Rubidium-82 ($^{82}$Rb), Yttrium-86 ($^{86}$Y), Technetium-94m ($^{94m}$Tc), Zirconium-89 ($^{89}$Zr), and Iodine-124 ($^{124}$I) [8,9]. Recently published papers confirm the investigational trend underlined in this paper [10–12]. The correspondent physical properties are presented in Table 1.

**Table 1.** Some unconventional positron emission tomography (PET) radionuclides and corresponding physical properties [13].

| Radionuclide | Half-Life | $\beta^+$ Endpoint Energy (MeV) |
|:---:|:---:|:---:|
| $^{44}$Sc | 3.927 h | 1.47 |
| $^{45}$Ti | 3.1 h | 1.04 |
| $^{55}$Co | 17.53 h | 1.50 |
| $^{60}$Cu | 23.7 min | 3.77 |
| $^{61}$Cu | 3.333 h | 1.21 |
| $^{64}$Cu | 12.700 h | 0.653 |
| $^{76}$Br | 16.2 h | 3.94 |
| $^{82}$Rb | 1.273 min | 3.15 |
| $^{86}$Y | 14.7 h | 3.14 |
| $^{89}$Zr | 3.3 day | 0.902 |
| $^{94m}$Tc | 52.0 min | 2.44 |
| $^{124}$I | 4.176 day | 2.14 |

As already cited, conventional applications of PET imaging are based on organic elements with short half-lives (minutes to hours). A careful analysis of the radionuclides listed in Table 1 shows a tendency to explore radionuclides with higher physical half-lives and chemical characteristics associated with metallic elements (inorganic elements). In fact, the use of nuclides of inorganic elements has been studied since the discovery of radioactivity, but nuclear properties are principal factors that justify the development of the new compounds, while the availability of radionuclides in high specific activities and inorganic chemistry issues at the radiotracer level are the most frequent challenges [14]. Even so, unconventional radiometals are gradually being introduced, diversifying the available tools and procedures in nuclear medicine, particularly in PET imaging.

## 2. Cyclotrons and the Production of Unconventional Radionuclides

Today, more than 2700 radionuclides have been produced artificially with particle accelerators [15]. In addition, a few examples are also being obtained *in loco* using radionuclide generators through radioactive decay.

PET radionuclides, in particular, can be produced in cyclotrons, especially using inducing *(p,n)* nuclear reactions in the targets of stable isotopes.

Routine cyclotron production processes are made possible by the actual dissemination of these devices. As a matter of fact, by the end of 2005, there were 262 cyclotrons operating in the 39 member states of International Atomic Energy Agency (IAEA); however, it was believed that around 350 cyclotrons were operating in the whole world, according to a database of the agency [16]. Unfortunately, there is no official update of this report, because it is not easy to correctly estimate the number of cyclotrons operating nowadays all over the world.

The commercially available cyclotrons can be classified with respect to the particle type and maximum energy reached, the method of ion production, the technique of beam extraction from the cyclotron (or absence of extraction), the intensity of the accelerated ion beams, and other specific properties or features [17,18]. There are different classifications based on the type and energy of the accelerated particles [17,19]. Independently from the classification used, an important aspect to mention is that near 70% of the cyclotrons disseminated over the world are low-energy cyclotrons ($\leq$20 MeV) [16].

According to empirical and practical evidence, the cyclotrons that typically have been applied worldwide in radionuclide production comprise properties such as: (i) the capability of accelerating negative ions ($H^-$); (ii) beam extraction using stripper foils; (iii) fixed beam energy between 10–18 MeV, or 10–24 MeV mainly if the installation is intended for the production of many radionuclides, large-scale production and/or research purposes; (iv) fixed frequency of the RF generator; (v) two or four *dees* placed in valleys; (vi) internal ion source(s); (vii) the possibility of adjusting the beam position on the target; (viii) possibility of multi-target irradiations; (ix) compact radiation shielding around the device ("self-shielded" cyclotron); and (x) a high level of automation and simplicity in maintenance.

Despite the existence of the formal classifications, cyclotrons that respect the criteria stated in the last paragraph are normally classified by professionals of the field as "small cyclotrons", "low-energy cyclotrons" or even as "medical cyclotrons", and are applied to induce *(p,n)* reactions, which are typically low-energy processes with a constant onset below 9 MeV [20].

After the careful optimization of technical details such as targetry and irradiation parameters, high production yields of some of the unconventional radionuclides covered in this paper have been obtained in cyclotrons [21–26]. Examples of such radionuclides and recommended nuclear reactions are given in Table 2.

**Table 2.** Examples of cyclotron-produced unconventional PET radionuclides.

| Radionuclide | Most Common Nuclear Reaction |
|---|---|
| $^{44}$Sc | $^{44}$Ca$(p,n)^{44}$Sc |
| $^{45}$Ti | $^{45}$Sc$(p,n)^{45}$Ti |
| $^{55}$Co | $^{58}$Ni$(p,\alpha)^{55}$Co |
| $^{60}$Cu | $^{60}$Ni$(p,n)^{60}$Cu |
| $^{61}$Cu | $^{61}$Ni$(p,n)^{61}$Cu |
| $^{64}$Cu | $^{64}$Ni$(p,n)^{64}$Cu |
| $^{76}$Br | $^{76}$Se$(p,n)^{76}$Br |
| $^{86}$Y | $^{86}$Sr$(p,n)^{86}$Y |
| $^{89}$Zr | $^{89}$Y$(p,n)^{89}$Zr |
| $^{94m}$Tc | $^{94}$Mo$(p,n)^{94m}$Tc |
| $^{124}$I | $^{124}$Te$(p,n)^{124}$I |

As more research studies in radionuclide production are delivered with successful results, which are conjugated with promising results in the preclinical and clinical trials that are currently being developed, these "unconventional" radionuclides are set to play an increasingly important role in nuclear medicine. These considerations highlight the pivotal role that cyclotrons have in solving the availability issues of interesting radionuclides for PET imaging [21].

## 3. Titanium-45 and Its Applications

Among the different examples that could be used to explore the paradigm of unconventional radionuclides for PET imaging in analysis, $^{45}$Ti is an interesting case study.

The nuclide $^{45}$Ti has a half-life of 3.09 h, which allows for distribution within a broader region than radiopharmaceuticals based on $^{18}$F or other shorter half-life radionuclides. On the other hand, its decay is predominantly based on positron emission (85%), with a maximum positron energy of 1040 keV and an average energy of 439 keV, which assures good image quality in PET images.

Preliminary results that were obtained in the experimental program of our group revealed the production feasibility of $^{45}$Ti in low-energy cyclotrons through the use of a $^{45}$Sc$(p,n)^{45}$Ti nuclear reaction, with proton beams higher than 3 MeV. The maximum cross-section values were determined for proton beams with energy in the range between 10–14 MeV, while significant impurities production was found for energies higher than 17 MeV (see Table 3). These results mean that the development of a routine production process is possible, and research on the subject is ongoing [27].

**Table 3.** Preliminary results of experimental cross-section values for $^{45}$Sc$(p,n)^{45}$Ti nuclear reaction.

| Mean Beam Energy (MeV) | Cross-Section (mbarn) | Uncertainty (mbarn) |
|---|---|---|
| $16.1 \pm 1.0$ | $1.2 \times 10^2$ | $\pm 0.4 \times 10^2$ |
| $15.3 \pm 1.0$ | $2.1 \times 10^2$ | $\pm 0.7 \times 10^2$ |
| $12.4 \pm 0.9$ | $4.4 \times 10^2$ | $\pm 1.4 \times 10^2$ |
| $8.9 \pm 0.8$ | $3.4 \times 10^2$ | $\pm 1.0 \times 10^2$ |
| $4.9 \pm 0.7$ | $1.2 \times 10^2$ | $\pm 0.4 \times 10^2$ |
| $3.4 \pm 0.7$ | $0.8 \times 10^1$ | $\pm 0.3 \times 10^1$ |

In addition, the chemical properties of Ti allow two different scenarios for developing radiopharmaceuticals: (i) the "direct" labeling of ligands using chelation chemistry and/or simple covalent bounds with ligands of interest; or (ii) radiolabeled nanoparticles (either to label nanoparticles that are already explored in the field of nuclear medicine, or to label titanium oxide nanoparticles that could be relevant for clinical purposes).

The first reference to a potential application of $^{45}$Ti belongs to a Japanese research team, and was published in 1981 by the *Cyclotron and Radionuclide Center* (CYRIC) of Tokohu University [28].

This group simultaneously evaluated the production, target chemistry (extraction and purification of $^{45}$Ti), and radiochemistry (labeling), and conducted some animal experiments. The main conclusion was centered on the potential of $^{45}$Ti-compounds to become *"a new series of the positron-emitting radiopharmaceutical"*, even considering some problems related to the stability of $^{45}$TiCl$_4$ [28]. Examples of $^{45}$Ti-phytate and $^{45}$Ti-DTPA images are shown in Figure 1.

Some years later, the same group reported the use of other $^{45}$Ti-labeled compounds, namely $^{45}$TiOCl$_2$ and $^{45}$TiO-phytate, as potential colloid agents for imaging the reticuloendothelial system. They also indicated $^{45}$Ti-DTPA (diethylenetriaminepentaacetic), $^{45}$Ti-citrate, and $^{45}$Ti-HSA (human serum albumin) as possible agents for estimating blood volume and indicators of the breakdown of the blood–brain barrier [29].

Considering the beginning of the study on the pathway of incorporation of titanium-based anticancer drugs such as budotitane($C_{24}H_{28}O_6$Ti), $^{45}$Ti-budotitane was prepared and incorporated into liposomes to provide optimal tumor targeting [30]. Apart from this work, no further experimental studies were found on biodistribution or imaging using this compound.

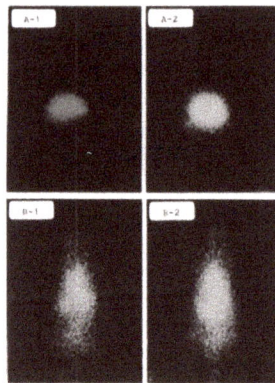

**Figure 1.** Whole body images of $^{45}$Ti-phytate (**A**) and $^{45}$Ti-DTPA (diethylenetriaminepentaacetic) (**B**) acquired 10 minutes (**A1** and **B1**) and 60 minutes (**A2** and **B2**) after injection in rats. Images published by Ishiwata et al. in 1981 [28].

After a gap of some years, Vavere et al. [31] reported an evaluation study of $^{45}$Ti as a radionuclide for PET imaging. Despite other considerations, it was reported that the cyclotron production and purification of $^{45}$Ti is feasible. The work also demonstrated a clear spatial resolution observed down to a rod diameter of 1.25 mm using a Derenzo phantom, i.e., a resolution comparable with $^{18}$F, only with a slight degradation due to the higher energy endpoint and consequent range widening of positrons emitted by $^{45}$Ti (see Figure 2) [31].

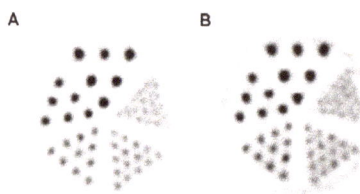

**Figure 2.** Comparison of image quality of $^{18}$F (**A**) and $^{45}$Ti (**B**) using a Derenzo phantom. Images published by Vavere et al. [31].

Another study of *Vavere and Welch* [32] produced [45]Ti and used small animal imaging with [45]Ti-transferrin, which provided new insights about the mechanism of action of a new class of cytostatic agents based on titanium complexes. This research performed a direct labeling of apotransferrin with [45]Ti and studied the resultant biodistribution. The microPET images provided indications on the increased uptake of [45]Ti-transferrin in tumors, with retention up to 24 h, demonstrating that titanium radiopharmaceuticals could be explored as new tools for the in vivo imaging of tumors [31]. This data also supports the use of [45]Ti-labeled compounds in theranostics and personalized medicine, taking the ability to select patients for a specific type of treatment into account.

However, Price et al. [33] published a work reporting some problems regarding the purification of [45]Ti to be used in the radiolabeling process of transferrin. They reported losses of up to 53% of the [45]Ti activity in the waste fractions during the separation process. However, it seems to be a problem with this particular experiment, because other authors have achieved good radiochemical yields in the labeling of proteins such as transferrin or Df-antibody [34].

In general, the reappearance of titanium-based drugs to treat cancer has been met with some success. The several chemical steps that are needed to stabilize these new titanium-based antineoplastics are being developed. In this research, the ligand salan seems to play a special role and lead to a significant investment in the development of these titanium-based drugs [35,36]. However, it is also known that the phase of drug development represents a highly complex, inefficient and costly process that usually takes huge amounts of time and money. Again, nuclear medicine imaging can be decisive, enhancing the efficiency of selecting the candidate drugs that should move forward into clinical trials or be abandoned [37]. The conjugation of the need for more information about drug pharmacokinetics, and the added value of nuclear medicine in the drug development process, can be used strategically; [45]Ti labeling of compounds of interest could boost the research on titanium-based drugs. In fact, this conceptual idea is not totally original, because already published data shows the use of [45]Ti-salan compounds as translational tools to help these drugs in the passage from fundamental research to clinical applications (example in Figure 3) [38].

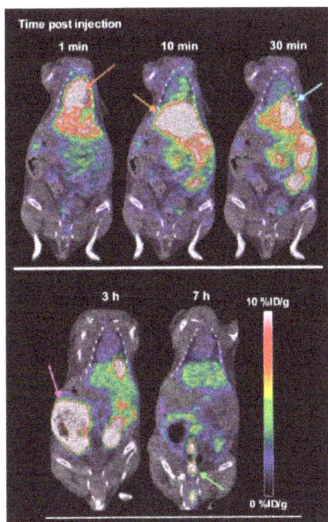

**Figure 3.** Example of acquired PET/computerized tomography (CT) images of a [45]Ti-compound in mice. Images published by Severin et al. [38].

However, the concept can be even taken further, if one hypothesizes that $^{45}$Ti compounds could act as diagnostic imaging agents of the same cancers that could be hereafter treated with titanium-based drugs. This once again suggests a role for $^{45}$Ti in theranostics and personalized medicine.

In the paradigm of personalized medicine, nanoparticles are being studied as drug delivery systems, and their application as vectors for radionuclide-based molecular imaging is a powerful tool of growing interest [39]. Indeed, in another context, recent data suggests that functionalized titanium dioxide nanoparticles ($TiO_2$) can be surface-engineered to target cancer cells [40–44]. Thus, the radiolabeling of $TiO_2$ nanoparticles with $^{45}$Ti could constitute a tool to provide in vitro biodistribution studies, which can also allow the in vivo monitoring of its biological distribution and the pharmacokinetics of such drug delivery systems. The nanoparticle radiolabeling could be achieved by different methodologies, including activation methods and synthesis based on radioactive precursors [45]. However, apart from other advantages related to the specific activity obtained, synthesis based on radioactive precursors appears to be the simplest and most favorable option, mainly due to the coincidence between the nanoparticle chemical nature (titanium) and the radionuclide that is aimed to be used for radiolabeling ($^{45}$Ti). Figure 4 presents an overview of the processing steps to radiolabel $TiO_2$ nanoparticles by the method suggested above.

To implement the process presented in Figure 4, once the cyclotron production of $^{45}$Ti seems to be feasible, our group looks for optimizing all of the technical aspects. Then, the extraction of $^{45}$Ti from the solid target constitutes a formidable challenge due to pharmaceutical requirements and technical issues, such as the specific activity needed. The task will require, for sure, wet chemistry techniques to fulfill the predefined requisites. The use of radioactive precursors for the labeling of $TiO_2$ nanoparticles was already mentioned and briefly justified, but special attention should be given to the reduction of preparation time, because the operation is actually very time-consuming. Finally, $^{45}TiO_2$ nanoparticles should be purified, filtered by size, and controlled regarding the surface properties for the appropriate targeting of biological processes.

As a final note, our literature review made on $^{45}$Ti found only one non-oncological application where labeled cations were used for the investigation of cerebral neurodegeneration [46].

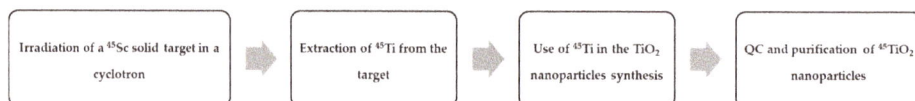

| Irradiation of a $^{45}$Sc solid target in a cyclotron | Extraction of $^{45}$Ti from the target | Use of $^{45}$Ti in the $TiO_2$ nanoparticles synthesis | QC and purification of $^{45}TiO_2$ nanoparticles |

**Figure 4.** Possible approach to obtain $^{45}TiO_2$ nanoparticles using the cyclotron irradiation of a solid target of $^{45}$Sc.

## 4. Conclusion and Future Perspectives

Over the last decades, significant efforts have been made in the fundamental study, testing, development, and optimization of radionuclide production processes. The achievements have benefited the field of medical imaging and, particularly, radionuclide imaging using nuclear medicine methods and techniques. PET presents some advantages that could be used in accordance with the actual paradigms of medicine, in such a way that truthfully modern and personalized medical care could be conducted on patients. The wider use of low-energy cyclotrons is playing a crucial role in increasing the availability of radionuclides for preclinical or clinical trials, diversifying the set of radionuclides available for biomedical applications.

Several unconventional radionuclides for PET are being suggested to be introduced in clinical practice, because of the diversity of pathophysiological processes that could be studied. $^{55}$Co, $^{60/61/64}$Cu, $^{76}$Br, $^{82}$Rb, $^{86}$Y, $^{89}$Zr or $^{124}$I represent examples of this trend. This review focused on the exploration of $^{45}$Ti as a reliable candidate for PET imaging, emphasizing again the importance of cyclotrons to provide the production of the radionuclide. The titanium-45 half-life of 3.09 h, together with low energy positrons, were key factors for its selection as an interesting candidate. Potential

*Instruments* **2018**, *2*, 8

applications of $^{45}$Ti include some possibilities that have been proposed and briefly studied by other groups, such as imaging with $^{45}$Ti-transferrin to study titanium-based chemotherapy agents or with $^{45}$Ti-salan compounds to be applied in theranostics of cancer. The use of $^{45}$TiO$_2$ nanoparticles here suggested is an innovative proposal that should be developed in the near future.

**Acknowledgments:** The authors would like to thank the support of all the institutions involved in this work, namely School of Health of the Polytechnic Institute of Porto, IsoPor, S.A. and ICNAS.

**Conflicts of Interest:** The authors declare no conflict of interest.

## References

1. Acharya, R.; Wasserman, R.; Stevens, J.; Hinojosa, C. Biomedical imaging modalities: A tutorial. *Comput. Med. Imaging Graph.* **1995**, *19*, 3–25. [CrossRef]
2. Cherry, S.R. Multimodality in vivo imaging systems: Twice the power or double the trouble? *Annu. Rev. Biomed. Eng.* **2006**, *8*, 35–62. [CrossRef] [PubMed]
3. World Health Organization. *Technical Report Series No 492*; World Health Organization: Geneva, Switzerland, 1972; pp. 34–50.
4. Nuclear Medicine and Radiopharmaceuticals Market 2012–2017. Available online: https://healthmanagement.org/c/imaging/news/nuclear-medicine-and-radiopharmaceuticals-market-2012-2017 (accessed on 20 April 2018).
5. Saha, G.B. Diagnostic Uses of Radiopharmaceuticals in Nuclear Medicine. In *Fundamentals of Nuclear Pharmacy*; Saha, G.B., Ed.; Springer: Berlin, Germany, 2005.
6. Saha, G.B. Cyclotron and Production of PET Radionuclides. In *Basics of PET Imaging: Physics, Chemistry and Regulations*; Saha, G.B., Ed.; Springer: New York, NY, USA, 2010; pp. 257–339.
7. Conti, M.; Eriksson, L. Physics of pure and non-pure positron emitters for PET: A review and a discussion. *EJNMMI Phys.* **2016**, *3*, 8. [CrossRef] [PubMed]
8. Jalilian, A.R. The Application of Unconventional PET Tracers in Nuclear Medicine. *Iran. J. Nucl. Med.* **2009**, *17*, 1–11.
9. Welch, M.J.; Laforest, R.; Lewis, J.S. Production of Non-Standard PET Radionuclides and Application of Radiopharmaceuticals Labeled with these Nuclides. In *PET Chemistry: The Driving Force in Molecular Imaging Series*; Schubiger, P.A., Lehmann, L., Friebe, M., Eds.; Springer: Berlin, Germany, 2007.
10. Arasaratnam, P.; Sadreddini1, M.; Yam, Y.; Kansal, V.; Dorbala, S.; di Carli, M.F.; Beanlands, R.S.; Merhige, M.E.; Williams, B.A.; Veledar, E.; et al. Prognostic value of vasodilator response using rubidium-82 positron emission tomography myocardial perfusion imaging in patients with coronary artery disease. *Eur. J. Nucl. Med. Mol. Imaging* **2018**, *45*, 538–548. [CrossRef] [PubMed]
11. Rossi, S.; Toschi, L.; Castello, A.; Grizzi, F.; Mansi, L.; Lopci, E. Clinical characteristics of patient selection and imaging predictors of outcome in solid tumors treated with checkpoint-inhibitors. *Eur. J. Nucl. Med. Mol. Imaging* **2017**, *44*, 2310–2325. [CrossRef] [PubMed]
12. England, C.G.; Jiang, D.; Ehlerding, E.B.; Rekoske, B.T.; Ellison, P.A.; Hernandez, R.; Barnhart, T.E.; McNeel, D.G.; Huang, P.; Cai, W. $^{89}$Zr-labeled nivolumab for imaging of T-cell infiltration in a humanized murine model of lung cancer. *Eur. J. Nucl. Med. Mol. Imaging* **2018**, *45*, 110–120. [CrossRef] [PubMed]
13. Chu, S.Y.F.; Ekström, L.P.; Firestone, R.B. WWW Table of Radioactive Isotopes, database version 1999-02-28. Available online: http://nucleardata.nuclear.lu.se/nucleardata/toi/ (accessed on 23 April 2018).
14. Carrol, V.; Demoin, D.W.; Hoffman, T.J.; Jurisson, S.S. Inorganic chemistry in nuclear imaging and radiotherapy: Current and future directions. *Radiochim. Acta* **2012**, *100*, 653–667. [CrossRef] [PubMed]
15. Saha, G.B. *Production of Radionuclides*, 6th ed.; Springer: Berlin, Germany, 2010.
16. International Atomic Energy Agency (IAEA). *Directory of Cyclotrons used for Radionuclide Production in Member States—2006 Update*; IAEA: Vienna, Austria, 2006.
17. International Atomic Energy Agency (IAEA). *Technical Report Series No465—Cyclotron Produced Radionuclides: Principles and Practice*; IAEA: Vienna, Austria, 2008; pp. 59–72.
18. Papash, A.I.; Alenitskii, Y.G. Commercial Cyclotrons. Part I: Commercial Cyclotrons in the Energy Range 10–30 MeV for Isotope Production. *Phys. Part. Nucl.* **2008**, *39*, 597–631. [CrossRef]

19. Qaim, S.M. Nuclear data relevant to the production and application of diagnostic radionuclides. *Radiochim. Acta* **2001**, *89*, 223–232. [CrossRef]
20. Jensen, M. Particle accelerators for PET radionuclides. *Nucl. Med. Rev.* **2012**, *15*, C9–C12.
21. Holland, J.; Williamson, M.; Lewis, J. Unconventional Nuclides for Radiopharmaceuticals. *Mol. Imaging* **2010**, *9*, 1–20. [CrossRef] [PubMed]
22. Sadeghi, M.; Enferadi, M.; Aref, M.; Jafari, H. Nuclear data for the cyclotron production of $^{66}$Ga, $^{86}$Y, $^{76}$Br, $^{64}$Cu and $^{43}$Sc. *Nucleonika* **2010**, *55*, 293–302.
23. Link, J.; Krohn, K.A.; O'Hara, M.J. A simple thick target for production of $^{89}$Zr in an 11 MeV cyclotron. *Appl. Radiat. Isot.* **2017**, *122*, 211–214. [CrossRef] [PubMed]
24. Qaim, S. Production of High Purity $^{94m}$Tc for Positron Emission Tomographic Studies. *Nucl. Med. Biol.* **2000**, *27*, 323–328. [CrossRef]
25. Asabella, A.; Cascini, G.; Altini, C.; Paparella, D.; Notaristefano, A.; Rubini, G. The Copper Radioisotopes: A systematic review with special interest to $^{64}$Cu. *Biomed. Res. Int.* **2014**, *2014*, 786463.
26. Schmitz, J. The production of [124I]iodine and [86Y]yttrium. *Eur. J. Nucl. Med. Mol. Imaging* **2011**, *38* (Suppl. 1), S4–S9. [CrossRef] [PubMed]
27. Costa, P.; Metello, L.F.; do Carmo, S.J.C.; Alves, F.; Duarte Naia, M. Titanium-45 as an innovative radionuclide for PET imaging: From cyclotron production to potential biomedical applications. In Proceedings of the 29th Annual Congress of the European Association of Nuclear Medicine 2016, Barcelona, Spain, 15–19 October 2016; Oral Presentation 447.
28. Ishiwata, K.; Ido, T.; Monma, M.; Murakami, M.; Fukuda, H.; Yamada, K.; Endo, S.; Yoshioka, H.; Sato, T.; Matsuzawa, T. *Preparation and Medical Application of* $^{45}$*Ti*; CYRIC Annual Report; CYRIC: Tohoku, Japan, 1981.
29. Ishiwata, K.; Ido, T.; Monma, M.; Murakami, M.; Fukuda, H.; Kameyama, M.; Yamada, K.; Endo, S.; Yoshioka, H.; Sato, T.; et al. Potential radiopharmaceuticals labeled with titanium-45. *Int. J. Radiat. Appl. Instrum. Part A Appl. Radiat. Isot.* **1991**, *42*, 707–712. [CrossRef]
30. Waterhouse, R.N.; Mattner, F.; Najdovski, L.; Collier, T.L.; Fallon, J. Synthesis and Characterisation of [111In]-Liposome Encapsulated [$^{45}$Ti]-Budotitane. In Proceedings of the Eleventh International Symposium on Radiopharmaceutical Chemistry, Vancouver, BC, Canada, 13–17 August 1995.
31. Vavere, A.L.; Laforest, R.; Welch, M.J. Production, processing and small animal PET imaging of titanium-45. *Nucl. Med. Biol.* **2005**, *32*, 117–122. [CrossRef] [PubMed]
32. Vavere, A.L.; Welch, M.J. Preparation, biodistribution, and small animal PET of $^{45}$Ti-transferrin. *J. Nucl. Med.* **2005**, *46*, 683–690. [PubMed]
33. Price, R.I.; Sheil, R.W.; Scharli, R.K.; Chan, S.; Gibbons, P.; Jeffery, C.; Morandeau, L. Titanium-45 as a Candidate for PET Imaging: Production, Processing & Applications. In Proceedings of the 15th International Workshop on Targetry and Target Chemistry, Prague, Czech Republic, 18–21 August 2015.
34. Siikanen, J.; Hong, H.; Valdovinos, H.; Hernandez, R.; Zhang, Y.; Barnhart, T.; Cai, W.; Nickles, R. Production, separation and labeling of $^{45}$Ti. In Proceedings of the SNMMI Annual Meeting, Vancouver, BC, Canada, 8–12 June 2013.
35. Tshuva, E.Y.; Ashenhurst, J.A. Cytotoxic Titanium(IV) Complexes: Renaissance. *Eur. J. Inorg. Chem.* **2009**, *2009*, 2203–2218. [CrossRef]
36. Immel, T.A.; Groth, U.; Huhn, T. Cytotoxic Titanium Salan Complexes: Surprising Interaction of Salan and Alkoxy Ligands. *Chem. Eur. J.* **2010**, *16*, 2775–2789. [CrossRef] [PubMed]
37. Cunha, L.; Szigeti, K.; Mathé, D.; Metello, L.F. The role of molecular imaging in modern drug development. *Drug Discov. Today* **2014**, *19*, 936–948. [CrossRef] [PubMed]
38. Severin, G.W.; Nielsen, C.H.; Jensen, A.I.; Fonslet, J.; Kjær, A.; Zhuravlev, F. Bringing Radiotracing to Titanium-Based Antineoplastics: Solid Phase Radiosynthesis, PET and ex Vivo Evaluation of Antitumor Agent [$^{45}$Ti](salan)Ti(dipic). *J. Med. Chem.* **2015**, *58*, 7591–7595. [CrossRef] [PubMed]
39. Assadi, M.; Afrasiabi, K.; Nabipour, I.; Seyedabadi, M. Nanotechnology and nuclear medicine; research and preclinical applications. *Hell. J. Nucl. Med.* **2011**, *14*, 149–159. [PubMed]
40. Zarytova, V.F.; Zinov'ev, V.V.; Ismagilov, Z.R.; Levina, A.S.; Repkova, M.N.; Shikina, N.V.; Evdokimov, A.A.; Belanov, E.F.; Balakhnin, S.M.; Serova, O.A.; et al. An Examination of the Ability of Titanium Dioxide Nanoparticles and Its Conjugates with Oligonucleotides to Penetrate into Eucariotis Cells. *Nanotechnol. Russia* **2009**, *4*, 732–735. [CrossRef]

*Instruments* **2018**, *2*, 8

41. Stefanou, E.; Evangelou, A.; Falaras, P. Effects of UV-irradiated titania nanoparticles on cell proliferation, cancer metastasis and promotion. *Catal. Today* **2010**, *151*, 58–63. [CrossRef]

42. El-Said, K.S.; Ali, E.M.; Kanehira, K.; Taniguchi, A. Molecular mechanism of DNA damage induced by titanium dioxide nanoparticles in toll-like receptor 3 or 4 expressing human hepatocarcinoma cell lines. *J. Nanobiotechnol.* **2014**, *12*, 48. [CrossRef] [PubMed]

43. Mund, R.; Panda, N.; Nimesh, S.; Biswas, A. Novel titanium oxide nanoparticles for effective delivery of paclitaxel to human breast cancer cells. *J. Nanopart. Res.* **2014**, *16*, 2739. [CrossRef]

44. Chen, Y.; Wan, Y.; Wang, Y.; Zhang, H.; Jiao, Z. Anticancer efficacy enhancement and attenuation of side effects of doxorubicin with titanium dioxide nanoparticles. *Int. J. Nanomed.* **2011**, *6*, 2321–2326.

45. Gibson, N.; Holzwarth, U.; Abbas, K.; Simonelli, F.; Kozempel, J.; Cydzik, I.; Cotogno, G.; Bulgheroni, A.; Gilliland, D.; Ponti, J.; et al. Radiolabelling of engineered nanoparticles for in vitro and in vivo tracing applications using cyclotron accelerators. *Arch. Toxicol.* **2011**, *85*, 751–773. [CrossRef] [PubMed]

46. Salber, D.; Manuvelpillai, J.; Spahn, I.; Klein, S.; Uhlenbruck, F.; Palm, C.; Matusch, A.; Becker, S.; Langen, K.J.; Coenen, H.H. [45]Ti-cations as potential PET-tracers for cerebral neurodegeneration. In Proceedings of the International Symposium on Technetium and Other Radiometals in Chemistry and Medicine, Bressanone, Italy, 8–11 September 2010.

*instruments*

MDPI

Article

# New Cross-Sections for $^{nat}$Mo($\alpha$,x) Reactions and Medical $^{97}$Ru Production Estimations with Radionuclide Yield Calculator

Mateusz Sitarz [1,2,3,*], Etienne Nigron [4], Arnaud Guertin [4], Férid Haddad [1,4] and Tomasz Matulewicz [2]

1   Groupement d'Intérêt Public ARRONAX, 44817 Saint-Herblain CEDEX, France; haddad@subatech.in2p3.fr
2   Faculty of Physics, University of Warsaw, 02-093 Warszawa, Poland; Tomasz.Matulewicz@fuw.edu.pl
3   Heavy Ion Laboratory, University of Warsaw, 02-093 Warszawa, Poland
4   Subatech, CNRS/IN2P3, IMT Atlantique, Université de Nantes, CS 20722 44307 Nantes CEDEX, France;
    nigron@subatech.in2p3.fr (E.N.); Arnaud.Guertin@subatech.in2p3.fr (A.G.)
*   Correspondence: mateusz.sitarz@univ-nantes.fr

Received: 17 December 2018; Accepted: 18 January 2019; Published: 22 January 2019

**Abstract:** The production of $^{97}$Ru, a potential Single Photon Emission Computed Tomography (SPECT) radioisotope, was studied at ARRONAX. The cross-section of $^{nat}$Mo($\alpha$,x)$^{97}$Ru reaction was investigated in the range of 40–67 MeV irradiating the $^{nat}$Mo and Al stacked-foils. The activities of $^{97}$Ru and other radioactive contaminants were measured via gamma spectroscopy technique. A global good agreement is observed between obtained cross-section results, previously reported values and TENDL-2017 predictions. Additionally, Radionuclide Yield Calculator, a software that we made available for free, dedicated to quickly calculate yields and plan the irradiation for any radioisotope production, was introduced. The yield of investigated nuclear reactions indicated the feasibility of $^{97}$Ru production for medical applications with the use of $\alpha$ beam and Mo targets opening the way to a theranostic approach with $^{97}$Ru and $^{103}$Ru.

**Keywords:** SPECT; cyclotron; medical radioisotope production; radioactive impurities; cross-section; stacked-foils; gamma spectroscopy; thick target yield; Radionuclide Yield Calculator

## 1. Introduction

The $^{97}$Ru radioisotope was first acknowledged as medically interesting in 1970 [1] and is even studied in recent measurements [2,3]. It has a half-life of 2.9 d allowing non-local production and emits low-energy high-intensity gamma lines (see Table 1) which have favorable characteristics for prolonged Single Photon Emission Computed Tomography (SPECT) examinations. It decays only by electron capture (EC) which lowers the contribution to the dose as compared to $\beta^+$ decays. It has a theranostic matched pair in the form of $^{103}$Ru ($T_{1/2}$ = 39.26 d) that decays to the short-lived Auger emitter $^{103m}$Rh ($T_{1/2}$ = 56.12 min), a promising gamma-free therapeutic agent. Moreover, ruthenium element has a rich chemistry associated with its various oxidation states (II, III, IV and VIII) and forms more stable compounds compared to the SPECT-standard $^{99m}$Tc [4]. Many radioactive Ru-labeled compounds have been studied and found applications as summarized recently by [5], in particular as the chemotherapy agents [6,7].

Due to these interesting characteristics, many studies on production of $^{97}$Ru have been conducted. The reactor route via $^{96}$Ru(n,$\gamma$)$^{97}$Ru was reported by [1] but it yields very low specific activity which may limit its use for some applications such as molecular imaging. To obtain high specific activity product, one can use charged projectile from accelerators. In case of cyclotron routes, the first and most used reaction is $^{103}$Rh(p,spall)$^{97}$Ru with 200 MeV proton beam and natural rhodium target, as suggested by [8]. While producing high amount of activity of no-carrier-added (NCA) $^{97}$Ru, this method requires high energy protons but no details about the impurity levels were reported. Another reaction route is the $^{103}$Rh(p,x)$^{97}$Ru reaction using 60 MeV proton beam [9]; the $^{97}$Ru production yield is very high but accompanied by Tc radioactive impurities which are difficult to discard even after the chemical separation step. A very feasible option is the $^{99}$Tc(p,3n)$^{97}$Ru reaction suggested by [10] and studied later up to 100 MeV by [4,11,12] as it produces significant amounts of $^{97}$Ru with very small amount of radioactive impurities. However, the availability of $^{99}$Tc radioactive target is an issue. Later, experimental excitation functions were reported for $^{nat}$Ag(p,x)$^{97}$Ru up to 80 MeV by [13] and for $^{nat}$Pd(p,x)$^{97}$Ru up to 70 MeV by [14]. These two production routes have much smaller cross-section, hence $^{97}$Ru production would require long irradiation time and would contain a substantial amount of radioactive impurities. In case of deuteron beam, the available reaction $^{96}$Ru(d,x)$^{97}$Ru studied by [15] is favorable but would produce low specific activity as the target material is an isotope of the nuclide of interest. Some groups have also investigated more exotic projectiles such as helium-3 through $^{nat}$Mo($^3$He,x)$^{97}$Ru [16], $^{93}$Nb($^7$Li,3n)$^{97}$Ru [17] and $^{89}$Y($^{12}$C,p3n)$^{97}$Ru [2,18]. In these cases, after chemical separation, low level of radioactive impurities can be achieved but the availability of these beams is scarce making these processes not suitable to launched clinical trials. Finally, the cross-sections for α-induced reactions on Mo were investigated by [19]. $^{nat}$Mo(α,x)$^{97}$Ru production and impurities up to 40 MeV were thoroughly studied in [3,20].

In this work, we investigate the optimization of $^{nat}$Mo(α,x)$^{97}$Ru production route and extend the available cross-section data to higher energy in coherence with commercially available cyclotrons, [21] which are able to deliver up to about 70 MeV alpha beam. We also report on the coproduction of the measured radioactive impurities (listed in Table 1) via $^{nat}$Mo(α,x) and explore the possible commercial production of $^{97}$Ru with the α beam on Mo target using the software Radionuclide Yield Calculator (RYC) that we developed and made freely available to the community.

**Table 1.** Nuclear data [22] of $^{97}$Ru and observed radionuclidic contaminants as well as reactions contributing to their formation during the irradiation of $^{nat}$Mo target*.

| Radionuclide | $T_{1/2}$ | Decay Mode (%) | γ-Lines [keV] and Intensities** (%) | Contributing Reactions*** | Q-Value [MeV] |
|---|---|---|---|---|---|
| $^{97}$Ru | 2.83 d | EC (100) | 215.7 (85.8) 324.5 (10.8) | $^{94}$Mo(α,n)$^{97}$Ru | −7.9 |
| | | | | $^{95}$Mo(α,2n)$^{97}$Ru | −15.3 |
| | | | | $^{96}$Mo(α,3n)$^{97}$Ru | −24.5 |
| | | | | $^{97}$Mo(α,4n)$^{97}$Ru | −31.3 |
| | | | | $^{98}$Mo(α,5n)$^{97}$Ru | −41.6 |
| | | | | $^{100}$Mo(α,7n)$^{97}$Ru | −54.1 |
| $^{89g}$Zr | 78.4 h | β$^+$ (23), EC (77) | 908.96 (100) | $^{92}$Mo(α,x)$^{89tot}$Zr | −16.7 |
| | | | | $^{94}$Mo(α,x)$^{89tot}$Zr | −14.0 |
| | | | | $^{95}$Mo(α,x)$^{89tot}$Zr | −21.4 |
| | | | | $^{96}$Mo(α,x)$^{89tot}$Zr | −30.6 |
| | | | | $^{97}$Mo(α,x)$^{89tot}$Zr | −37.4 |
| | | | | $^{98}$Mo(α,x)$^{89tot}$Zr | −46.0 |
| | | | | $^{100}$Mo(α,x)$^{89tot}$Zr | −60.2 |
| | | | | $^{92}$Mo(α,x)$^{89tot}$Nb→$^{89tot}$Zr | −21.1 |
| | | | | $^{94}$Mo(α,x)$^{89tot}$Nb→$^{89tot}$Zr | −38.9 |
| | | | | $^{95}$Mo(α,x)$^{89tot}$Nb→$^{89tot}$Zr | −46.2 |
| | | | | $^{96}$Mo(α,x)$^{89tot}$Nb→$^{89tot}$Zr | −55.4 |
| | | | | $^{97}$Mo(α,x)$^{89tot}$Nb→$^{89tot}$Zr | −62.2 |
| | | | | $^{98}$Mo(α,x)$^{89tot}$Nb→$^{89tot}$Zr | −70.9 |
| | | | | $^{100}$Mo(α,x)$^{89tot}$Nb→$^{89tot}$Zr | −85.1 |

Table 1. Cont.

| Radionuclide | $T_{1/2}$ | Decay Mode (%) | $\gamma$-Lines [keV] and Intensities** (%) | Contributing Reactions*** | Q-Value [MeV] |
|---|---|---|---|---|---|
| $^{96g}$Tc | 4.28 d | EC (100) | 778.22 (100) 812.58 (82) 849.93 (98) 1126.97 (15.2) | $^{94}$Mo($\alpha$,x)$^{96tot}$Tc $^{95}$Mo($\alpha$,x)$^{96tot}$Tc $^{96}$Mo($\alpha$,x)$^{96tot}$Tc $^{97}$Mo($\alpha$,x)$^{96tot}$Tc $^{98}$Mo($\alpha$,x)$^{96tot}$Tc $^{100}$Mo($\alpha$,x)$^{96tot}$Tc | −13.3 −14.4 −23.7 −30.4 −39.0 −53.3 |
| $^{99}$Mo | 65.9 h | $\beta^-$ (100) | 140.51 (89.43) 739.50 (12.13) | $^{97}$Mo($\alpha$,2p)$^{99}$Mo $^{98}$Mo($\alpha$,x)$^{99}$Mo $^{100}$Mo($\alpha$,x)$^{99}$Mo | −13.7 −14.7 −8.3 |
| $^{95g}$Tc | 20.0 h | EC (100) | 765.8 (93.82) | $^{92}$Mo($\alpha$,n)$^{95g}$Tc $^{94}$Mo($\alpha$,x)$^{95g}$Tc $^{95}$Mo($\alpha$,x)$^{95g}$Tc $^{96}$Mo($\alpha$,x)$^{95g}$Tc $^{97}$Mo($\alpha$,x)$^{95g}$Tc $^{98}$Mo($\alpha$,x)$^{95g}$Tc $^{100}$Mo($\alpha$,x)$^{95g}$Tc $^{92}$Mo($\alpha$,n)$^{95}$Ru→$^{95g}$Tc $^{94}$Mo($\alpha$,3n)$^{95}$Ru→$^{95g}$Tc $^{95}$Mo($\alpha$,4n)$^{95}$Ru→$^{95g}$Tc $^{96}$Mo($\alpha$,5n)$^{95}$Ru→$^{95g}$Tc $^{97}$Mo($\alpha$,6n)$^{95}$Ru→$^{95g}$Tc $^{98}$Mo($\alpha$,7n)$^{95}$Ru→$^{95g}$Tc $^{100}$Mo($\alpha$,9n)$^{95}$Ru→$^{95g}$Tc | −5.7 −14.9 −22.3 −31.4 −38.3 −46.9 −61.1 −9.0 −26.7 −34.1 −43.3 −50.1 −58.7 −73.0 |

* $^{nat}$Mo composition: $^{92}$Mo (14.6%), $^{94}$Mo (9.2%), $^{95}$Mo (15.9%), $^{96}$Mo (16.7%), $^{97}$Mo (9.6%), $^{98}$Mo (24.3%), $^{100}$Mo (9.7%); ** lines with less than 10% intensities are not included; *** "tot"—the reaction produces the radionuclide directly and via decay of its metastable state.

## 2. Materials and Methods

### 2.1. Stacked-Foils Irradiations

Three experiments were performed at the ARRONAX facility [23], irradiating stacked-foils targets in vacuum with $\alpha$ beam of 67.4(5) MeV for about 1 h with beam currents of 40–60 nA. The stacked-foils technique and set-up in our facility have been described most recently in [24–26]. A typical stacked-foil target consisted of an Al monitor foil (~10 μm thick) in front, followed by the set of multiple $^{nat}$Mo foils (~10 μm thick) and Al degraders (50–500 μm thick), arranged alternately. The order of the foils in the stacks were planned so that each $^{nat}$Mo foil is activated with a different energy, all covering the energy range from 40 MeV to 67 MeV in about 3 MeV intervals (the projectile stopping-power in the stacks was calculated using SRIM software [27]). Certain foils were also used as catchers of the recoil atoms.

All foils were purchased from the GoodFellow© company with a purity of 99% for Al and 99.9% for $^{nat}$Mo. Each foil was weighed before irradiation using an accurate scale ($10^{-5}$ g) and scanned for area determination, allowing the precise thickness calculation (assuming the homogeneity over the whole surface).

As recommended by the International Atomic Energy Agency [28], the activity of the $^{24}$Na radioisotope formed in Al monitor foil was used to calculate the beam current impinging the stack. Additionally, during the irradiations, the online beam current monitoring was performed using a Faraday's Cup with an electron suppressor for precise measurement and located behind the stack. The two measurements were consistent with each other.

## 2.2. Gamma Spectroscopy and Data Analysis

After about 14 h of cooling time, the gamma ray spectra of irradiated samples were collected using a HPGe Canberra detector with efficiency 20% at 1.33 MeV equipped with low-background lead and copper shielding. Each foil was placed at a height of 19 cm from the detector to ensure the dead-time below 10%. The detector was calibrated in energy and efficiency at 19 cm with $^{57}$Co, $^{60}$Co and $^{152}$Eu calibrated sources from LEA-CERCA (France) prior to the measurements. Gamma spectra were recorded using the LVis software from Ortec© while the activity of the radionuclides produced at the End of Bombardment (EOB) were derived using the FitzPeaks Gamma Analysis and Calibration Software (JF Computing Services). For the identification and activity estimation we used the γ-line and associated branching presented in Table 1. Knowing the activity of each isotope and the thickness of the foil in which they were observed, it was possible to calculate their production cross-section σ with the following formula:

$$\sigma = \frac{A_{EOB} \, M \, Z \, e}{H \, N_A \, I \, \rho \, x \, (1 - exp\{-\lambda \, t\})}$$

where: $A_{EOB}$—activity of the radioisotope at the EOB, $M$—atomic mass of the target, $Z$—ionization number of the projectile, $e$—elementary charge, $H$—enrichment and purity of the foil, $N_A$—Avogadro's number, $I$—beam current, $\rho$—target material density, $x$—thickness of the foil, $\lambda$—decay constant of the radioisotope, $t$—time of the irradiation. The similar formula, solved for $I$ and with cross-section values from [28], was used to calculate the beam current from the monitor foils. The projectile energy in the middle of the foils was adopted to the corresponding cross-section value.

The errors of cross-section values were propagated from the uncertainty of thickness measurements (around 1%), uncertainty of the counts in the γ-line peaks in the spectroscopy measurements (around 5–10%) and the error of the calculated beam current (around 5–10%) while the corresponding energy errors were propagated with SRIM software [27] considering the beam energy straggling through the foils (the initial energy error estimated by the cyclotron operators was 0.5 MeV).

The obtained cross-section values are then compared with TENDL-2017 (TALYS-based evaluated nuclear data library) [29] and the experimental results from other research groups.

## 2.3. Radionuclide Yield Calculator

Given the cross-section values, one can calculate the Thick Target Yield (TTY) with the following formula [30,31]:

$$TTY(E) = \frac{H \, N_A \, \lambda}{Z \, e \, M} \int_{E_{min}}^{E_{max}} \frac{\sigma(E)}{dE/dx(E)} dE$$

where: $E_{max}$ and $E_{min}$—maximal and minimal energy of the projectile penetrating the target (in case of TTY, $E_{min} \leq$ reaction threshold), $dE/dx$—stopping-power of the projectile in the irradiated target. To facilitate this calculation for $^{97}$Ru, as well as any other radioisotope and cross-section, we developed a Radionuclide Yield Calculator, later named RYC.

RYC is graphical user interface software written in python programming language (version 2.7) [32] using the TKinter module and compiled with PyInstaller software (version 3.4) [33]. It uses the cross-section and basic target data inputs to instantly calculate TTY and activity produced in any irradiation scenario. Data points can be fitted using different type of function, gaussian-like and polynomial functions, using the least-squares method. Excitation functions from TENDL [29] can be easily imported to compare with experimental data and to look for potential radioactive impurities. RYC with its detailed documentation can be downloaded from the ARRONAX website [34]. In particular, RYC uses implemented SRIM module [27] for stopping-power calculation.

The validation of this software was performed using data from the literature. On Figure 1, we compare the RYC-calculated TTY with the values published by IAEA [28,35] based on the same cross-section values for $^{127}$I(p,3n)$^{125}$Xe, $^{64}$Ni(d,2n)$^{64}$Cu and $^{209}$Bi(α,2n)$^{211}$At reactions on metallic targets. Data calculated by RYC are presented as points whereas the curve published by IAEA correspond to the lines. As can be seen, for the 3 types of projectiles and for the different target masses, a very good agreement is obtained. The same good results have been obtained for all our tests.

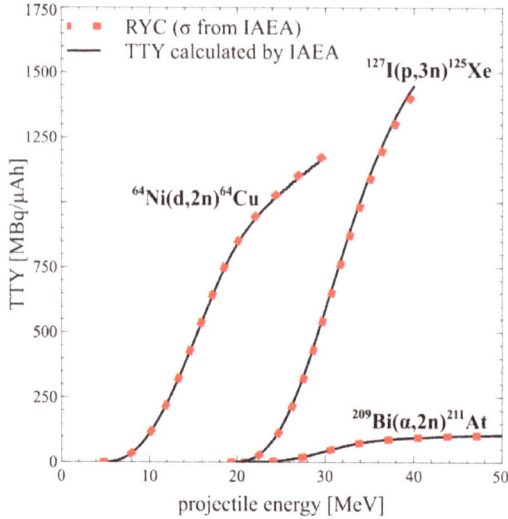

**Figure 1.** Comparison of TTY for selected nuclear reactions on metallic targets calculated with RYC and adapted from [28,35] based on the same cross-section values from IAEA.

## 3. Results and Discussion

### 3.1. Cross-Section Measurements

On Figures 2–6 we present the measured cross-sections for $^{nat}$Mo(α,x) reactions producing the $^{97}$Ru radioisotope as well as observed radioactive impurities: $^{89g}$Zr, $^{95g}$Tc, $^{96g}$Tc, and $^{99}$Mo. The contributing reactions forming these radioisotopes are shown in Table 1. The experimental data are compared with previous experiments reported in literature [3,19,20] and the values from TENDL-2017 library [29]. Measured cross-section values are also listed in Table 2 (with corresponding energy errors, not visible on the graphs).

**Table 2.** Measured cross-sections for $^{nat}$Mo(α,x) reactions (with the uncertainties in the parenthesis).

| E [MeV] | $^{nat}$Mo(α,x) Cross-Section [mb] | | | | |
|---|---|---|---|---|---|
| | $^{97}$Ru | $^{89g}$Zr | $^{95g}$Tc | $^{96tot}$Tc | $^{99}$Mo |
| 41.80(75) | 237(20) | ND* | 81(11) | 73(7) | 7.5(1.0) |
| 46.03(68) | 225(20) | ND | 127(14) | 89(8) | 10.1(1.2) |
| 50.00(64) | 199(18) | ND | 163(17) | 100(9) | 11.4(1.3) |
| 51.93(62) | 166(14) | ND | 170(16) | 101(9) | 12.8(1.3) |
| 55.30(60) | 159(13) | 3.6(9) | 177(17) | 109(9) | 13.5(1.4) |
| 58.51(56) | 176(15) | 11.7(1.6) | 205(17) | 119(10) | ND |
| 59.97(55) | 176(15) | 18(2) | 174(24) | 116(10) | 14.0(1.5) |
| 63.47(53) | 180(16) | 30(3) | 188(16) | 118(10) | 15.0(1.7) |
| 66.84(50) | 173(14) | 40(3) | 203(17) | 122(10) | 15.6(1.3) |

* ND = not detected.

In the case of $^{97}$Ru production (Figure 2), our measurements correspond well to the data at lower energies. Compared to the experimental data, TENDL shows similar structure but underestimates the cross-section by about 30 mb in the region 20–40 MeV. The subsequent fall of the excitation function and a bump seem to be shifted by 5–10 MeV with respect to the experimental data.

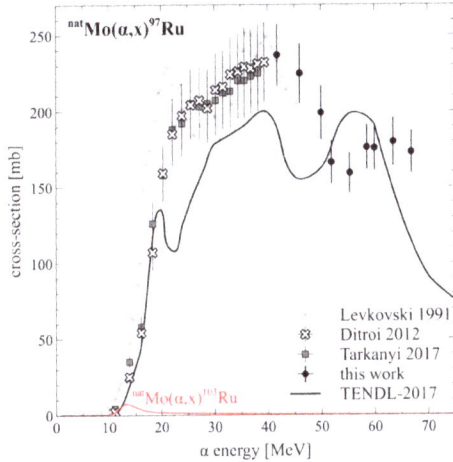

**Figure 2.** Measured cross-section for $^{nat}$Mo(α,x)$^{97}$Ru reaction compared with data available in the literature. The coproduction of $^{103}$Ru via $^{100}$Mo(α,n)$^{103}$Ru and $^{100}$Mo(α,p)$^{103}$Tc→$^{103}$Ru reactions was not observed in the investigated energy range, but the cross-section for $^{nat}$Mo(α,x)$^{103}$Ru from TENDL-2017 is plotted (red line) to complement the discussion from the text.

The $^{89g}$Zr excitation function (Figure 3) is measured for the first time. The predictions of TENDL shows a similar trend as our measurements but with a slightly shifted toward lower energies (5 MeV).

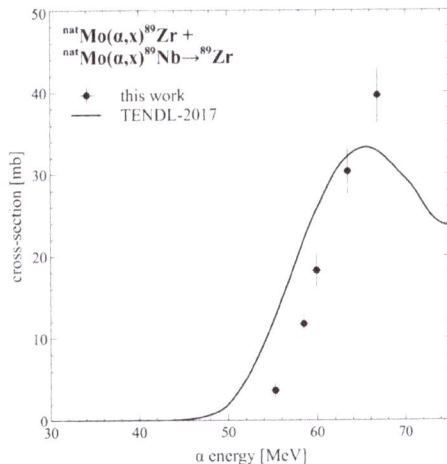

**Figure 3.** Measured cross-section for $^{nat}$Mo(α,x)$^{89g}$Zr reaction compared with TENDL.

For $^{95g}$Tc (Figure 4), our measurements are consistent with the previously measured data at lower energies. The shape of TENDL calculations seems to be the same as obtained in the measurements, but again a shift in energy is observed. This shift is probably related to the code since 3 different sets of data acquired at different time, different laboratories and overlapping energy range are consistent with each other and shows the same shift with respect to TENDL.

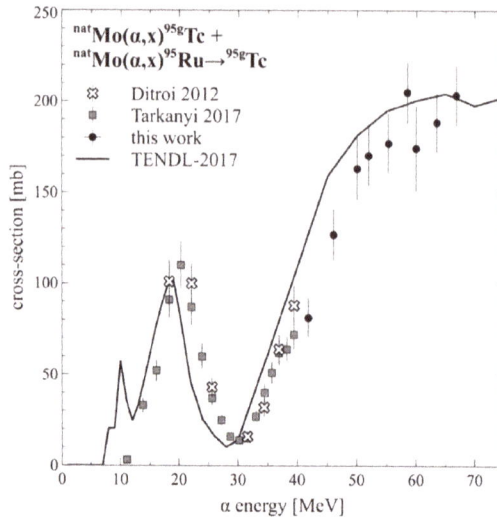

**Figure 4.** Measured cross-section for $^{nat}Mo(\alpha,x)^{95g}Tc$ reaction compared with the literature.

The experimental data describes well the excitation function for $^{96tot}Tc$ (Figure 5). Additionally, our measurements preserve the trend of the ones reported earlier for lower energies. Here we do not observe any shift with respect to the TENDL calculations, as seen in the previous reactions.

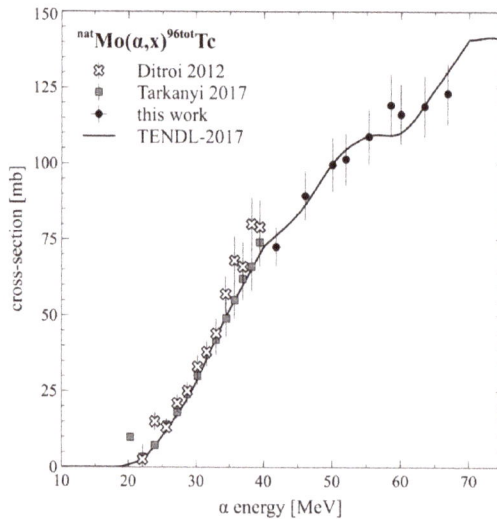

**Figure 5.** Measured cross-section for $^{nat}Mo(\alpha,x)^{96tot}Tc$ reaction compared with the literature. This is a cumulative cross-section of $^{nat}Mo(\alpha,x)^{96g}Tc$ and $^{nat}Mo(\alpha,x)^{96m}Tc$ reactions.

The experimental cross-sections for $^{99}Mo$ production (Figure 6) are consistent. Our results show a continuous rise of the excitation function up to the maximum energy of our measurements. This is in obvious contrast to TENDL which predicts a maximum at around 30 MeV and then a decrease of the excitation function. Slight shift between TENDL and experimental results is observed at low energies.

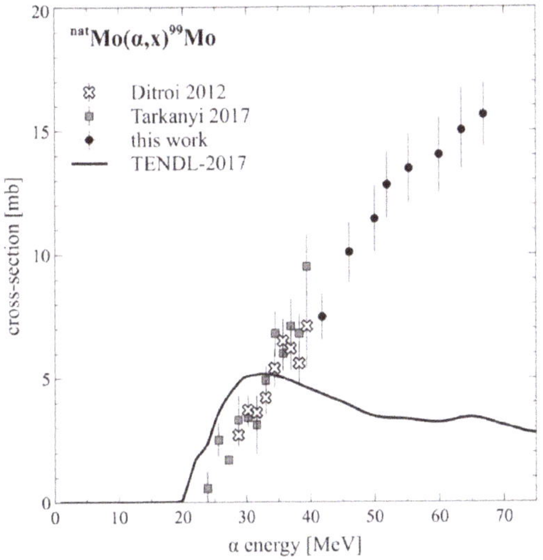

**Figure 6.** Measured cross-section for $^{nat}Mo(\alpha,x)^{99}Mo$ reaction compared with the literature.

*3.2. Calculated Yield and Production*

Using RYC, we calculated TTY for $^{nat}Mo(\alpha,x)^{97}Ru$ reaction on metallic $^{nat}Mo$ target, based on our cross-section measurements above 40 MeV and the values reported by [3,20] below 40 MeV (Figure 7). The TTY values for other radioisotopes were also calculated in a similar way (not shown) to estimate the radioactive impurities.

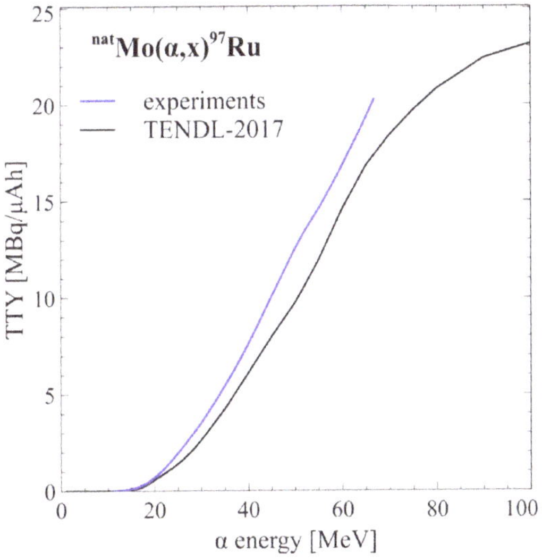

**Figure 7.** TTY for $^{97}Ru$ production via $^{nat}Mo(\alpha,x)$ on metallic $^{nat}Mo$ target. The experimental curve (blue) is calculated using the cross-section from this work (above 40 MeV) and the data provided by [3,20] below 40 MeV.

The obtained experimental TTY values for $^{97}$Ru and radioactive impurities were used to estimate the possible production of $^{97}$Ru (Table 3) with $^{nat}$Mo target and for two energies: 30 MeV and 67 MeV, which are the most common in commercially available cyclotrons. The $^{97}$Ru production yields are 3.5 MBq/μAh and 20 MBq/μAh respectively. Although the yield is almost 6 times larger at 67 MeV than at 30 MeV, the latter energy of α beam offer for example the optimal production of $^{211}$At (summarized recently by [36]) and $^{43}$Sc [37] medical radioisotopes. Additionally, in Table 3 we show the possible production of 50 MBq as this amount was proven SPECT-applicable in several clinical trials [38]. We have also calculated the yield and the number of produced stable Ru atoms ($^{96}$Ru, $^{98}$Ru, $^{99}$Ru, $^{100}$Ru, $^{101}$Ru, $^{102}$Ru) based on the TENDL cross-sections to estimate the specific activity (SA) of $^{97}$Ru. The SA presented here assumes 100% successful chemical extraction of Ru isotopes from Mo target at EOB and hence is just an estimation used to compare different production routes.

It is worth mentioning that from the diagnostics point of view, the most dangerous impurity is $^{103}$Ru. It is the only other radioactive Ru element with long half-life ($T_{1/2}$ = 39.26 d), which will contribute to the patient's dose via high-intensity gamma-line (497 keV with 90.9% intensity) and Auger electrons from its daughter ($^{103m}$Rh). During the irradiation of $^{nat}$Mo with α beam it can be only formed via $^{100}$Mo(α,x) reactions marked as the red line on Figure 2. Its contribution is rather small in our energy range and its activity was below our detection limit but we address it nevertheless (based on the measurements of [3,20]).

For the completeness of this study, we show the alternative production of $^{97}$Ru with the use of 100% enriched $^{95}$Mo and $^{96}$Mo targets and α beams of 30–15 MeV and 67–15 MeV, respectively.

**Table 3.** Estimation of $^{97}$Ru activity produced via the irradiation of $^{nat}$Mo (based on experimental data) and enriched $^{95,96}$Mo targets (based on TENDL-2017 [29]) with α beam in two energy ranges. The list of radioactive impurities is narrowed down to the long-lived ones and shows their activity relative to activity of $^{97}$Ru at EOB.

| α energy | | 30–15 MeV | | 67–15 MeV | |
|---|---|---|---|---|---|
| target | | $^{nat}$Mo | $^{95}$Mo (100%) | $^{nat}$Mo | $^{96}$Mo (100%) |
| thickness | | 100 mg/cm$^2$ | 100 mg/cm$^2$ | 540 mg/cm$^2$ | 540 mg/cm$^2$ |
| $^{97}$Ru yield | | 3.5 MBq/μAh | 14 MBq/μAh | 20 MBq/μAh | 31 MBq/μAh |
| irradiation | | 1 h, 15 μA | 1 h, 15 μA | 1 h, 2.5 μA | 1 h, 2.5 μA |
| $^{97}$Ru A$_{EOB}$ | | 50 MBq (1.4 mCi) | 200 MBq (5.4 mCi) | 50 MBq (1.4 mCi) | 80 MBq (2.2 mCi) |
| SA at EOB | | 350 GBq/μmol (9 kCi/mmol) | 1300 GBq/μmol (36 kCi/mmol) | 420 GBq/μmol (11 kCi/mmol) | 630 GBq/μmol (17 kCi/mmol) |
| relative activity [%] | $^{97}$Ru | 100 | 100 | 100 | 100 |
| | $^{89g}$Zr | 0 | 0 | 3 | 0.04 |
| | $^{95g}$Tc | 95 | 1E−3 | 200 | 150 |
| | $^{96g}$Tc | 4 | 0.2 | 25 | 34 |
| | $^{103}$Ru | 0.12 | 0 | 0.02 | 0 |
| reference | | [3], [20] | TENDL-2017 | [3], [20] this work | TENDL-2017 |

Further chemical separation would be required to extract Ru element from Mo target and separate it from formed radioactive and stable elements of Tc, Nb, and Zr. This can be done for example with either the solvent extraction or distillation methods with an efficacy better than 80% [16]. The SA should also be considered in further chemical research as each production route form additional stable atoms of Ru, which would chelate the labeling compound.

## 4. Conclusions and Summary

We have extended the available cross-section measurements of selected $^{nat}Mo(\alpha,x)$ reactions up to 67 MeV. Our measurements preserve well the trend of the cross-section values reported previously below 40 MeV and are consistent in overlapping energy ranges. A reasonable agreement with TENDL is observed however in certain cases the shift of 5–10 MeV is visible with respect to the experimental data.

We have shown the feasibility of no-carrier-added $^{97}Ru$ production with $\alpha$ beam up to 67 MeV and thick $^{nat}Mo$ targets. The impurity of the only long-lived radioactive Ru radioisotope ($^{103}Ru$) is small, around 0.1%. An irradiation of 1 h with few $\mu A$ $\alpha$-beam should satisfy the need for SPECT imaging for the patient. Several doses could be produced with longer irradiations at higher currents or using enriched $^{95,96}Mo$ targets which will substantially increase the produced activity and SA.

The use of RYC [34] to calculate TTY based on cross-section data was also demonstrated.

**Author Contributions:** M.S.: Investigation, formal analysis, software, writing—original draft, writing—review and editing; E.N.: Investigation, resources, writing—review and editing; A.G.: Investigation, resources, writing—review and editing; F.H.: Investigation, conceptualization, supervision, funding acquisition, writing—review and editing; T.M.: supervision, writing—review and editing.

**Funding:** The cyclotron Arronax is supported by CNRS, Inserm, INCa, the Nantes University, the Regional Council of Pays de la Loire, local authorities, the French government, and the European Union. This work has been, in part, supported by a grant from the French National Agency for Research called "Investissements d'Avenir", Equipex Arronax-Plus noANR-11-EQPX-0004 and Labex IRON noANR-11-LABX-18-01.

**Acknowledgments:** The PhD cotutelle scholarship from French Government and 17$^{th}$ WTTC bursary for Mateusz Sitarz are acknowledged. Special thanks to RYC beta testers: Roberto Formento, Julio Panama and Katarzyna Szkliniarz.

**Conflicts of Interest:** The authors declare no conflict of interest.

## References

1. Subramanian, G.; McAfee, J.G.; Poggenburg, J.K. Ruthenium-97: A preliminary evaluation of a new radionuclide for use in nuclear medicine. *J. Nucl. Med.* **1970**, *11*, 365.
2. Maiti, M.; Lahiri, S. Measurement of yield of residues produced in $^{12}C+^{nat}Y$ reaction and subsequent separation of $^{97}Ru$ from Y target using cation exchange resin. *Radiochim. Acta* **2015**, *103*, 7–13. [CrossRef]
3. Tárkányi, F.; Hermanne, A.; Ditrói, F.; Takács, S.; Ignatyuk, A. Investigation of activation cross section data of alpha particle induced nuclear reaction on molybdenum up to 40 MeV: Review of production routes of medically relevant $^{97,103}Ru$. *Nucl. Inst. Meth. B* **2017**, *399*, 83–100. [CrossRef]
4. Zaitseva, N.G.; Stegailov, V.I.; Khalkin, V.A.; Shakun, N.G.; Shishlyannikow, P.T.; Bukow, K.G. Metal Technetium Target and Target Chemistry for the Production of $^{97}Ru$ via the $^{99}Tc(p,3n)^{97}Ru$ Reaction. *Appl. Radiat. Isot.* **1996**, *47*, 145–151. [CrossRef]
5. Mukhopadhyay, B.; Mukhopadhyay, K. Applications of the Carrier Free Radioisotopes of Second Transition Series Elements in the Field of Nuclear Medicine. *J. Nucl. Med. Radiat. Ther.* **2011**, *2*, 1000115. [CrossRef]
6. Shao, H.S.; Meinken, G.E.; Srivastava, S.C.; Slosman, D.; Sacker, D.F.; Sore, P.; Brill, A.B. In vitro and in vivo characterization of ruthenium bleomycin compared to cobalt- and copper-bleomycin. *J. Nucl. Med.* **1986**, *27*, 1044.
7. Clarke, M.J. Ruthenium in Cancer Chemotherapy. *Platin. Met. Rev.* **1988**, *32*, 198.
8. Ku, T.H.; Richards, P.; Srivastava, S.C.; Prach, T.; Stang, L.G., Jr. Production of ruthenium-97 for medical applications. In Proceedings of the 2nd International Congress of the World Federation of Nuclear Medicine and Biology, Washington, DC, USA, 17–21 September 1978.
9. Lagunas-Solar, M.C.; Avila, M.J.; Navarro, N.L.; Johnson, P.C. Cyclotron Production of No-carrier-added $^{97}Ru$ by Proton Bombardment of $^{103}Rh$ Targets. *J. Appl. Radiat. Isot.* **1983**, *34*, 915–922. [CrossRef]
10. Lebowitz, E.; Kinsley, M.; Klotz, P.; Bachsmith, C.; Ansari, A.; Richards, P.; Atkins, H.L. Development of $^{97}Ru$ and $^{67}Cu$ for medical use. *J. Nucl. Med.* **1974**, *15*, 511.
11. Zaitseva, N.G.; Rurarz, E.; Vobecký, M.; Hwan, K.H.; Nowak, K.; Téthal, T.; Khalkin, V.A.; Popinenkova, L.M. Excitation function and yield for $^{97}Ru$ production in $^{99}Tc(p,3n)^{97}Ru$ reaction in 20–100 MeV proton energy range. *Radiochim. Acta* **1992**, *56*, 59–68. [CrossRef]

12. Dmitriev, S.N.; Zaitseva, N.G.; Starodub, G.Y.; Maslov, O.D.; Shishkin, S.V.; Shishkina, T.V. High-purity radionuclide production: Material, construction, target chemistry for $^{26}$Al, $^{97}$Ru, $^{178}$W, $^{235}$Np, $^{236,237}$P. *Nucl. Inst. Meth. A* **1997**, *397*, 125–130. [CrossRef]

13. Uddin, M.S.; Hagiwara, M.; Baba, M.; Tarkanyi, F. Experimental studies on excitation functions of the proton-induced activation reactions on silver. *Appl. Radiat. Isot.* **2005**, *62*, 533–540. [CrossRef] [PubMed]

14. Ditrói, F.; Tárkányi, F.; Takács, S.; Mahunka, I.; Csikai, J.; Hermanne, A.; Uddin, M.S.; Hagiwara, M.; Baba, M.; Ido, T.; et al. Measurement of activation cross sections of the proton induced nuclear reactions on palladium. *J. Radioanal. Nucl. Chem.* **2007**, *272*, 231–235. [CrossRef]

15. Mito, A.; Komura, K.; Mitsugashira, T.; Otozai, K. Excitation functions for the (d, p) reactions on $^{96}$Ru, $^{102}$Ru and $^{104}$Ru. *Nucl. Phys. A* **1969**, *129*, 165–171. [CrossRef]

16. Comparetto, G.; Qaim, S. A Comparative Study of the Production of Short-Lived Neutron Deficient Isotopes $^{94,95,97}$Ru in α- and $^3$He-Particle Induced Nuclear Reactions on Natural Molybdenum. *Radiochim. Acta* **1980**, *27*, 177–180. [CrossRef]

17. Maiti, M.; Lahiri, S. Production and separation of $^{97}$Ru from $^7$Li activated natural niobium. *Radiochim. Acta* **2011**, *99*, 359–364. [CrossRef]

18. Maiti, M. Production and separation of $^{97}$Ru and coproduced $^{95}$Tc from $^{12}$C-induced reaction on yttrium target. *Radiochim. Acta* **2013**, *101*, 437–444. [CrossRef]

19. Levkovski, V.N. *Cross-Section of Medium Mass Nuclide Activation (A = 40–100) by Medium Energy Protons and Alpha-Particles (E = 10–50 MeV)*; Inter-Vesi: Moscow, Russia, 1991.

20. Ditrói, F.; Hermanne, A.; Tárkányi, F.; Takács, S.; Ignatyuk, A.V. Investigation of the α-particle induced nuclear reactions on natural molybdenum. *Nucl. Instr. Meth. B* **2012**, *285*, 125–141. [CrossRef]

21. Poirier, F.; Girault, F.; Auduc, S.; Huet, C.; Mace, E.; Delvaux, J.I.; Haddad, F. The C70 Arronax and beam lines status. In Proceedings of the IPAC2011, San Sebastián, Spain, 4–9 September 2011.

22. International Atomic Energy Agency, Live Chart of Nuclides. 2018. Available online: https://www-nds.iaea.org/relnsd/vcharthtml/VChartHTML.html (accessed on 21 January 2019).

23. Haddad, F.; Ferrer, L.; Guertin, A.; Carlier, T.; Michel, N.; Barbet, J.; Chatal, J.F. ARRONAX, a high-energy and high-intensity cyclotron for nuclear medicine. *Eur. J. Nucl. Med. Mol. Imaging* **2008**, *35*, 1377–1387. [CrossRef] [PubMed]

24. Garrido, E.; Duchemin, C.; Guertin, A.; Haddad, F.; Michel, N.; Métivier, V. New excitation functions for proton induced reactions on natural titanium, nickel and copper up to 70 MeV. *Nucl. Inst. Meth. B* **2016**, *383*, 191–212. [CrossRef]

25. Guertin, A.; Duchemin, C.; Fardin, A.; Guigot, C.; Nigron, E.; Remy, C.; Haddad, F.; Michel, N.; Métivier, V. How nuclear data collected for medical radionuclides production could constrain nuclear codes. In *EPJ Web of Conferences*; EDP Sciences: Les Ulis, France, 2017; p. 146.

26. Pupillo, G.; Sounalet, T.; Michel, N.; Mou, L.; Esposito, J.; Haddad, F. New production cross sections for the theranostic radionuclide $^{67}$Cu. *Nucl. Inst. Meth. B* **2018**, *415*, 41–47. [CrossRef]

27. Ziegler, J.F.; Ziegler, M.D.; Biersack, J.P. SRIM Code, Version 2008.04. Available online: http://www.srim.org/ (accessed on 21 January 2019).

28. International Atomic Energy Agency. Charged-Particle cross Section Database for Medical Radioisotope Production: Diagnostic Radioisotopes and Monitor Reactions. 2017. Available online: https://www-nds.iaea.org/medical (accessed on 21 January 2019).

29. Koning, A.J.; Rochman, D. Modern Nuclear Data Evaluation with The TALYS Code System. *Nucl. Data Sheets* **2012**, *113*, 2841–2934. [CrossRef]

30. Phelps, M.E. *PET: Molecular Imaging and Its Biological Applications*; Springer: New York, NY, USA, 2004.

31. de Lima, J.J.P. *Nuclear Medicine Physics*; CRC Press: Boca Raton, FL, USA, 2011.

32. Python Software Foundation. Python Programming Language, Version 2.7. 2010. Available online: https://www.python.org (accessed on 21 January 2019).

33. Cortesi, D. PyInstaller Documentation, Release 3.4. 2018. Available online: https://www.pyinstaller.org/documentation.html (accessed on 21 January 2019).

34. ARRONAX, Radionuclide Yield Calculator. 2018. Available online: http://www.cyclotron-nantes.fr/spip.php?article373 (accessed on 21 January 2019).

35. International Atomic Energy Agency. Cross Section Database for Medical Radioisotope Production: Production of Therapeutic Radionuclides. 2011. Available online: https://www-nds.iaea.org/radionuclides (accessed on 21 January 2019).

36. Kim, G.; Chun, K.; Park, S.H.; Kim, B. Production of α-particle emitting $^{211}$At using 45 MeV α-beam. *Phys. Med. Biol.* **2014**, *59*, 2849. [CrossRef] [PubMed]

37. Szkliniarz, K.; Sitarz, M.; Walczak, R.; Jastrzębski, J.; Bilewicz, A.; Choiński, J.; Jakubowski, A.; Majkowska, A.; Stolarz, A.; Trzcińska, A.; et al. Production of medical Sc radioisotopes with an alpha particle beam. *Appl. Radiat. Isot.* **2016**, *118*, 182–189. [CrossRef]

38. Zanzi, I.; Srivastava, S.C.; Meinken, G.E.; Robeson, W.; Mausner, L.F.; Fairchild, R.G.; Margouleff, D. A New Cholescintigraphic Agent: Ruthenium-97-DISIDA. *Nucl. Med. Biol.* **1989**, *16*, 397–403. [CrossRef]

MDPI

St. Alban-Anlage 66

4052 Basel

Switzerland

Tel. +41 61 683 77 34

Fax +41 61 302 89 18

www.mdpi.com

*Instruments* Editorial Office

E-mail: instruments@mdpi.com

www.mdpi.com/journal/instruments

www.ingramcontent.com/pod-product-compliance
Lightning Source LLC
Chambersburg PA
CBHW051839210326

41597CB00033B/5707